ORDERED PHASES IN METALLIC SYSTEMS

ORDERED PHASES IN METALLIC SYSTEMS

N.M. Matveeva and E.V. Kozlov

NOVA SCIENCE PUBLISHERS, INC.

Art Director: Maria Ester Hawrys
Assistant Director: Elenor Kallberg
Graphics: Susan A. Boriotti and Frank Grucci
Manuscript Coordinator: Phylis Gaynor
Book Production: Gavin Aghamore, Joanne Bennette, Michele Keller, Christine Mathosian and Tammy Sauter
Circulation: Iyatunde Abdullah, Cathy DeGregory and Annette Hellinger

Library of Congress Cataloging–in–Publication Data available upon request

ISBN 1-56072-357-2

6080 Jericho Turnpike, Suite 207
Commack, New York 11725
Tele. 516-499-3103 Fax 516-499-3146
E Mail Novasci1@aol.com

Printed in the United States of America

Contents

PREFACE

This monograph summarizes, to a certain extent, the results of studying the structure and regularities in the formation of ordered phases in binary metallic systems.

The ordered phases form on the base of the main crystal structures, i.e. face-centered cubic (f.c.c.), body-centered cubic (b.c.c.) and hexagonal close-packed (h.c.p.) lattices owing to the regular arrangement of atoms over lattice sites (i.e. due to the formation of a long-range order). They may form by crystallization from the liquid phase or arise in the solid state. Specific conditions for their formation and stability, stoichiometry and position in the phase diagram as well as the properties different from those of solid solutions enable the ordered phases to be singled out as an individual class of metallic compounds.

The authors of this book have attempted to specify the ordered phases as a separate class from the positions of physical metallurgy, physics and chemistry, and to show the progress in the understanding of their nature achieved during the last two decades as a result of experimental and theoretical studies of such phases. The monograph gives a consistent description of the crystal and electron structure of ordered phases, of the laws governing the order-disorder phase transitions and the conditions for stability of ordered phases.

The results of the study of ordered phases have been discussed at several international (New York, 1969; Tubingen, 1973; Colorado, 1986) and all-union (Kiev, 1969; Tomsk, 1972 and 1976; Kiev, 1978; Sverdlovsk, 1983) conferences and symposia. The monograph takes account of a world experience in this field but, at the same time, is based primarily on the original authors' investigations published, at least partly, in the papers cited in the reference list. These investigations have been carried out during the last two decades in the laboratory of metal chemistry in the Baikov

Institute of Metallurgy, Moscow and in the physical department of the Institute of Construction Engineers, Tomsk.

Authors are greatly indebted to Professor I.I. Kornilov and Professor L.E. Popov for substantial support in setting up and generalizing the results of these investigations. They also acknowledge with thanks the contribution of their coworkers V.M.Dementyev, N.M. Kormin, S.V. Starenchenko, A.S. Tailashev, D.M. Stern, and others whose results, together with the authors', have been employed in writing this book. A great assistance in the technical preparation of the manuscript made by Professor N.A.Koneva and senior research associates A.V. Paul, V.P. Podkovka, and M.A. Popova should also be appreciated. Chapter 1 was written by the authors jointly, Chapter 2 by N.M. Matveeva, Chapters 3, 4, 5 by E.V. Kozlov.

INTRODUCTION

Phases with the ordered arrangement of atoms, i.e. superstructures, are present in the phase diagrams of binary metallic systems in the range of the existence of primary solid solutions. In the solid state, they form by means of the order-disorder phase transformations. In the conventional classifications of metallic phases, the superstructures are usually referred to the primary solid solutions (see, for example, Hume-Rothery's classification). This point of view cannot be recognized as satisfactory because phases possessing the structure of the ordered solid solution more frequently form by direct solidification of the melt instead of being formed by successive crystallochemical reactions passing through a series of intermediate phases (e.g. Fe, Ga, Fe_3Ge, Ni_3Si, BeCu, etc.). To date, over 1000 ordered phases have been identified just in the binary systems.

Binary ordered phases show a stoichiometric (or nearly stoichiometric) ratio of the component atoms and arise, as a rule, near the AB, AB_2, AB_3, and AB_4 compositions. Irrespective of the origin from the liquid or solid state, the crystal structure of an ordered phase is based on a slightly distorted structure of the basic metallic lattices, i.e., f.c.c., b.c.c., or h.c.p. The distinctions manifest themselves in the completely or partially ordered arrangement of atoms over the sites of the crystal lattice subdivided into the different sublattices. As a rule, the formation of a long-range order (which distinguishes the ordered phases from the primary solid solutions) is accompanied, along with an increase in the lattice constant, by a small distortion (e.g., tetragonal one) of the original unit cell.

As a result, the following superstructures arise: B_2, B19, B32, $C11_b$, DO_a, DO_3, DO_{19}, DO_{22}, DO_{24}, $D1_a$, $L1_0$(M), $L1_0$, $L1_1$, $L1_2$, $L1_2$(M), $L1_2$(MM), $L1_3$, $L6_0$, and a Pt_2Mo-type superstructure. This is a comprehensive list of the superstructures, i.e., the crystal structures of ordered phases that will be described in detail in Chapter 1. Other intermediate phases, such as most electron compounds, the Lavez phases, μ-, δ-, ω-phases, though possessing a certain stoichiometry and an ordered arrangement of atoms, cannot be referred to the ordered phases because the f.c.c., b.c.c. and h.c.p. lattices do not represent the basic ones for these phases.

As an individual class, the ordered phases should be characterized by specific stability conditions. For this reason, the authors have analyzed, from different positions, the conditions for their existence. Empirical analysis has shown that the existence of the ordered phases and the superstructure type are controlled by a set of factors, primarily by the atomic properties of the components forming these phases. Various ordered phases are characterized by specific ranges of the electron concentration, size factor and electronegativity. As a consequence, the approaches developed by Hume-Rothery and Darken and Gurry have proved to be applicable to the ordered phases. The stability of a crystal lattice of an ordered phase with respect to the martensitic transformation is also connected with the aforementioned characteristics. Analysis of the role of the size factor and applicability of the Vegard rule allowed one to discover a new phenomenon inherent in the ordered phases, namely, the superstructure contraction. Specific volumes per one atom of the ordered phase were found to be less, as a rule, than those of the melted components.

Quantitative parameters defining stability of ordered phases depend upon the interaction between electrons of atoms on the first coordination shell. Such characteristics as the melting temperature depend on the overlap of the electron clouds of neighboring ions.

It is known that atomic ordering is always accompanied by a decrease in the symmetry of a crystal lattice. The symmetry of positions occupied by atoms of different kinds in ordered phases is lower than that in the parent crystal lattice. This effect entails an additional split in the energy levels of a crystal which may give, at certain electron concentrations, the gain in energy needed for stabilization of a given superstructure.

Approach to the consideration of symmetry effects developed in this monograph permitted the authors to single out the possible ranges of stability in terms of the initial atomic configurations. At the same time, analysis of different types of known superstructures revealed the important role of these effects in the problem of stability. Also of importance is the consideration of the band structure effects. Upon ordering, the collectivized electron states corresponding to different wave vectors in the Brillouin zones may increase or decrease their symmetry. Depending on the Fermi surface parameters for particular phases, these factors may facilitate the stabilization of various superstructures.

Thus, a detailed analysis of stability conditions for particular superstructures shows the necessity of considering not only atomic properties of the components, but also the electron structure of disordered solid solutions. An empirical approach has proved to be insufficient. There arises the need for a rigorous quantum-mechanical treatment of the stability conditions. The pseudopotential method confers such a possibility. With the aid of this method, significant results have been obtained during the last decades in describing the properties of pure metals and alloys of simple and polyvalent metals, including ordered alloys.

Because most ordered phases are based on transition elements, the monograph suggests an up-to-date quantum-mechanical approach to the theory of ordering alloys formed by transition and noble metals. Using this approach, a series of quantum-mechanical calculations of the stability conditions and the properties of the ordered alloys has been carried out, especially of the alloys of the Cu-Au system which has become a "classical" one.

When saying about the important role of the Fermi surface, we should stress, above all, the substantial part of its shape, particularly the presence of plane and cylindrical sections, for the stabilization of a given superstructure. The shape of the Fermi surface determines, in particular, the antiphase period of the long-period superstructures appearing in the Cu-Au system. At the same time, the quantum-mechanical calculations have shown that the short-range order structure existing in this system above the transformation temperature at different component concentrations is determined chiefly by the atomic properties of Cu and Au. Energy effects related to the phase transitions in the Cu-Au system are also primarily connected with the atomic properties of the component atoms.

A multifactor character of the problem of the ordered phase stability should also be stressed. The atomic properties and various band-structure effects are concurrent with the size and symmetry effects. The quantum theory of ordered phases is capable, even at the present time, of explaining many of the experimental observations. It shows a limited role of pair interactions in alloys and the need to take account of the three-particle or more complex contributions into the total energy, as well as the role of purely volume factors. Although in some cases one can formulate a set of limited rules concerning the stability conditions for ordered phases in the interaction space, the stability of a given superstructure in a particular system can be ultimately established only by allowing for all contributions into the Gibbs energy of the considered superstructure and the competing phases.

Also considered in the monograph are the modes of formation of superstructures from disordered solid solutions or, in other words, the laws governing the order-disorder transformation. The order-disorder phase transition represents a classical example of the phase transformation. Some of the characteristic features of phase transitions have been established by studying the order-disorder transformation. Tamman was first to set forward the hypothesis on the formation of a long-range order in binary metallic alloys. The proof of the existence of the ordered arrangement of atoms based on X-ray diffraction data was obtained by Bain in the 1920s. An important contribution to the study of the order-disorder transformations has been made by Russian scientists Kurnakov, Komar, Ageev, Smirnov, Panin, Kornilov, Zhdanov, Iveronova, Krivoglaz, Katsnel'son, Popov, Gladyshevskii, Matysina, Golosov, Dekhtyar, and others.

The order-disorder phase transitions are predominantly of the first order. Transformations close to the second-order type occur just in a few alloys, mostly with the B2 superstructure. During the last decade a significant progress has been achieved in the understanding of the nature of the second-order transitions. As regards the origin of the first-order transitions, the theory seems at present not to be markedly advanced. This situation makes the order-disorder transition a convenient object for investigating the general characteristics of the first-order transformations owing to the possibility of a considerable decrease in the equilibrium value of the long-range order parameter in this transformation. The stability of some superstructures with respect to the internal disordering makes them quite interesting for studying the states arising in the vicinity of the first-order phase transition and close to the critical states.

The order-disorder phase transition is a typical configurational transformation because it produces the change in the mutual arrangement of atoms over the sites of the parent crystal lattice and, respectively, the change in its symmetry. The configurational effects in atomic ordering, namely in changing from the state with a short-range order to the state with a long-range order, involve the changes in the long-range order parameter, the pair, three-particle and many-particle correlation functions and their dependence on the distance, and in the numbers of different neighbors in the first, second and more distant coordination shells which depend on a particular type of the superstructure.

In other words, the configurational effects consist in specific changes of the order parameters in ordering within a given superstructure; these changes depend, however, on a particular superstructure. The atomic configuration in an alloy changes not only with changing the order parameters and correlation functions, and/or the superstructure type but also with the appearance of the antiphase boundaries. The configurational effects determine the regularities in the atomic ordering via the values of the pair and many-particle ordering energies. The knowledge of the order parameters and the ordering energy enables one to determine the configurational energy and the entropy of an alloy, i.e., the fraction of the free energy which explicitly depends on the order parameters and correlations.

The order-disorder transition is accompanied by changing in the volume of the unit cell and interatomic spacings. These effects are not configurational. The nonconfigurational effects may be only indirectly expressed in terms of the order parameters whereas energy and entropy are strongly dependent on them. For this reason, the study of the order- disorder transformation will not be comprehensive if the configurational and nonconfigurational effects are treated separately. Only the combined treatment of these effects would permit one to understand the regularities and the nature of the stability of the ordered state. This is the conception the authors of this monograph hold. Such approach is especially important because some effects exist which cannot be strictly classified as configurational or nonconfigurational; their dscription depends on the degree of the approximation adopted.

To such effects we refer the symmetry and the bond-structure factors, low-symmetry lattice distortions and some others. The configurational and nonconfigurational effects determine the Gibbs energy of an alloy, the possibility of forming a superstructure, the superstructure type, the temperature of its existence, the heat of transformation, and the internal and external stabilities of the ordered state.

In the monograph we give a detailed description of the results of our experimental study of the order-disorder transformation in a variety of alloys. The generalization of these results enabled us to establish the relation between the superstructure type, interatomic interactions, size effects, the instability with regard to martensitic transformations, and the observed trends in the order-disorder transformations. It was found that all superstructures can be divided into the three specific groups differing in the characteristic features of the order-disorder transformation.

Very important for any phase transition is the problem of the order parameter. It is shown in the monograph that the long-range order parameter η is a unique parameter of the phase transition and the behavior of all order parameters is correlated in a definite way with η. To describe quantitatively the order-disorder transformation, an efficient formalism of the statistical theory of transformations based on the modern methods of statistical physics has been developed.

All previous versions of the theory are shown to represent the partial cases of the newly developed formalism. With the aid of this method, the behavior of order parameters, interatomic correlation and heat capacity in the order-disorder phase transition has been assessed quantitatively. The results of the calculation and comparison with experimental data confirmed a high efficiency of the new variant of the theory and led us to a series of interesting physical conclusions. The quantitative trends in the order-disorder transformation proved to be dependent on such factors as the interatomic long-range interaction, the size of the cluster specifying the interatomic correlations, etc.

The thermal effects in the order-disorder transformation are important both for verifying the statistical theories and for a deeper insight into the nature of the stabilization of the ordered state. We describe the results of our measurements of the temperature dependence of heat capacity in several ordered alloys. The comparison of the entropy of transformation with the configurational entropy allowed us to distinguish a group of alloys in which the lattice properties either facilitate the order-disorder transition or impede it.

Based on the experimental evidence, the thermodynamical classification of the order-disorder phase transitions has been suggested. It has urged, in a closed form, to incorporate not only the temperature dependence of the internal energy and heat capacity but also the dependence of the long-range and short-range order parameters on temperature. It was found that it would be more convenient to classify the order-disorder transitions taking into consideration the properties of the disordered phase. For this aim, a parameter was required in order to describe, in a unified scheme, both the ordered and disordered phases. This was found to be the Cowley parameter.

The introduction of the Cowley parameter into the treatment of the order-disorder transition underlines the substantial role of correlation effects. It means that the transformation may be affected by the peculiarities of the fine structure of the solid solution: the variation in the unit cell parameter, the distortion of the unit cell, directed chaotic displacements, etc. Thus, as in the problem of the ordered phase stability, the specific features of the order-disorder transformation depend on a large variety of factors such as atomic and crystal structure, etc.

1. CRYSTAL STRUCTURE OF ORDERED PHASES

1.1 Symmetry types, symmetry and the problem of stability of ordered phases

An order-disorder transformation leads to the formation of a new crystal structure, the superstructure, below the transformation temperature T_c. As a result of ordering, a series of superstructures (see Tables 1 and 2) form in the primary solid solutions based on the f.c.c. (A1), b.c.c. (A2), or h.c.p. (A3) lattices. The unit cells corresponding to the most symmetric representation of the above superstructures are shown in Fig. 1. We use and give, along with the international classification, the Schönfliss notation for each symmetry group. The Schönfliss notations will be, however, most commonly used henceforth.

We must point out some specific features of the superstructures. The $L1_1$ and $L1_3$ superstructures, by contrast to all the others, are virtually tetragonal and monoclinic analogs of the B2 superstructure, respectively. The close-packed planes in the $L1_2$, DO_{19}, and DO_{24} superstructures are identical, differing only in the succession of atomic layers (ABCABC in the $L1_2$ superstructure, ABAB in the DO_{19}, ABACABAC in the DO_{24}). In the DO_{23}, $C11_b$, B19, and in the $MoPt_2$-type superstructures, some of the atoms occupy the lattice sites on symmetry axes, while in the DO_{19}, DO_{24}, $D1_a$, and DO_a structures a certain fraction of atoms is situated even in symmetry planes. These atoms are frequently shifted away from their positions in the original disordered crystal lattice by various distances, so that the ordering in these structures may be accompanied by the modulation of interatomic spacings (i.e., by the appearance of static displacement waves).

Table 1. Crystallography of superstructures.

Structure type of disordered solid solution;	Composition;	Structure type and space group;			Characteristic stars;	Number of formular units (unit cell contents);	Example of phases;
		Strukturbericht	Schönfliess;	International			
$A1-O_h^5$	AB	$L1_0$	D_{4h}^1	P4/mmm	001	1	CuAuI
		$L1_0$(M)	C_{2v}^{20}	Imm2	$10\frac{1}{2M}$	4M	CuAuII
		$L1_1$	D_{3d}^5	$R\bar{3}m$	$\frac{1}{2}\frac{1}{2}\frac{1}{2}$	1	PtCu
	AB_2	–	D_{2h}^{25}	Immm	$\frac{2}{3}\frac{2}{3}0$	2	$MoPt_2$
	AB_3	$L1_2$	O_h^1	Pm3m	100	1	$AuCu_3$
		$L6_0$	D_{4h}^1	P4/mmm	001	1	$SrPb_2$
		$D0_{22}$	D_{4h}^{17}	I4/mmm	$100; 10\frac{1}{2}$	2	$TiAl_3$
		$D0_{23}$	D_{4h}^{17}	I4/mmm	$100; 10\frac{1}{4}$	4	$ZrAl_3$
		$L1_2$(M)	D_{4h}^{17}	I4/mmm	$100; 10\frac{1}{2M}$	2M	$AuCu_3$ II
		$L1_2$(MM)	D_{2h}^{13}	Pmnn	$10\frac{1}{2M_1}$; $10\frac{1}{2M_2}$	$4M_1 \times M_2$	$CuAu_3$ II
		$L1_3$	D_{2h}^{19}	Cmmm	$001; \frac{1}{2}\frac{1}{2}\frac{1}{2}$	2	$CuPt_3$
	AB_4	$D1_a$	C_{4h}^5	I4/m	$\frac{2}{5}\frac{4}{5}0$	2	$MoNi_4$
$A2-O_h^9$	AB	B2	O_h^1	Pm3m	100	1	βCuZn
		B32	O_h^7	Fd3m	111	8	NaTl
	AB_2	$C11_b$	D_{4h}^{17}	I4/mmm	$00\frac{2}{3}$	2	$AlCr_2$
	AB_3	$D0_3$	O_h^5	Fm3m	$\frac{1}{2}\frac{1}{2}\frac{1}{2}$	4	$AlFe_3$
$A3-D_{6h}^4$	AB	B19	D_{2h}^5	Pmna	010	2	AuCd
	AB_3	$D0_{19}$	D_{6h}^1	$P6_3$/mmc	$\frac{1}{2}00$	2	$CdMg_3$
		$D0_a$	D_{2h}^{13}	Pmmn	$\frac{1}{2}00; 01\frac{1}{2}$	2	$\beta TiCu_3$
		$D0_{24}$	D_{6h}^1	$P6_3$/mmc	$\frac{1}{2}00$	4	$TiNi_3$

Table 2. Crystallography of superstructures.

Structure type;	2. Site coordinates;	Multiplicity;	Point symmetry group of lattice sites;	Number of order parameters;	Change in number of bonds in ordering (per atom); I	II	III
1	2	3	4	5	6	7	8
B2	A: 000	1	m3m	1	+2	$-\frac{3}{2}$	−3
	B: $\frac{1}{2}\frac{1}{2}\frac{1}{2}$	1	m3m				
B32	A: $\left.\begin{matrix}000\\ \frac{1}{4}\frac{1}{4}\frac{1}{4}\end{matrix}\right\} + \left\{\begin{matrix}000\\ \frac{1}{2}\frac{1}{2}0\\ 0\frac{1}{2}\frac{1}{2}\\ \frac{1}{2}0\frac{1}{2}\end{matrix}\right\}$	8	$4\bar{3}m$	1	0	$+\frac{3}{2}$	−3
	B: $\left.\begin{matrix}\frac{1}{2}\frac{1}{2}\frac{1}{2}\\ \frac{3}{4}\frac{3}{4}\frac{3}{4}\end{matrix}\right\}$ + (same set)	8					
$C11_b$	A: 000	2	4/mmm				
	B: $\left.\begin{matrix}00\bar{z}\\ 00z\end{matrix}\right\} + \left\{\frac{1}{2}\frac{1}{2}\frac{1}{2}\right\}$	4	4 mm	1	$+\frac{8}{9}$	$-\frac{2}{3}$	0
	A:000	4	m3m				
	$B(1)$: $\left.\begin{matrix}\frac{1}{2}\frac{1}{2}\frac{1}{2}\\ \frac{1}{4}\frac{1}{4}\frac{1}{4}\end{matrix}\right\} + \left\{\begin{matrix}000\\ \frac{1}{2}\frac{1}{2}0\\ \frac{1}{2}0\frac{1}{2}\\ 0\frac{1}{2}\frac{1}{2}\end{matrix}\right\}$	4	m3m	2	$+\frac{1}{2}$	$-\frac{9}{8}$	$-\frac{9}{4}$
$D0_3$	$B(2)$: $\frac{1}{4}\frac{1}{4}\frac{1}{4}$	8	$\bar{4}3m$				

table 2 continued

1	2	3	4	5	6	7	8
$L1_0$	A: 000	1	4/mmm	1	+1	$-\frac{3}{2}$	+2
	B: $\frac{1}{2}\frac{1}{2}\frac{1}{2}$	1					
$L1_0(M)$ M = 1	$A(2)$: 000	2					
	$A(2)$: $0\frac{1}{2}z_2$	2	mm	4m − 1	+1	$-(\frac{3}{2}-\frac{1}{2M})$	$+(2-\frac{2}{M})$
	$B(1)$: $00z_3$	2					
	$B(2)$: $0\frac{1}{2}z_4$ $\Bigg\} + \left\{\frac{1}{2}\frac{1}{2}\frac{1}{2}\right\}$	2					
$L1_1$	$A(1)$: 000	1	$\bar{3}$m				
	$A(2)$: $\frac{1}{2}\frac{1}{2}0$; $\frac{1}{2}0\frac{1}{2}$; $0\frac{1}{2}\frac{1}{2}$	3	2/m	1	0	$+\frac{3}{2}$	0
	$B(1)$: $\frac{1}{2}\frac{1}{2}\frac{1}{2}$	1	$\bar{3}$m				
	$B(2)$: $\frac{1}{2}00$; $0\frac{1}{2}0$; $00\frac{1}{2}$	3	2/m				
type $MoPt_2$	A: 000	2	mmm	1	$+\frac{2}{3}$	0	$-\frac{4}{3}$
	B: $0y0$, $0\bar{y}0$ $\Big\} + \left\{\frac{1}{2}\frac{1}{2}\frac{1}{2}\right\}$	4	mm				
B19	A: $\frac{1}{4}\frac{1}{2}z$; $\frac{3}{4}\frac{1}{2}\bar{z}$	2	mm	1	$-\frac{3}{2}$	$-\frac{3}{2}$	$+\frac{3}{2}$
	B: $\frac{1}{4}0z$; $\frac{3}{4}0\bar{z}$	2	mm				

table 2 continued

$D0_{19}$	$A: \pm(\frac{1}{3}\frac{2}{3}\frac{1}{4})$	2	$\bar{6}m2$				
	$B: \pm(x, 2x, \frac{1}{4}); \pm(2\bar{x}; \bar{x}, \frac{1}{4});$ $\pm(x, \bar{x}, \frac{1}{4})$	6	mm	1	$+\frac{3}{4}$	$-\frac{9}{8}$	$-\frac{9}{8}$
$D0_3$	$A: 00z; \frac{1}{2}\frac{1}{2}\bar{z}$	2	mm				
	$B(1): 0\frac{1}{2}z; \frac{1}{2}0\bar{z};$	2	mm	2	$+\frac{3}{4}$	$-\frac{5}{8}$	$-\frac{3}{8}$
	$B(2): \begin{cases} \frac{1}{2}-x, \frac{1}{2}, z; x0z \\ \frac{1}{2}+x, \frac{1}{2}, z; x0\bar{z} \end{cases}$	4	m				
$L1_2(M)$ $M \geqslant 3$	$A: 00z_i \; (z_i = \frac{1}{4M}, \ldots, \frac{M-1}{4M})$	2M	4mm				
	$B: 00z_j \; (z_j = \frac{M+1}{4M}, \ldots, \frac{2M-1}{4M})$	2M	4mm	$\leqslant 3M/2$ (theory)			
	$B: \frac{1}{2}00$ $+\left\{\frac{1}{2}\frac{1}{2}\frac{1}{2}\right\}$	4	mmm	2 (experiment)	$+\frac{3}{4}$	$-(\frac{9}{8} - \frac{1}{2M})$	$+(\frac{3}{2} - \frac{2}{M})$
	$B: 0\frac{1}{2}\frac{1}{4}$	4	4m2				
	$B: \frac{1}{2}0z_k \; (z_k = \frac{1}{2M}, \ldots, \frac{M-2}{4M})$	4M – 8	mm				
$L1_2(MM)$	Coordinates of right points depend on M_1 and M_2 values;	–	–	2 (experiment);	$+(\frac{3}{4} - \frac{1}{8M_1M_2})$	$-(\frac{9}{8} - \frac{M_1+M_2}{4M_1M_2})$	$+(\frac{3}{8} - \frac{4M_1+4M_2-3}{4M_1M_2})$

table 2 continued

1	2	3	4	5	6	7	8
$L1_3$	A: 000	2	mmm				
	B: $\frac{1}{2}00$ $\Big\}+\left\{\frac{1}{2}\frac{1}{2}0\right\}$	2	mmm	2	$+\frac{1}{4}$	$+\frac{3}{8}$	$+\frac{1}{2}$
	B: $\frac{1}{4}\frac{1}{4}\frac{1}{2};\ \frac{1}{4}\frac{\bar{1}}{4}\frac{1}{2}$	4	2/m				
$D1_a$	A: 000 $\Big\}+\left\{\frac{1}{2}\frac{1}{2}\frac{1}{2}\right\}$	2	4/m	1	$+\frac{12}{25}$	$-\frac{4}{25}$	$-\frac{16}{25}$
	B: $xy0;\ y\bar{x}0;\ x\bar{y}0;\ \bar{y}x0$	8	m				
$L1_2$	A: 000	1	m3m	1	$+\frac{3}{4}$	$-\frac{9}{8}$	$+\frac{3}{2}$
	B: $\frac{1}{2}\frac{1}{2}0;\ \frac{1}{2}0\frac{1}{2};\ 0\frac{1}{2}\frac{1}{2}$	3	4/mmm				
$L6_0$	A: 000	1	4/mmm				
	$B(1)$: $\frac{1}{2}\frac{1}{2}0$	1	4/mmm	2	$+\frac{3}{4}$	$-\frac{9}{8}$	$+\frac{3}{2}$
	$B(2)$: $\frac{1}{2}0\frac{1}{2};\ 0\frac{1}{2}\frac{1}{2}$	2	mmm				
$D0_{22}$	A: 00	2	4/mmm				
	$B(1)$: $\frac{1}{2}\frac{1}{2}\frac{1}{2}$ $\Big\}+\left\{\frac{1}{2}\frac{1}{2}\frac{1}{2}\right\}$	2	4/mmm	2	$+\frac{3}{4}$	$-\frac{5}{8}$	$-\frac{1}{2}$
	$B(2)$: $\frac{1}{2}0\frac{1}{4};\ 0\frac{1}{2}\frac{1}{4}$	4	$\bar{4}$m2				

table 2 continued

DO_{23}	A: $00z_1$; $00\bar{z}_1$	4	4mm	2 (experiment);			
	$B(1)$: $00z_2$; $00z_2$	4	4mm				
	$B(2)$: $\frac{1}{2}00$; $0\frac{1}{2}0$;	4	mmm	3 (theory).	$+\frac{3}{4}$	$-\frac{7}{8}$	$+\frac{1}{2}$
	$B(3)$: $\frac{1}{2}0\frac{1}{4}$; $0\frac{1}{2}\frac{1}{4}$	4	$\bar{4}$m2				
DO_{24}	$A(1)$: 000; $00\frac{1}{2}$	2	$\bar{3}$m	3			
	$A(2)$: $\pm(\frac{1}{3}\frac{2}{3}\frac{1}{4})$	2	$\bar{6}$m2				
	$B(1)$: $\frac{1}{2}00$; $0\frac{1}{2}0$; $\frac{1}{2}\frac{1}{2}0$; $\frac{1}{2}0\frac{1}{2}$; $0\frac{1}{2}\frac{1}{2}$; $\frac{1}{2}\frac{1}{2}\frac{1}{2}$	6	2/m		$+\frac{3}{4}$	$-\frac{9}{8}$	$+\frac{9}{8}$
	$B(2)$: $\pm[x, 2x, \frac{1}{4}$; $x\bar{x}\ \frac{1}{4}$; $2\bar{x}, x, \frac{1}{4}]$	6	mm				

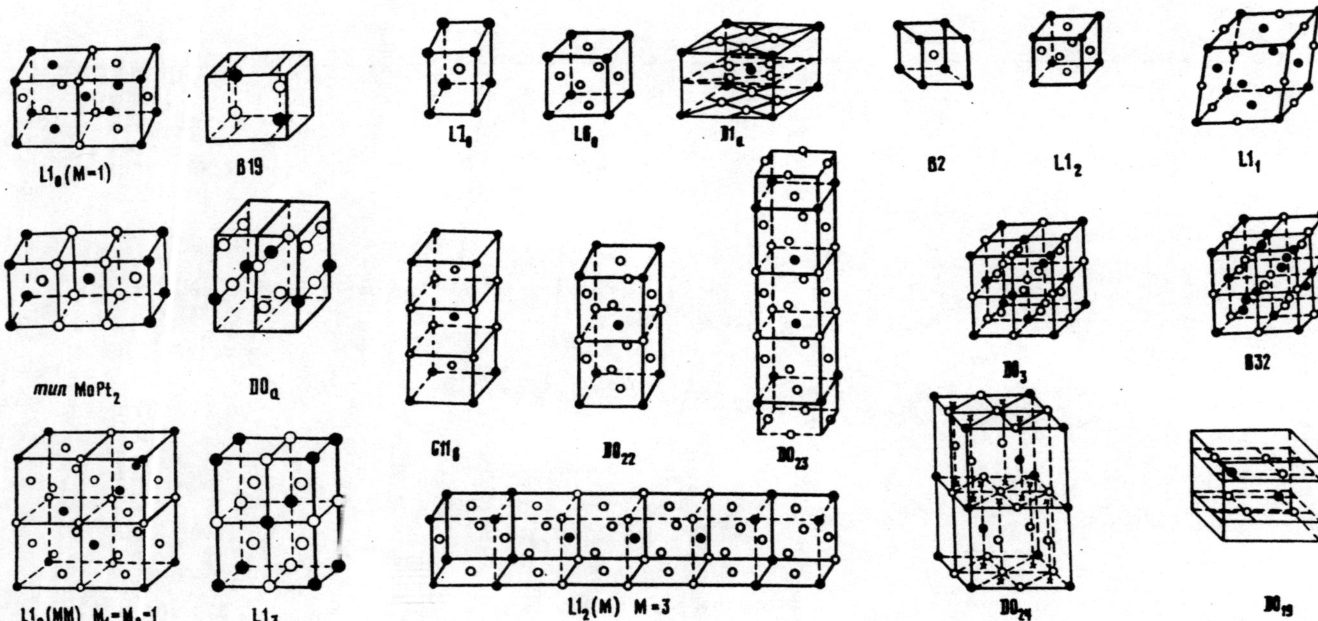

Fig. 1. Unit cells of superstructures.

The $L1_2(M)$ and $L1_2(MM)$ superstructures are usually called the long-period ones because they can be obtained from the $L1_2$ superstructure by introducing the periodic antiphase boundaries. As will be shown in Section 1.3, this is somewhat arbitrary definition since the majority of superstructures can be transformed to one another with the aid of periodic antiphase shifts, whereas the superstructures originated from the same disordered solid solution are always converted to one another in this way, irrespective of their type.

Prediction of superstructures has its own history. Several methods for their theoretical derivation have been suggested so far. A geometric method is based on a crystallogeometric listing of the possible arrangements of the two kinds of atoms over the sites of the original crystal lattice. This labor-consuming method allows one to find, in principle, all possible superstructures. The development of this method is based on the studies of Belov [1], Hume-Rothery and Reiner [2], Schubert [3], Smirnova [4 - 8] and Gladyshevskii [9 - 11]. Its main disadvantage is the lack of the criterion that could help one to establish the completeness of the superstructures list.

The second approach based on the group theory analysis suffers from the same drawback. In this method, one obtains all possible structures for the given phase transition by distributing them over the symmetry subgroups of disordered solid solution [12, 13]. Gufan et al. [14 - 16] applied this method directly to atomic ordering. Although the approach enables, in principle, all possible superstructures to be predicted, this work has not yet been completed.

The third approach is the method of the static concentration waves. The basic ideas have been put forward by Lifshits [17]. The up-to-date formulation was developed in the works [18 - 21]. By contrast to the two above methods, this approach is specially developed to deal with the atomic ordering problems. The method includes the criterion of the completeness of the superstructures list for a given number of the order parameters and operates with the long-range order parameters. Regretfully, the theory, as formulated by Khachaturyan [19, 21], has proved to be incapable of describing some of the observed superstructures due to the redundant limitations imposed on the function of the distribution of atoms over the lattice sites. We proposed a modification to the static concentration-wave method, free of the above disadvantages. Stern and coworkers used this method to predict all possible superstructures based on the f.c.c., b.c.c. and h.c.p. lattices. In addition, the superstructures of nonstoichiometric alloys have been deduced as well as the structures corresponding to the quasibinary sections of ternary alloys. We have presented here merely the results obtained for the f.c.c. and b.c.c. lattices.

The crystal structure of a binary solid solution will be defined if the probability f(R) of finding a certain kind of atom, say A, at any lattice site R is known. The function f(R) can be expressed as the Fourier superposition of the concentration plane waves

$$f(\mathbf{R}) = c + \frac{1}{2}\sum_{s,j}\eta_s\left[A(\mathbf{k}_s^j)e^{i\mathbf{k}_s^j\mathbf{R}} + A^*(\mathbf{k}_s^j)e^{-i\mathbf{k}_s^j\mathbf{R}}\right]. \qquad (1.1.1)$$

The summation is over the first Brillouin zone in the reciprocal lattice k{R}; s is the index of the star of the superstructure vectors, j denotes the vector in the star. The $A(\mathbf{k}_s^j)$ values are chosen so that ηs be the long-range order parameters.

An obvious constraint should be imposed on the f($\overline{\mathrm{R}}$) function concerning the number of its values in the set {$\overline{\mathrm{R}}$}. This number must be equal to the number of different sublattices, or to that of the long-range order parameters plus one. Khachaturyan [19 - 21] imposes on the f($\overline{\mathrm{R}}$) the complementary condition of translational invariance of the superstructure vectors. Let us consider this matter in more detail. The coefficients in series (1.1.1) may be real or complex. If A is real, formula (1.1.1) takes the form:

$$f(\mathbf{R}) = c + \sum_{s,j}\eta_s A(\mathbf{k}_s^j)\cos(\mathbf{k}_s^j\mathbf{R}). \qquad (1.1.2)$$

For the perfectly ordered alloy of stoichiometric composition, f($\overline{\mathrm{R}}$) is equal to unity or zero. Then, for any $\overline{\mathrm{R}}$

$$f^{(2)}(\mathbf{R}) = f^{(3)}(\mathbf{R}) = \ldots = f(\mathbf{R}) \qquad (1.1.3)$$

Combining (1.1.2) and (1.1.3) makes the condition of translational invariance necessary. Khachaturyan obtained this condition in other way restricting his consideration to the case of small η values. Meanwhile, $A(\mathbf{k}_s^j)$ may be complex, so that $A(\mathbf{k}_s^j) = a(\mathbf{k}_s^j) + ib(\mathbf{k}_s^j)$, and we obtain

$$f(\mathbf{R}) = c + \sum_{s,j} \eta_s [a(\mathbf{k}_s^j)\cos(\mathbf{k}_s^j\mathbf{R}) + b(\mathbf{k}_s^j)\sin(\mathbf{k}_s^j\mathbf{R})]. \tag{1.1.4}$$

If $a(\mathbf{k}_s^j) \neq b(\mathbf{k}_s^j)$, combining (1.1.3) and (1.1.4) makes the translational invariance condition necessary. But if $a(\mathbf{k}_s^j) = b(\mathbf{k}_s^j)$, the combination of (1.1.3) and (1.1.4) leads to another possibility that has not been considered in reference [19], namely,

$$\sin(2\mathbf{k}_s^j\mathbf{R}) = 0 \tag{1.1.5}$$

for any $\overline{\mathrm{R}}$. If, in addition, $a(\mathbf{k}_{s'}^{j'}) = 0$, yet another way of breaking the translation invariance condition is possible:

$$\sin[(\mathbf{k}_s^j + \mathbf{k}_{s'}^{j'})\mathbf{R}] = 0 \tag{1.1.6}$$

for any $\mathbf{k}_s^j$ and $\overline{\mathrm{R}}$. And finally, even in the case of the real Fourier coefficients, the terms in series (1.1.2) may have the alternate signs. Upon reducing similar terms, the terms like $\cos[(\mathbf{k}_s^j + \mathbf{k}_{s'}^{j'})\overline{\mathrm{R}}]$ and $\cos(2\mathbf{k}_s^j\mathbf{R})$ may be canceled out in the series, again making the translational condition superfluous.

Therefore, the above consideration eliminates, by contrast to [19], the requirement that the reciprocal lattices of superstructures be the Bravais lattices. This extends the range of possible superstructures. As a result, the necessary superstructure can be obtained from the theory [22].

The second limitation in the Khachaturyan theory is connected to the use for $\mathbf{k}_s^j$ of only high-symmetry (Lifshits) points in the Brillouin zone. This constraint is based on the assumption that the Lifshits assertion [17] is valid for all ordered phases. This criterion implies that the ordering energy has a minimum in the points of the reciprocal space corresponding to the superstructure vectors. However, this is really valid just in the vicinity of the points corresponding to the second-order phase transition. An attempt to extend this criterion to the first-order transition [19] is not correct due to the use of the wrong expression for the Fourier transform of the ordering energy. As a consequence, Khachaturyan [21] came to an erroneous inference that all the superstructures for which $\mathbf{k}_s^j$ do not correspond to the extremum positions, would be unstable with regard to the formation of the periodic antiphase boundaries with the temperature-dependent antiphase period. First, any superstructure can be represented as a long-period one relative to any other (see Section 1.3). Second, the minimum of $V(\mathbf{k})$ is not required at all for the stability of a superstructure (see Section 3.2).

It follows from the above reasoning that all possible superstructures with a given number of order parameters can be predicted by the theory. We shall further consider the ordering based on f.c.c. and b.c.c. lattices in the order of the increasing number of order parameters. The results are given in Tables 3 - 6.

Face-centered cubic lattice. One order parameter. Twelve possible superstructures are described in Table 3. In addition to the $L1_0$, $L1_0$ (M=1) and $L1_1$ superstructures obtained in [22], Table 3 also contains the $D1_a$ (of the Ni_4Mo type) and the Pt_2Mo-type superstructures, i.e., those which could not be predicted by Khachaturyan's scheme. Besides, the present theory predicts the formation of a set of relatively simple superstructures that are to be found experimentally (or, if not, it should be understandable why they do not form).

Face-centered lattice. Two order parameters. The theory has made it possible to derive one more well-known superstructure, DO_{23} (of the Ag_3Mg type). This deserves special saying. The distribution function does not include the star {0 0 1/2} which could lead to the occurrence of three order parameters. Nevertheless, owing to the complementary restriction placed on the number of different $f(\overline{R})$

Table 3. Predicted f.c.c. superstructures with one order parameter

Com-position;	Star	Super-structure	Distribution function;
AB	(001)	$L1_0$	$c+\frac{1}{2}\eta\cos 2\pi z$
AB	$(\frac{1}{2}\frac{1}{2}\frac{1}{2})$	$L1_1$	$c+\frac{1}{2}\eta\cos\pi(x+y+z)$
AB	$(\frac{1}{2}\frac{1}{2}\frac{1}{2})$	*	$c+\frac{1}{4}\eta[\cos\pi(x+y+z)+\cos\pi(x-y+z)+$ $+\cos\pi(-x+y-z)-\cos\pi(x+y+z)]$
AB	$(00\frac{1}{2})$	*	$c+\frac{1}{2}\eta(\cos\pi z+\sin\pi z)$
AB	$(10\frac{1}{2})$	$L1_0(M)$; $M=1$	$c+\frac{1}{2}\eta[\cos\pi(z+2x)+\sin\pi(z+2x)]$
AB	$(0\frac{1}{2}\frac{1}{2})$	*	$c+\frac{1}{2}\eta[\cos\pi(x+y)+\sin\pi(x+y)]$
AB	$(\frac{1}{4}\frac{1}{4}\frac{1}{4})$	*	$c+\frac{1}{2}\eta[\cos\frac{\pi}{2}(x+y+z)+\sin\frac{\pi}{2}(x+y+z)]$
AB_2	$(00\frac{2}{3})$	*	$c+\frac{2}{3}\eta\cos\frac{4}{3}\pi z$
AB_2	$(0\frac{2}{3}\frac{2}{3})$	тип Pt_2M_0	$c+\frac{2}{3}\eta\cos\frac{4}{3}\pi(x+y)$
AB_2	$(\frac{1}{3}\frac{1}{3}\frac{1}{3})$	*	$c+\frac{2}{3}\eta\cos\frac{2}{3}\pi(x+y+z)$
AB_3	(001)	$L1_2$	$c+\frac{1}{4}\eta(\cos 2\pi x+\cos 2\pi y+\cos 2\pi z)$
AB_4	$(\frac{2}{5}\frac{4}{5}0)$	$D1_a$	$c+\frac{2}{5}\eta[\cos\frac{4}{5}\pi(x+2y)+\cos\frac{4}{5}\pi(2x+y)]$

*Superstructure is predicted but not found.

values, the coefficient of the order parameter $\eta_{0\ 0\ 1/2}$ has vanished. In addition to the DO_{23} superstructure, there may exist several superstructures not observed experimentally (Table 4).

Thus, the theory predicts all simple superstructures based on f.c.c. lattice which have been experimentally observed. Upon ordering, the period may increase by 2 to 6 times. The increase by 2 to 4 or even 6 times is connected to the Lifshits stars

Table 4. Predicted f.c.c. superstructures with two order parameters

Comp-osition;	Star	Super-structure	Distribution function;
AB_2	$\left(0\,0\,\frac{2}{3}\right)\left(0\,0\,\frac{1}{3}\right)$	*	$c+\frac{1}{3}\eta_1\cos\frac{4}{3}\pi z+\frac{1}{\sqrt{3}}\eta_2\sin\frac{2}{3}\pi z$
AB_2	$\left(0\,\frac{2}{3}\,\frac{2}{3}\right)\left(0\,\frac{1}{3}\,\frac{1}{3}\right)$	*	$c+\frac{1}{3}\eta_1\cos\frac{4}{3}\pi(x+y)+\frac{1}{\sqrt{3}}\eta_2\sin\frac{2}{3}\pi(x+y)$
AB_2	$\left(\frac{1}{3}\,\frac{1}{3}\,\frac{1}{3}\right)\left(\frac{1}{6}\,\frac{1}{6}\,\frac{1}{6}\right)$	*	$c+\frac{1}{3}\eta_1\cos\frac{2}{3}\pi(x+y+z)+\frac{1}{\sqrt{3}}\eta_2\sin\pi(x+y+z)$
AB_2	$\left(0\,0\,\frac{2}{3}\right)\left(0\,1\,\frac{1}{3}\right)$	*	$c+\frac{2}{3}\eta_1\cos\frac{4}{3}\pi z+\frac{1}{\sqrt{3}}\eta_2\sin 2\pi\left(x+\frac{1}{3}z\right)$
AB_2	$\left(\frac{1}{3}\,\frac{1}{3}\,\frac{1}{3}\right)\left(\frac{1}{6}\,\frac{1}{6}\,\frac{5}{6}\right)$	*	$c+\frac{1}{3}\eta_1\cos\frac{2}{3}\pi(x+y+z)+$ $+\frac{1}{\sqrt{3}}\eta_2\sin\frac{1}{3}\pi(-5x+y+z)$.
AB_3	$(0\,0\,1)\left(\frac{1}{2}\,\frac{1}{2}\,\frac{1}{2}\right)$	$L1_3$	$c+\frac{1}{4}\eta_1\cos 2\pi z+\frac{1}{2}\eta_2[\cos\pi(x+y+z)+$ $+\cos\pi(x+y+z)]$
AB_3	$(0\,0\,1)\left(0\,1\,\frac{1}{2}\right)$	$D0_{22}$	$c+\frac{1}{4}\eta_1\cos 2\pi z+\frac{1}{2}\eta_2\cos\pi(2x+z)$
AB_3	$(0\,0\,1)\left(0\,0\,\frac{1}{2}\right)$	*	$c+\frac{1}{4}\eta_1\cos 2\pi z+\frac{1}{2}\eta_2\cos\pi z$
AB_3	$(0\,0\,1)\left(\frac{1}{2}\,\frac{1}{2}\,0\right)$	*	$c+\frac{1}{4}\eta_1\cos\pi(x+y+z)+\frac{1}{2}\eta_2\cos\frac{\pi}{2}(x+y+z)$
AB_3	$\left(\frac{1}{2}\,\frac{1}{2}\,\frac{1}{2}\right)\left(\frac{1}{4}\,\frac{1}{4}\,\frac{1}{4}\right)$	*	$c+\frac{1}{4}\eta_1\cos\pi(x+y+z)+\frac{1}{2}\eta_2\cos\frac{\pi}{2}(x+y+z)$
AB_3	$\left(\frac{1}{2}\,\frac{1}{2}\,\frac{1}{2}\right)\left(\frac{1}{4}\,\frac{1}{4}\,\frac{3}{4}\right)$	*	$c+\frac{1}{4}\eta_1\cos\pi(x+y+z)+\frac{1}{2}\eta_2\cos\frac{\pi}{2}(x+y-3z)$
AB_4	$\left(0\,\frac{4}{5}\,\frac{4}{5}\right)\left(0\,0\,\frac{2}{5}\right)$	*	$c+\frac{2}{5}\eta_1\cos\frac{8}{5}\pi z+\frac{2}{5}\eta_2\cos\frac{4}{5}\pi z$
AB_4	$\left(0\,\frac{4}{5}\,\frac{4}{5}\right)\left(0\,\frac{2}{5}\,\frac{2}{5}\right)$	*	$c+\frac{2}{5}\eta_1\cos\frac{8}{5}\pi(y+z)+\frac{2}{5}\eta_2\cos\frac{4}{5}\pi(y+z)$
AB_4	$\left(\frac{2}{5}\,\frac{2}{5}\,\frac{2}{5}\right)\left(\frac{1}{5}\,-\frac{1}{5}\,\frac{1}{5}\right)$	*	$c+\frac{2}{5}\eta_1\cos\frac{4}{5}\pi(x+y+z)+\frac{2}{5}\eta_2\cos\frac{\pi}{5}(x+y+z)$
AB_7	$(0\,0\,1)\left(\frac{1}{2}\,\frac{1}{2}\,\frac{1}{2}\right)$	*	$c+\frac{1}{8}\eta_1\cos 2\pi(x+y+z)+\frac{1}{8}\eta_2[\cos\pi(x+y+z)+$ $+\cos\pi(x+y-z)+\cos\pi(x-y+z)+$ $+\cos\pi(-x+y+z)]$.

*See Table 3.

Table 5. Predicted b.c.c. superstructures with one order parameter

Composition;	Star	Superstructure;	Distribution function;
AB	(0 0 1)	B2	$c+\frac{1}{2}\eta\cos 2\pi z$
AB	$\left(\frac{1}{2}\frac{1}{2}0\right)$	*	$c+\frac{1}{2}\eta\cos\pi(x+y)$
AB	$\left(0\,0\,\frac{1}{2}\right)$	*	$c+\frac{1}{2}\eta(\cos\pi z+\sin\pi z)$
AB	$\left(\frac{1}{2}\frac{1}{2}\frac{1}{2}\right)$	B32	$c+\frac{1}{2}\eta[\cos\pi(x+y+z)+\sin\pi(x+y+z)]$
AB	$\left(\frac{1}{2}\frac{1}{4}\frac{1}{4}\right)$	*	$c+\frac{1}{2}\eta\left[\cos\frac{\pi}{2}(2x+y+z)+\sin\frac{\pi}{2}(2x+y+z)\right]$
AB	$\left(0\,\frac{1}{4}\frac{3}{4}\right)$	*	$c+\frac{1}{2}\eta\left[\cos\frac{\pi}{2}(x+3y)+\sin\frac{\pi}{2}(x+3y)\right]$
AB	$\left(0\,\frac{1}{4}\frac{1}{4}\right)$	*	$c+\frac{1}{2}\eta\left[\cos\frac{\pi}{2}(x+y)+\sin\frac{\pi}{2}(x+y)\right]$
AB_2	$\left(0\,0\,\frac{2}{3}\right)$	$C11_b$	$c+\frac{2}{3}\eta\cos\frac{4}{3}\pi z$
AB_2	$\left(0\,\frac{1}{3}\frac{1}{3}\right)$	*	$c+\frac{2}{3}\eta\cos\frac{2}{3}\pi(y+z)$
AB_2	$\left(\frac{2}{3}\frac{1}{3}\frac{1}{3}\right)$	*	$c+\frac{2}{3}\eta\cos\frac{2}{3}\pi(2x+y+z)$
AB_4	$\left(0\,\frac{1}{5}\frac{3}{5}\right)$	*	$c+\frac{2}{3}\eta\left[\cos\frac{2}{5}\pi(x+3y)+\cos\frac{2}{5}\pi(3x-y)\right]$

*See Table 3.

$(\partial V/\partial k_x \neq \partial V/\partial k_y = \partial V/\partial k_z = 0)$. The fivefold increase of the period is connected with the star of lower symmetry $(\partial V/\partial k_x \neq 0,\ \partial V/\partial k_y \neq 0,\ \partial V/\partial k_z = 0)$ [23]. The ordering related to the stars of much lower symmetry has proved to be impossible. In particular, the problem of the formation of long-period superstructures based on the $L1_2$ structure remained unsolved. Perhaps, the gradient terms of the $\partial\eta/\partial x$ and $\partial c/\partial x$ type must be introduced into the theoretical scheme to improve the matter.

Body-centered cubic lattice. One order parameter. The relevant superstructures are listed in Table 5. For the first time, the theory proved to be able to predict the

Table 6. Predicted b.c.c. superstructures with two order parameters

Composition	Star	Superstructure	Distribution function;
AB_2	$\left(0\,0\,\frac{2}{3}\right)\left(0\,0\,\frac{1}{3}\right)$	*	$c+\frac{1}{3}\eta_1\cos\frac{4}{5}\pi z+\frac{1}{\sqrt{3}}\eta_2\sin\frac{2}{3}\pi z$
AB_2	$\left(0\,\frac{1}{3}\,\frac{1}{3}\right)\left(0\,\frac{1}{6}\,\frac{1}{6}\right)$	*	$c+\frac{1}{3}\eta_1\cos\frac{2}{3}\pi(x+y)+\frac{1}{\sqrt{3}}\eta_2\sin\frac{1}{3}\pi(x+y)$
AB_2	$\left(\frac{1}{3}\,\frac{1}{3}\,\frac{2}{3}\right)\left(\frac{1}{6}\,\frac{1}{6}\,\frac{1}{3}\right)$	*	$c+\frac{1}{3}\eta_1\cos\frac{2}{3}\pi(x+y+2z)+\frac{1}{\sqrt{3}}\eta_2\sin\frac{1}{3}\pi(x+y+2z)$
AB_2	$\left(\frac{1}{3}\,\frac{1}{3}\,0\right)\left(\frac{5}{6}\,\frac{1}{6}\,0\right)$	*	$c+\frac{1}{3}\eta_1\cos\frac{2}{3}\pi(x+y)+\frac{1}{\sqrt{3}}\eta_2\sin\frac{1}{3}\pi(x-5y)$
AB_3	$(0\,0\,1)\left(0\,0\,\frac{1}{2}\right)$	*	$c+\frac{1}{4}\eta_1\cos 2\pi z+\frac{1}{2}\eta_2\cos\pi z$
AB_3	$\left(0\,\frac{1}{2}\,\frac{1}{2}\right)\left(0\,\frac{1}{4}\,\frac{1}{4}\right)$	*	$c+\frac{1}{4}\eta_1\cos\pi(x+y)+\frac{1}{2}\eta_2\cos\frac{\pi}{2}(x+y)$
AB_3	$\left(0\,\frac{1}{2}\,\frac{1}{2}\right)\left(0\,\frac{1}{4}\,\frac{1}{4}\right)$	DO_3	$c+\frac{1}{4}\eta_1\cos 2\pi z+\frac{1}{2}\eta_2\cos\pi(x+y+z)$
AB_3	$(0\,0\,1)\left(0\,\frac{1}{2}\,\frac{1}{2}\right)$	*	$c+\frac{1}{4}\eta_1\cos 2\pi z+\frac{1}{4}\eta_2[\cos\pi(x+y)+\cos\pi(x-y)]$
AB_3	$\left(0\,\frac{1}{2}\,\frac{1}{2}\right)\left(\frac{1}{2}\,\frac{1}{4}\,\frac{1}{4}\right)$	*	$c+\frac{1}{4}\eta_1\cos\pi(x-y)+\frac{1}{2}\eta_2\cos\frac{\pi}{2}(x+y+2z)$
AB_4	$\left(0\,0\,\frac{4}{5}\right)\left(0\,0\,\frac{2}{5}\right)$	*	$c+\frac{2}{5}\eta_1\cos\frac{8}{5}\pi z+\frac{2}{5}\eta_2\cos\frac{4}{5}\pi z$
AB_4	$\left(\frac{2}{5}\,\frac{2}{5}\,0\right)\left(\frac{1}{5}\,\frac{1}{5}\,0\right)$	*	$c+\frac{2}{5}\eta_1\cos\frac{4}{5}\pi(x+y)+\frac{2}{5}\eta_2\cos\frac{2}{5}\pi(x+y)$
AB_4	$\left(\frac{2}{5}\,\frac{2}{5}\,\frac{2}{5}\right)\left(\frac{1}{5}\,\frac{1}{5}\,\frac{4}{5}\right)$	*	$c+\frac{2}{5}\eta_1\cos\frac{4}{5}\pi(x+y-z)+\frac{2}{5}\eta_2\cos\frac{2}{5}\pi(x+y+4z)$
AB_7	$(0\,0\,1)\left(0\,\frac{1}{2}\,\frac{1}{2}\right)$	*	$c+\frac{1}{8}\eta_1\cos 2\pi z+\eta_2[\cos\pi(x+y)+\cos\pi(x-y)+$ $+\cos\pi(x+z)+\cos\pi(x-z)+\cos\pi(y+z)+$ $+\cos\pi(y-z)]$

*See Table 3:

existence, among the others, of the C11g superstructure (of the $AlCr_2$ type) as well as a series of superstructures not observed earlier.

Body-centered cubic lattice. Two order parameters. In this case, the theory was capable of predicting several superstructures (Table 6). We will not consider here the superstructures specified by more than two order parameters. In general, all superstructures based on the b.c.c. lattice and observed experimentally have been predicted by the theory. The lattice period may increase by 2 to 6 times. In the case of a b.c.c. lattice, in addition to Lifshits' stars and those of the mean symmetry, there occur more frequently than for the f.c.c. lattice the stars showing no extremum of the ordering energy along two directions in the reciprocal lattice.

We can summarize the theoretical treatment of possible structures of ordered alloys as follows. First, we were able to include into consideration the ordering corresponding to the low-symmetry stars and derive theoretically all existing simple superstructures based on f.c.c. and b.c.c. lattices with one or two order parameters. Secondly, a set of new possible superstructures has been predicted. The total number of superstructures with one or two order parameters formed on the base of a f.c.c. lattice is equal to 27 and of those based on a b.c.c. lattice -- 24.

Atomic ordering in binary metallic alloys lowers the symmetry of a crystal lattice or increases the lattice constant [24]. In the last case, the symmetry of the right points occupied by atoms is also lowered. The effect is illustrated in Tables 1, 2 and 7. The decrease in symmetry due to atomic ordering may remove degeneracy and may serve, respectively, as an energy factor favoring the order-disorder transformation. To a certain approximation depending largely on hybridization, the electronic states in metallic solid solutions can be divided into the core, quasi-localized or collectivized states [25]. The core states are completely filled and, therefore, are not interesting for the treatment of symmetry effects.

In ordering, the symmetry of collectivized states changes in accord with the change in the structure and size of the Brillouin zone. The comparison of the symmetry of the electronic states lying on planes, lines and points of the Brillouin zone of a disordered solid solution (the structures of the Brillouin zones and the data on symmetry can be found in [23, 26, 27]) with those of the relevant superstructures shows that, despite the reduction in symmetry of the space group and in the crystal class upon ordering, situation may be different for some particular points of the Brillouin zone. Let us consider, for example, the phase transition $A1 \rightarrow L1_2$ [28].

Table 7. Symmetry characteristics of disordered solid solutions

Syngony	Space group			Coordinates of right points occupied by atoms	Point symmetry group.
	Struc-tur-bericht	Sch@onfliss	Internatio-nal		
Cubic	A1	O_h^5	Fm3m	$000;\ \frac{1}{2}\frac{1}{2}0;$ $0\frac{1}{2}\frac{1}{2};\ \frac{1}{2}0\frac{1}{2}$	m3m
Cubic	A2	O_h^9	Im3m	$000;\ \frac{1}{2}\frac{1}{2}\frac{1}{2};$	m3m
Hexagonal	A3	D_{6h}^4	$P6_3/mmc$	$\frac{1}{3}\frac{2}{3}\frac{1}{4};\ \frac{2}{3}\frac{1}{3}\frac{3}{4}$	$\bar{6}m2$
Hexagonal	A3′	D_{6h}^4	$P6_3/mmc$	$000;\ 00\frac{1}{2}$	$\bar{3}m$
				$\frac{1}{3}\frac{2}{3}\frac{1}{4};\ \frac{2}{3}\frac{1}{3}\frac{1}{4}$	$\bar{6}m2$

In ordering, the electronic states corresponding to some particular points and directions in the Brillouin zone tend to stick together. The change of topology of the electronic spectrum is shown schematically in Fig. 2. Sticking leads to enhanced effective jump in energy at the Brillouin zone boundary compared to the disordered state. If the Fermi level is close to these states, then the energy of the system will decrease or increase depending on its actual position. Such treatment has been performed for some other superstructures and for transitions between them. It was shown that in some cases the change in symmetry of the collectivized states can promote the formation of superstructures of different type. In each particular situation, the prediction should be based on the calculation of the electron energy spectrum.

The symmetry of atomic positions in a superstructure is equivalent or lower than that in a disordered solid solution. Atomic s, p, d or f states are split in a certain way by the crystal field of a solid solution. The formation of a superstructure entails an

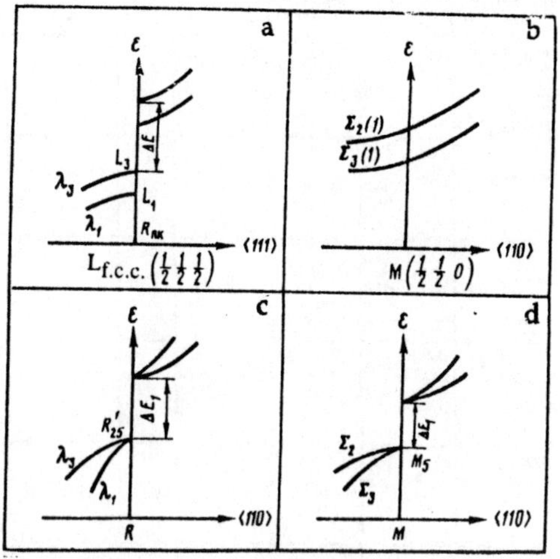

Fig. 2. Variation of the spectrum of electronic states at the A1 → $L1_2$ transition: disordered state (a, c); ordered state (b, d).

additional split of the part of the energy sublevels. In the case of the incompletely occupied shells and for some electron concentrations, the reduction in symmetry may result in the decrease in energy. The quasi-localized, mainly outer d and f shells may be regarded as being transformed in accordance with the point group symmetry of the lattice site occupied by the given ion [29]. The comparison of the data in Tables 2 and 7 shows that the decrease in energy connected with the removal of degeneracy of the quasi-localized states may be an important factor in the appearance of a long-range order and the stabilization of particular types of superstructures.

For all superstructures, the tables showing the split of the p and d terms have been constructed and the electronic configurations found for which a decrease in energy is possible. The data were used then to determine the preferable type of the superstructure, i.e., the type which is being stabilized by the crystal field acting upon the quasi-localized states. The analysis is based on the one-electron strong crystal field approximation [30, 31] taking into account only the Stark effect. The choice of such model is dictated by the level of the approximation in the band theory of solids [29, 32]

B \ A	Mn	Tc	Re	Fe	Ru	Os	Co	Rh	Ir	Ni	Pd	Pt	Cu	Ag	Au	Zn	Cd	Hg	Li	Cs	Be	Mg	Ca	Sr	Al	Ga	In	Tl	Si	Pb	Sb	Bi
Ti		B2	B2	B2	B2	B2	B2	B2 $L1_0$	B2 $L1_0$	B2	B2 B19	B2 B19	B2 $L1_0$	$L1_0$	$L1_0$ B19	B2		B2							$L1_0$	$L1_0$	$L1_0$			$L1_0$		
Zr					B2	B2	B2	B2				B2		$L1_0$		B2	$L1_0$	$L1_0$														
Hf		B2			B2	B2	B2	B2 $L1_0$	$L1_0$																							
V	B2	B2		B2	B2 $L1_0$	B2			$L1_0$			$L1_0$ B19																				
Nb								$L1_0$ B19	$L1_0$			B2 B19																				
Ta		B2			B2 $L1_0$ B19																											
Cr											$L1_0$	$L1_0$																				
Mo								B19	B19			B19					B19															
W																																
Mn								B2 $L1_0$	$L1_0$	B2 $L1_0$	B2 $L1_0$	$L1_0$			B2			B2 $L1_0$							B2 $L1_0$	$L1_0$						
Tc																																
Re																									B2							
Fe							B2	B2		$L1_0$	B2 $L1_0$	B2 $L1_0$													B2	B2						
Ru																									B2	B2			B2			
Os																									B2				B2			
Co												$L1_0$				B2					B2				B2	B2						
Rh																						B2			B2	B2	B2		B2			
Ir																									B2							
Ni												$L1_0$			$L1_0$	B2 $L1_0$					B2				B2	B2	B2					
Pd													B2			B2 $L1'_0$	B2 $L1_0$	$L1_0$			B2	B2 $L1_0$			B2		B2					
Pt													$L1_1$	B2	B2	$L1_0$	$L1_0$	$L1_0$														
Cu															$L1_0$	B2					B2											
Ag																B2	B2 B19		B2													
Au													$L1_0$(M)			B2	B2 B19		B2	B2		B2										
Zn																			B2													
Cd																						B19	B2	B2								
Hg																			B2			B2	B2	B2						$L1_0$	B2	B2
Be																																
Mg																								B2			$L1_0$	B2				
Ca																											B2	B2				
Sr																												B2				
Al																																
Ga																																
In																																$L1_0$
Tl																			B2												B2	B2
Si																																
Ge																																
Sn																																
Pb																			B2													

Fig. 3a. Types of forming superstructures. The phase type is indicated at the intercept of column and line. Compositions: AB (a); AB_2 (b); AB_3 (c); AB_4 (d).

B \ A	Ti	Zr	Hf	V	Nb	Ta	Cr	Mo	W	Re	Pd	Cu	Ag	Au	Zn	Cd	Al
Ti											$C11_b$	$C11_b$			$C11_b$	$C11_b$	
Zr											$C11_b$	$C11_b$	$C11_b$	$C11_b$	$C11_b$		
Hf											$C11_b$	$C11_b$	$C11_b$	$C11_b$	$C11_b$	$C11_b$	
Cr																	$C11_b$
Mn														$C11_b$			
Tc																	$C11_b$
Ni				type. $MoPt_2$			type. $MoPt_2$	type. $MoPt_2$									
Pd		$C11_b$	$C11_b$	type. $MoPt_2$	type. $MoPt_2$	type. $MoPt_2$		type. $MoPt_2$									
Pt				type. $MoPt_2$	type. $MoPt_2$			type. $MoPt_2$									
Au	$C11_b$	$C11_b$	$C11_b$														
Si								$C11_b$	$C11_b$	$C11_b$							
Ge								$C11_b$									

Fig. 3b.

and, in particular, by the assumptions made in the theory of transition metals based on the pseudopotential method [33, 34]. This method provides the basis for the electronic theory of atomic ordering developed in the present work. At present, two large branches of the theory of solids, the band theory and the crystal field theory, use the models strongly different in their structure. They can be matched only in the above mentioned approximation.

Consider, as a simple example, the possibility of realizing the $L1_2$ superstructures in the Cu_3Pt alloy [35]. When the 3d shell is filled, Cu atoms do not participate in the energy "play". We only need to consider the role of the Pt atoms. With the use of pure atomic configurations, nine d electrons of Pt atom do not produce the O_h symmetry providing only the D_{4h} symmetry [36]. On the other hand, such analysis for the AB_3 composition predicts the stabilization of the superstructure whose symmetry is lower than that of the $L1_2$ structure for any n_d values except $n_d = 5$ or 10.

Now we have two options. In the first, the pure states persist but the symmetry of the superstructure reduces in such a way that the Pt atoms occupy the positions

B \ A	Ti	Zr	Hf	V	Nb	Ta	Cr	Mo	W	Mn	Fe	Ru	Os	Co	Rh	Ir	Ni	Pd	Pt	Cu	Ag	Au	Zn	Cd	Hg	Li	Na	Mg	Ca	Sr	Al	Ga	In	Tl	Si	Ge	Sn	Pb	Sb
Ti																					L1₂				L1₂						D0₁₉	D0₁₉		D0₁₉ L1₂			D0₁₉	D0₁₉	D0₁₉
Zr																					L6₀										L1₂			L1₂					
Mn														L1₂	L1₂	L1₂			L1₂													D0₁₉			D0₁₉	D0₁₉	D0₁₉		
Re																																				L1₂	D0₁₉		
Fe																			L1₂												D0₃	L1₂			D0₃	L1₂	D0₁₉		
Co	L1₂ D0₂₄					L1₂		D0₁₉	D0₁₉	L1₂																					L1₂								
Rh	L1₂	L1₂	L1₂	L1₂	L1₂	L1₂		D0₁₉	D0₁₉																														
Ir	L1₂	L1₂	L1₂	L1₂	L1₂	L1₂	L1₂	D0₁₉	D0₁₉																										D0₂₂				
Ni	D0₂₄	D0₂₄	D0₂₄	D0₂₂	D0ₐ	D0ₐ D0₂₂		D0ₐ		L1₂	L1₂			L1₂*				L1₂	L1₂ L1₂M			L1₂*	L1₂								L1₂	L1₂	D0₁₉		L1₂	L1₂	D0₃ D0₁₉		D0₃ D0ₐ
Pd	D0₂₄	D0₂₄	D0₂₄	D0₂₂	D0₂₂ L1₂	D0₂₂	L1₂			L1₂	L1₂			L1₂								D0₂₂						L1₂		L6₀				D0₂₂ D0₂₃			L1₂	L1₂	
Pt	L1₂	D0₂₄	D0₂₄	D0₂₂	D0₂₂	D0₂₂	L1₂			L1₂	L1₂			L1₂		L1₂				L1₂	L1₂	L1₂	L1₂	L1₂				L1₂ L6₀			L1₂ L6₀	L1₂	L6₀		L1₂*	L1₂*	L1₂	L1₂	
Cu	D0ₐ									L1₂M		L1₂	L1₂					L1₂M	L1₂M			L1₂M	L1₂		D0ₐ										D0ₐ	D0ₐ	D0₃ D0ₐ		D0₃
Ag																		L1₂	L1₂				L1₂	L1₂				L1₂M					D0₁₉						D0ₐ
Au	D0ₐ	D0ₐ	D0ₐ							L1₂M							L1₂		L1₂	L1₂M			L1₂M	L1₂M	L1₂	L1₂							D0ₐ						D0ₐ
Zn	L1₂			L1₂	L1₂					L1₂															L1₂														
Cd																										L1₂		D0₁₉					L1₂						
Hg		L1₂																								D0₁₉								L1₂					
Mg																		L1₂	L6₀					D0₁₉		D0₃							L1₂						
Al	D0₂₂	D0₂₂ D0₂₃	D0₂₃	D0₂₂	D0₂₂	D0₂₂						D0₂₄																											
Ga	D0₂₂	D0₂₂	D0₂₂		D0₂₂																																		
In		D0₂₂																										L1₂											
Tl																													L1₂	L1₂						L1₂			
Sn																													L1₂	L1₂									
Pb																											L1₂		L1₂	L1₂ L6₀						L1₂			

Fig. 3c.

B \ A	Ti	Hf	V	Cr	Mo	W	Mn	Pd
Ni					$D1_a$	$D1_a$		
Cu								$P4_2/m$
Au	$D1_a$	$D1_a$	$D1_a$		$D1_a$		$D1_a$	

Fig. 3d.

characterized by the point group symmetry not higher than D_{4h}.. In this case $L1_2(M)$ superstructure may form. Another option suggests the mixing of the electronic configurations by the crystal field [29] thereby providing the O_h symmetry and retaining the $L1_2$ superstructure. The choice between the two possible superstructures can be made only after the calculation of the total internal energy is completed; it will not be considered at this point. Both the $L1_2$ and the $L1_2(M)$ superstructures are realized in the Cu_3Pt alloy.

The crystal field causes the split of the atomic p, d, and f terms; the character of splitting depends on the symmetry of atomic positions. In a disordered solid solution the symmetry is on the average the same for all atoms in the lattice. The ordering is accompanied, as a rule, by the reduction in symmetry of the entire crystal lattice and symmetry of sublattices occupied by atoms of different kind. The crystal fields acting upon different atoms after the formation of superstructures are not the same in many superstructures. These different fields may be energetically beneficial for the ions with different numbers of outer electrons providing the optimum removal of degeneracy. The split of the terms in a crystal field of lower symmetry may diminish the energy of the superstructures belonging to the lowest space group. Such treatment provided the formulation of the rules for the selection of superstructures based on the criterion of crystal field [35,37,38].

For the AB composition, the $L1_0$ and B19 superstructures (whose symmetry is lower than that of the B2 structure) are preferable when one of the components is a p element with $n_p \leq 5$ or d element with $n_d \leq 5$ or $6 < n_d < 10$. The choice between the

$L1_0$ and B19 superstructures can be made, at small numbers of d electrons, in favor of the latter. The B19 superstructure is preferable if n_d=3 - 5. The data concerning the influence of the electronic configuration upon the type of the superstructure formed are given in Fig. 3 [3, 35, 38 - 50]. We used the data for more than 800 phases. A significant role of the crystal field in stabilizing superstructures can be perceived from the comparison of the data in Fig. 3 and the above formulated rule.

For the AB_3 composition, the cubic superstructures DO_{19}, DO_{22}, DO_{23}, DO_{24}, DO_a are more preferable than $L1_2$ when the A component is the p element or d element with $n_d \leq 5$, or the B component is the d element with n_d = 7 - 9. With regard to the AB_2 and AB_4 compositions, the formation of known superstructures is energetically more beneficial in comparison to the disordered solid solution, if the A component has $n_d \leq 5$ and the B component has n_d = 7 - 9, or both of them are the p elements (Fig. 3c, d).

The quantitative evaluation of the contribution of the considered effects in the ordering energy is based on the crystal field theory [30,31]. This contribution can be expressed as

$$\Delta U = \sum_{L=1}^{Q} \sum^{\text{уп}} (n_L^{\downarrow} \Delta E_L^{\downarrow} - n_L^{\uparrow} \Delta E_L^{\uparrow}) - \sum_{L=1}^{Q} \sum^{\text{тв.р}} (n_L^{\downarrow} \Delta E_L^{\downarrow} - n_L^{\uparrow} \Delta E_L^{\uparrow}), \quad (1.1.7)$$

where the summation extends over the upper unoccupied d or f shell up to n_d or n_f ($n = n^{\downarrow} + n^{\uparrow}$), respectively, and over the different types of the sites in the ordered and disordered lattices, $n^{\downarrow}$ and $n^{\uparrow}$ being the numbers of electrons in the energy levels with the split parameters $\Delta E < 0$ and $\Delta E > 0$, respectively. The first and second sums in (1.1.7) describe the superstructure and the solid solution, respectively.

For the case of the Cu_3Pt alloy considered above, the $L1_2$ superstructure does not provide a gain in energy since the Cu atoms have completely filled d shell and the point symmetry for the Pt atoms is the position D_{4h} in comparison to a solid solution (for one lattice site the O_h symmetry changes for D_{4h}):

$$\Delta E = \Delta_s + 10\Delta_t, \quad (1.1.8)$$

where, according to [30]

$$\Delta_s = 2/7\,\mathrm{eq}[F_2(R_2) - F_2(R_1)]\,, \tag{1.1.9}$$

$$\Delta_t = 2/21\,\mathrm{eq}\,[F_4(R_4) - F_4(R_1)]\,, \tag{1.1.10}$$

$$F_n(R_i) = \int_0^\infty R_{4d}^2(r)\,\frac{1}{|r - R_i|^2}\,r^2\,dr. \tag{1.1.11}$$

From the diagram showing the split of the atomic d^9 level in the fields of the O_h and C_{4v} symmetry, it can be easily seen that the contribution into the stabilization of the $L1_2$(M) superstructure is due to the split $T_{2g} \rightarrow E_g + B_{2g}$. It is this contribution that is accounted for by formula (1.1.8).

The contribution to ordering energy considered above is of a cooperative or, at least, cluster nature. The minimum size of the cluster be not less than the radius of the first coordination shell. For the Pt atoms in the Cu_3Pt alloy with the $L1_2$(M) superstructure, the minimum cluster embraces the two nearest shells. Thus, although the contribution ΔU into the ordering energy can be formally separated into individual pair bonds for each coordination shell, this contribution is virtually of a multi-particle or even volume character.

1.2 Crystal chemistry and conditions for the existence of ordering phases

An important role of such factors as the atomic volume, the electronic concentration and their derivatives in the stabilization of a particular class of crystal structures of metallic phases is well known [2, 51]. These factors are referred to as the Hume-Rothery criterion, the Darken-Gurry diagrams, Vegard's rule and others. Analysis of their role, in many cases, allows one to single out the physical reasons for the predominant formation of a particular phase or structure and represents necessary and simultaneously underlying foundation for constructing the quantum theory of their stability.

Such analysis has not been comprehensively applied to ordered phases as to an individual class of metallic compounds. Just one fact has been set up firmly: the critical role of the electron concentration in stabilizing the long-period superstructures [3, 52]. It was shown that the stabilization of these superstructures is determined by the contact of the Fermi sphere with the planes of the Brillouin zone of the superstructure [52 - 5]. Although the significance of this fact for the stabilization of crystal superstructures raised doubts [56, 57], one may agree with the arguments of the opponents of this concept rather than with their conclusions. In fact, this concept was confirmed on a sound basis in [32, 58].

In this section the crystallophysical and crystallochemical stability criteria are applied to the ordered alloys as to a special class of metallic compounds [35, 37, 59, 60]. In the last time, besides our works, these conceptions have been partly developed by Pearson [61, 62].

The size factor. Consider, first, the superstructures of the AB composition. The dependence of the number of phases forming B2, B19 and $L1_0$ superstructures upon the size factor

$$\delta = \left| \frac{R_A - R_B}{R_B} \right| ,$$

is represented in Fig. 4 by the histograms with the step 0.05; R_A and R_B are the atomic radii of pure components [61].

It is evident that we have a series of superstructures B19 - $L1_0$ - B2 whose stability increases with the size factor (δ=0.15; 0.25; 0.40, respectively). The close-packed superstructures do not form at large size factors but more than 50 percent of phases with the B2 superstructure form at δ > 0.15. Notice, this phase was found to be absolutely stable with regard to the structure transformation if the δ value exceeds 0.25. At the same time, a number of phases with the $L1_0$ and B19 are unstable with respect to the structure transformation. It is worth noting that the increase in a tetragonal distortion in the $L1_0$ superstructure is distinctly related to the difference in the atomic radii. The greater this difference, the closer the $L1_0$ superstructure approaches that of the B2 type. In addition, the tendency takes place to form a non-close-packed structure with increasing discrepancy in atomic volumes. If $\delta > 0.2$, the ordered phase of the AB composition based on an f.c.c. lattice does not form. This finding demonstrates clearly the role of size effects in the stabilization of various superstructures.

Another factor, the electronegativity difference, for the $L1_0$ superstructure lies in the range from 0.3 to 0.35, which is less than that for the B2 superstructure. The $L1_0$ structure is largely formed by melting together copper or nickel group metals with transition and polyvalent metals. The value $\delta = 0.15$ is found to be critical. At $\delta > 0.15$, the B19 and $L1_0$ superstructures practically do not form, whereas for the B2 superstructure, the range $\delta = 0.15$ - 0.20 separates two stability peaks. The origin of these maxima on the B2 histogram is connected with the difference of the elements forming the alloy.

The first maximum corresponds to the phases whose components are, on the one hand, polyvalent metals and, on the other, noble and transition metals. This maximum is characterized by a broad uniform spectrum of electronic concentrations (1–8 e/a). The first maximum is also formed by elements not so different in electronegativity (0.35–0.5). The drop at $\delta = 0.15$ is due to the phases obtained by alloying transition elements. The existence of the second maximum owes rare earth elements forming the phases with noble and transition metals. In this case, the dominant electronic concentrations are 2 e/a and 6 e/a. The electronegativity difference in the second maximum is equal to 0.7 - 0.8. Evidently, the compounds corresponding to the second peak are more polar than those in the first peak. In other words, the electrostatic energy in the B2 superstructures increases with the difference in the atomic radii.

For the AB_2 composition (Fig. 5), the size factor ranging from 0.10 to 0.15 is again critical. The close-packed lattice of the $MoPt_2$ type does not form at $\delta > 0.12$; the second peak of stabilization can be observed at $\delta > 0.15$ for the non-close-packed phase $MoSi_2$.

For the superstructures of the AB_3 composition based on a f.c.c. lattice, the value of the size factor (δ = 0.15) is also critical (Fig. 6). The phases with DO_{22}, DO_{23}, $L1_2$(M), or $L1_2$(MM) superstructures, long-periodic with respect to the $L1_2$ structure, are formed only at $\delta > 0.15$. Most (about 2/3) of the phases having the $L1_2$ superstructure are formed at $\delta \leq 0.15$. Those formed at $\delta > 0.15$ have a specific electronic structure. A detailed analysis of the phases belonging to the L12 superstructure showed that the electronegativity difference ranges from 0.3 to 0.45 at $\delta < 0.15$ and extends from 0.45 to 1.0 at $\delta > 0.15$. In other words, compounds with a great size factor have an obvious tendency to form the polar bonds (which correspond to a greater contribution to the electrostatic energy). The phases with a greater δ value have, as a rule, the electronic concentration in the range 7.5 - 8.5. A broader spectrum of electronic concentrations occurs at small differences in the atomic sizes. Hexagonal superstructures are found to be more stable at greater values of the size factor. They are almost uniformly distributed over the range δ=0 - 0.35.

For the $D1_a$ superstructures of the AB_4 composition, the value δ = 0.15 is also critical (Fig. 5). Only the phases formed by gold with rare earth elements exist at $\delta >$ 0.15.

We must say, in conclusion, that the size factor, as defined by Hume-Rothery, substantially affects the stabilization of superstructures. The structures with lower atomic density are more stable at greater δ values. The critical value δ=0.15 bounds the region where more closely packed phases exist and separates the regions of lower closely packed phases of different electronic nature. The above consideration gives evidence of the important role the volume-dependent part of energy plays in the stabilization of superstructures. Clearly, a considerable size discrepancy may be compensated for by a more complicated band structure.

Vegard's rule. An ordered solid solution or an ordered phase cannot be assigned to ideal solid solutions. As a consequence, the deviation of their lattice parameters from Vegard's rule can be used to analyze the role of volume effects in the formation of superstructures. Vegard's rule for the atomic volumes may be written as

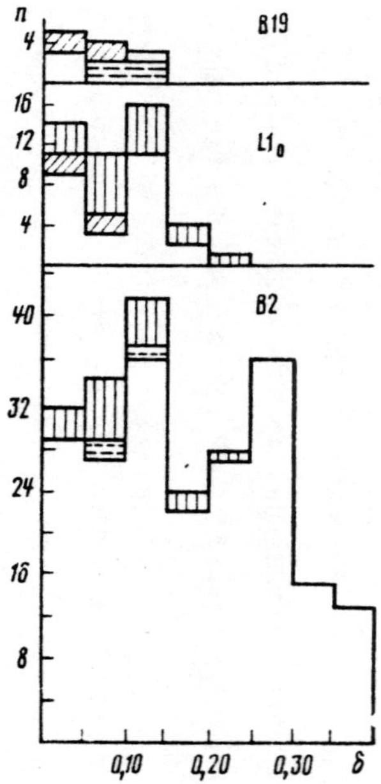

Fig. 4.

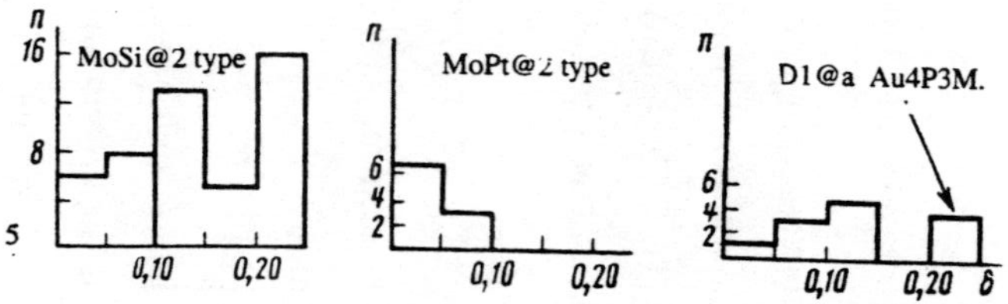

Fig. 5

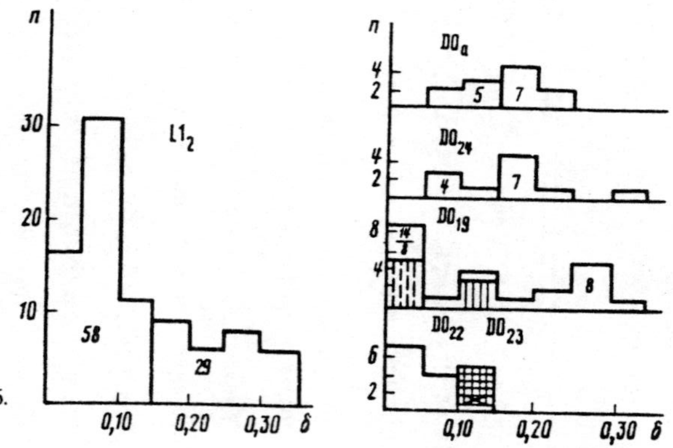

Fig. 6.

Fig. 4. Influence of size factor on the number of known superstructures. AB composition. Hatched are phases undergoing martensitic transformations.

Fig. 5. Influence of size factor on the number of superstructures. Compositions AB_2 and AB_4.

Fig. 6. Influence of size factor on the number of superstructures. AB_3 composition. Hatched are phases prone to martensitic transformations.

$$\Omega = c_A \Omega_A + c_B \Omega_B, \tag{1.2.1}$$

where Ω_A and Ω_B are the atomic volumes of pure components [63]. This formula can be rewritten as

$$\frac{\Omega}{\Omega_B} = (1 - c_A) + c_A \frac{\Omega_A}{\Omega_B}. \tag{1.2.2}$$

to represent it in the graphical form. The behavior of atomic volumes for various superstructures is shown in Figures 7 and 8. It can be seen that Vegard's rule holds for a small number of superstructures (Ir_3Cr, Pt_3Co, Cu_3Pd, Cu_3Au, Cu_3Ti, CoPt, AuCu). Evidently, there is no size contribution to the energy of stabilization of superstructure for these phases.

There are some superstructures, usually of the AB_3 composition, with the positive deviation from Vegard's rule. These are the superstructures based on the elements of the nickel, copper and cobalt subgroups. The volume-dependent contribution to the energy for these structures impedes the superstructure formation.

For most ordered phases the negative (sometimes very large) deviation is observed. Such behavior is typical of superstructures and is classified here as **superstructure contraction.** It should be noted that contraction is more pronounced for the AB_3 superstructures and increases with difference in the atomic volumes of the components making up the phase. For the AB compositions, superstructure contraction decreases as the d shell is filled over a long period, say from Ti and its analogs through Cu and its analogs. Simultaneously decreases (and disappears for the d^{10} configuration) the dependence of the value of superstructure contraction upon the volume difference of the components. The last effect is characteristic of the AB composition but not of AB_2 or AB_3.

The phenomenon of superstructure contraction represents one of the most important effects in the formation of long-range order. For this reason, we shall consider in more detail the influence of different crystallophysical effects upon superstructure contraction. For the elements of isoelectronic groups forming the ordered

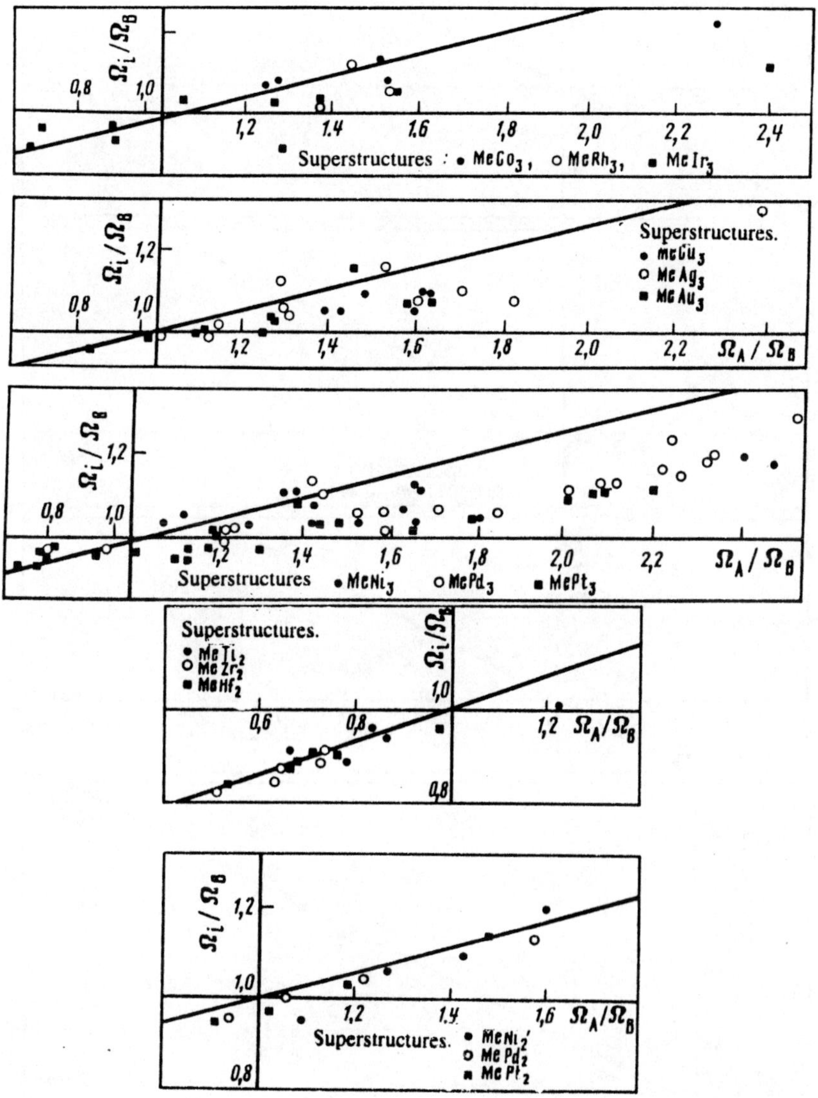

Fig. 7. Predominantly negative deviation from Vegard's rule. Compositions AB_2 and AB_3.

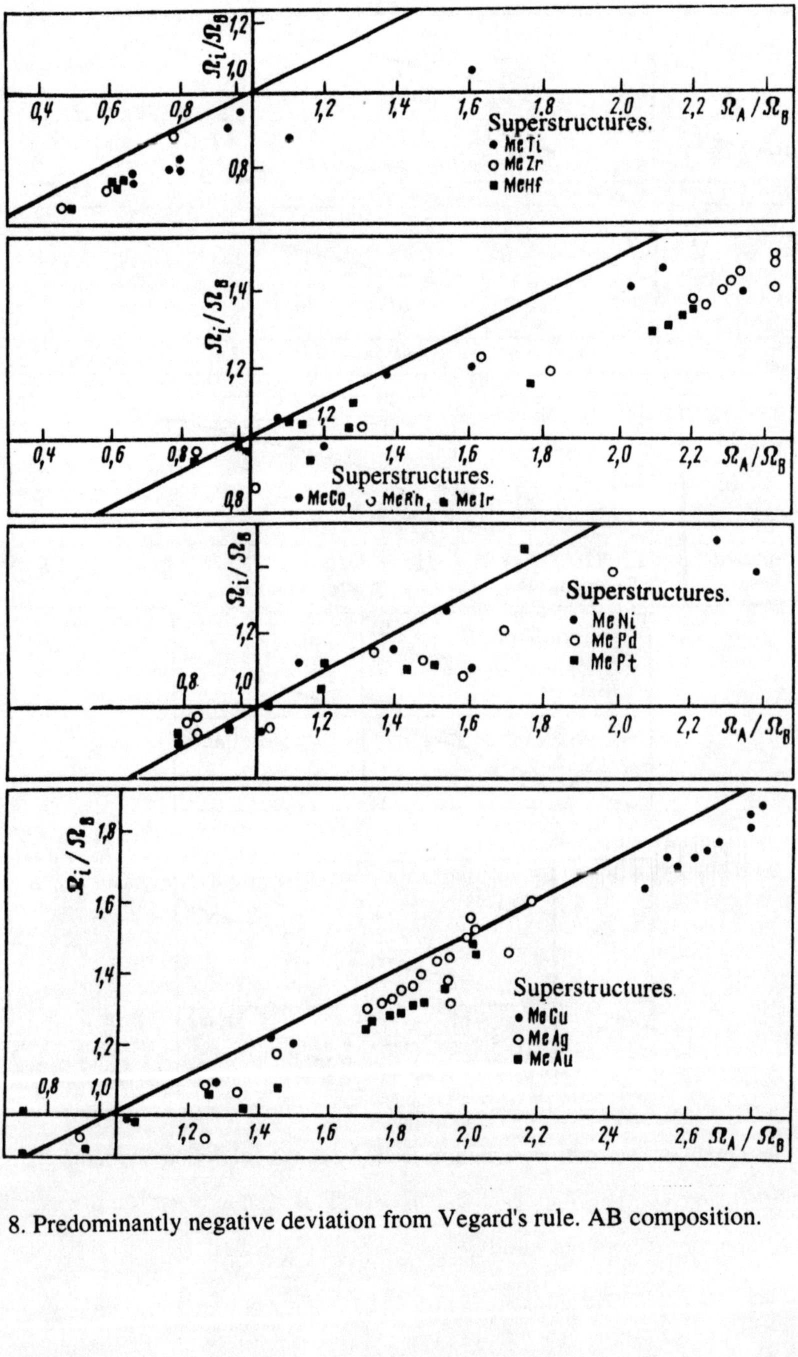

Fig. 8. Predominantly negative deviation from Vegard's rule. AB composition.

phases of the AB and AB_3 stoichiometry, there has been established the combined effect of the R_A/R_B ratio, the electronic concentration (spd/atom), and the electronegativity difference (Δx) on the parameter of superstructure contraction, $(\Omega_i/\Omega_\beta)_{exp}-(\Omega_i/\Omega_\beta)_T$, where i denotes the calculation based on Vegard's rule. Although all the factors affect superstructure contraction simultaneously, the above tendencies can be traced for each element forming the ordered phase.

The strongest is the effect of the size or atomic factor. When the atomic radii or volumes of the components forming the ordered phase are identical, the effect of superstructure contraction is usually small or completely absent. As the size effect increases, the value of superstructure contraction increases. For a small number of phases, the crystal lattice dilatates upon their formation, namely when the atomic radius or volume of the base element exceeds the radius or volume of the alloying element. Deviations from this behavior display the alloys based on the noble metals, where, along with the great size and volume effects, the ascending branch is observed, especially typical of Ag-based alloys.

Simultaneously, the new effect, not marked elsewhere earlier, has been discovered: the ordered phases of nonequiatomic stoichiometry are formed, as a rule, on the base of elements with smaller atomic radii and volumes, and this is just typical of superstructure contraction. This nonconfigurational effect is entirely caused by the change in the electronic states when the alloying elements enter the lattice of the base component. This effect of a small radius of the element forming the ordered phase can be explained by the fact that the ordering phases are mainly formed between elements of the Fe triad and noble metals. In other words, the ordered phases are dominantly formed on the base of the elements with completely or almost completely occupied d shell. This behavior is due to the comparatively high stabilization of the lattice involving elements at the end of long periods of the Periodic table with respect to alloying. Conversely, the elements of Ti, V, and Cr subgroups form, under such conditions (at the AB_3 composition) the phases of the β tungsten (A15) type changing their coordination number.

The growth of the electronic concentration decreases the effect of superstructure contraction. But this effect is virtually absent for the phases based on noble metals. A clear-cut dependence on the electronegativity difference is also not marked. It should be mentioned only the occurrence of the critical values of $|\Delta x|$ = 0.7 - 0.9 dividing the ordered phases into two groups.

The occurrence of superstructure contraction and its magnitude give evidence of the important role of the volume effects in stabilizing the ordered state. The volume-dependent contributions to the total energy may be small or large, hindering or favoring contraction. At the same time, the analysis of the data on superstructure contraction reveals a strong interaction of such factors as the electronic concentration, electronegativity, the volume and size factors.

The Darken-Gurry diagrams (the combined treatment of electronegativity and size factor). It is evident from the foregoing discussion that the Hume-Rothery and related crystallochemical factors should be taken in consideration jointly. This conclusion is supported by the quantum theory of atomic ordering (Chapter 5). The method of such combined approach has been proposed by Darken and Gurry [64].

Figure 9 gives, as an example, the diagrams for the Ni (AB_3 composition) and Ir (AB composition). Only Ni-based phases of the most high-symmetry $L1_2$ superstructure and one phase of the DO_{22} superstructure fall into the "favorability" ellipse. All the remaining low-symmetry phases lie outside this ellipse. Almost all phases in the ellipse (except for Ni_3Si and Ni_3Ge) undergo the order-disorder transition in the solid state. All phases outside the ellipse melt in the ordered state. In the case of Ir, the other correlation is observed: all close-packed and nearby phases are located inside the "favorability" ellipse whereas B2 phases lie outside.

Similar analysis has been accomplished for the superstructures of different stoichiometry based on several other elements [65]. The conclusion was drawn that the increase in the volume discrepancy and electronegativity difference leads to the rise in the critical ordering temperature and to the reduction in symmetry and the packing of a superstructure lattice. The critical is the Darken-Gurry ellipse in which the ratio of the size factor (the first half-axis) to the main component radius (the second half-axis) is equal to 0.85 - 1.15 and electronegativity x = 0.4 (sometimes ± 0.5 or even 0.55). Thus, electronegativity should also be regarded as an important parameter, which, like the size factor, affects the stabilization of superstructures. Electronegativity, as a parameter, determines the tendency for atoms to interact with an electron. There is no direct analog of this factor in the quantum theory of solid solutions. Its part in this theory takes on, in a certain sense, the effective valency. The ordering energy rises with the increase in the difference of both electronegativity and effective valency (Section 5.2).

Another inference is that a great value of the ordering energy caused by a great difference in electronegativity entails the low-symmetry distortion of the superstructure or even of the basic crystal structure toward a lower density.

The electronic concentration. The next important factor determining the number of phases with particular superstructures is the electronic concentration. This parameter is associated with the number of outer s, p, d, f electrons outside the upper, completely filled inert gas shell per one atom in a given superstructure. Numerous examples are known of the existence of the regions of stability and instability of different metallic phases depending on the electronic concentration parameter [2 ,3 ,51 ,63]. Various properties of phases such as electron heat capacity, magnetic characteristics or temperature of the transition to the superconducting state may depend strongly upon the electronic concentration.

The physical nature of the electronic concentration factor is not fully understood [57, 56]. Undoubtedly, this factor is closely related to the bond structure of the phase under consideration, namely, to the number of the electronic states in the Brillouin zone and the volume enclosed in the Fermi sphere (or, to be more accurate, the volume under the Fermi surface touching the Brillouin zone). Since all outer s, p, d electrons are more or less collectivized and occupy the states in the Brillouin zone in accordance with the Pauli principle [67], some specific values of the electronic concentration factor will correspond to the complete filling of the Brillouin zone or to the contact of the Fermi surface with the Brillouin zone boundaries.

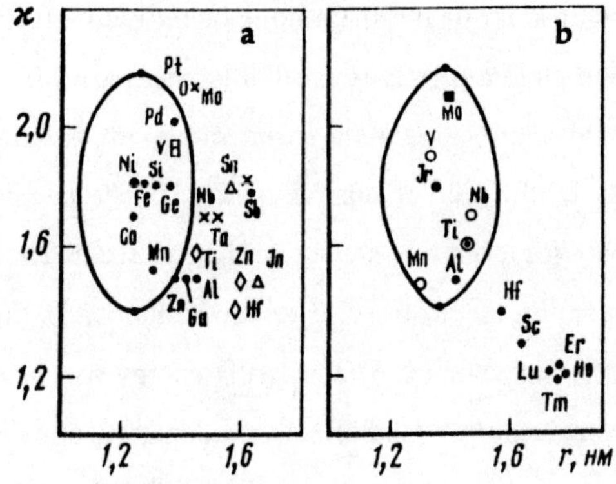

Fig. 9. Darken-Gurry diagrams. Compositions: Ni-base alloys, AB_3 (a); Ir-base alloys, AB (b).

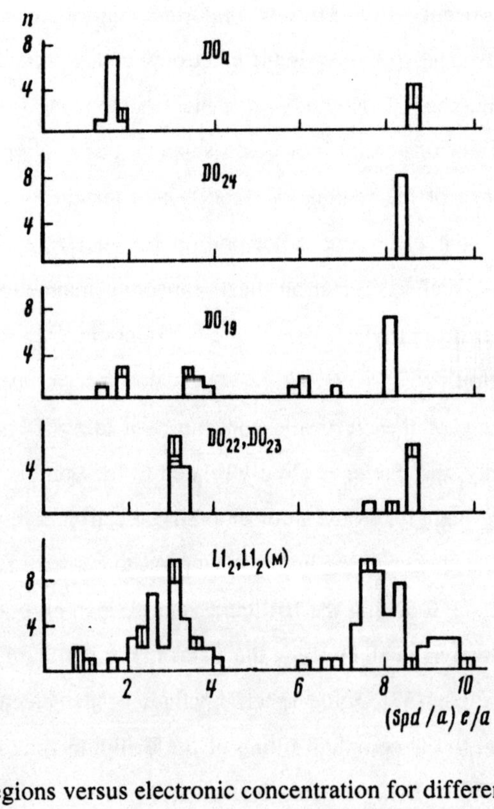

Fig. 10. Stability regions versus electronic concentration for different superstructures.

Also of importance may be the electronic concentration corresponding to one of the d resonances or to the singularity in the multiple scattering (associated with higher orders of the pseudopotential perturbation theory). All these factors are in turn connected with the specific contributions to the band-structure energy which promote the phase stabilization. In general, it would be not so simple to find the relation between a particular value of the electronic concentration and the fine features of the electronic structure: the preliminary calculation of the band energy spectrum is needed for each compound separately.

There exists another cause for the appearance of stabilization maxima that is due to the purely atomic properties of the components forming the superstructure. It can be easily understood if one expresses the stability criterion determined by the split of the terms in a crystal field in terms of the electronic concentration. As we have already emphasized, the choice between different explanations of the nature of the stability dependence on the electronic concentration could be made after a detailed calculation of the band structure.

The dependence of the number of phases forming different superstructures on the mean number of electrons per atom is shown in Fig.10 and 11. We can see that a small electronic concentration (less than 2 e/a) stabilizes the distorted (others than L12) superstructures, and the increase in the electronic concentration (up to 4 e/a) gives rise to the stabilization of the close-packed $L1_2$, DO_{22}, DO_{23} superstructures based on an f.c.c. lattice. And finally there exists a region of stability of ordered phases in the range from 7 e/a to 9 e/a where the phases with the $L1_2$ superstructure are stable, whereas those with the other superstructures (DO_{19}, Do_{22}, DO_{24}, DO_a) have narrow Δdelta-shaped distribution peaks. At e/a > 9 only the phases with the $L1_2$ superstructure may occur. Further, with the reduction in symmetry, the range of the electronic concentrations where the ordered phase exists will narrow. This fact correlates with the dependence of the structure types on the split of the p and d terms. As the symmetry reduces, the ordered phases are likely to approach more and more the electronic compounds.

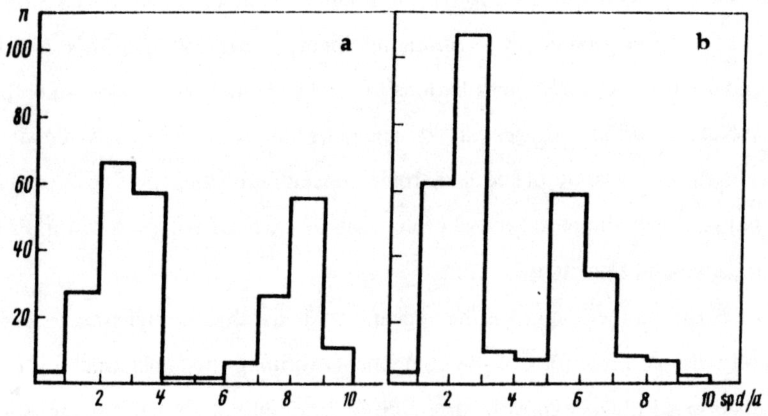

Fig. 11. Distribution of ordered phases with electronic concentration. Compositions: AB_3 (a); AB (b).

As we have noticed above, the ordered phases are intermediate between solid solutions and electronic compounds; the transition is accompanied by the reduction in symmetry. If one summarizes the data concerning the formation of the superstructures of the AB or AB_3 stoichiometry, irrespectively to the phase type, then the superstructure stability regions will be characterized by the presence of the two maxima (Fig.11). For the AB phases the highest stability corresponds to the ranges 1 - 3 e/a and 5 - 7 e/a; for the AB_3 phases -- to the ranges 1 - 4 e/a and 7 - 9 e/a. It is of interest that the superstructures do not virtually occur if the shell is half-filled. In this case, the electronic concentration factor interacts with the collective Jan-Teller effect.

Overlap integrals. This is another factor whose value must correlate with the stability of the ordered state. Qualitative substantiation may be found in the relation between the values of the overlap integrals $S_{dd'}$ for d elements of long periods and the melting temperatures. Qualitative explanation follows directly from the quantum theory of transition metal alloys [59] (see Chapter 5). The overlap integrals have been estimated for the superstructures of the AB_3 composition based on Ni, Cu and their analogs and compared to the temperature of the order-disorder transformation T_C. If the phase in question melts in the disordered state, the value of T_C is chosen to be equal to the melting temperature. The overlap integrals were calculated by the method developed in [68]. The observed correlation between T_C and the magnitude of the overlap integral is especially pronounced for the compounds based on the same element. The stability of the ordered state grows with the increase of the overlap integrals in accordance with the theoretical predictions (Chapter 5).

In forming the ordered phases, there may be another changes in the outer electron states, besides the removal of degeneracy, which also may be responsible for the stabilization of a superstructure. They may be linked with (1) the hybridization effects (involving mainly the s and d states), (2) the interatomic electron transitions, and (3) the Fermi energy variation. All these effects are closely interrelated so that the influence of one of them may cause the emergence of the other. The experimental data presented in this section shows that such effects take place in the ordering alloys and intermetallic compounds.

Information on the direction of the **Fermi energy change** can be obtained from the observed dependence of resistivity and thermo-electromotive force (t.e.m.f.) [32, 69]

$$Q = \frac{\pi^2}{3}\frac{k^2 T}{e}\left[\frac{\partial \ln \sigma(E)}{\partial E}\right]_{E = E_F}, \tag{1.2.3}$$

where k is the Boltzmann constant, T is temperature, $\sigma(E)$ is conductivity, E_F is the Fermi energy. Taking into account the dependence of t.e.m.f. and specific resistivity $\rho = 1/\sigma$ on the degree of long-range order η, formula (1.2.3) can be transformed to

$$\frac{\partial E_F}{\partial \eta} = \frac{\pi^2 k^2 T}{3|e|Q\rho}\frac{\partial \rho}{\partial \eta}. \tag{1.2.4}$$

Let us consider the Cu_3Au alloy as an example. The shape of the Fermi surface in this alloy changes significantly on ordering [70]. The corresponding changes in the sign and the magnitude of the Hall parameter are due to the change in the relative fraction of the electron and hole states. The one-electron density of states near the Fermi level increases, according to the photoeffect and the electron heat capacity data. Also changes, according to the t.e.m.f. data [71], the slope of the density of states curve near the Fermi level. Using formula (1.2.4) and the experimental dependences for conductivity and t.e.m.f. measured on the Cu_3Au alloy [71], it can be shown that E_F in this alloy increases with increasing η up to the value $\eta = 0.8$ and then drops in the range of η between 0.8 and 1.0 [72]. Interestingly, it is just the range where the ordered phase Cu_3Au is stable. One can show in a similar way that the Fermi energy changes on ordering in some other solid solutions. In particular, it increases with η in Ni_3Fe and Ni_3Mn. If the long-range order parameter in Cu_3Au alloy remains constant, the Fermi energy increases with the size of the antiphase domains.

Information on the presence in an alloy of **the hybridization effects, the overlap and the interatomic redistribution of the electron density** can be deduced in most straightforward manner from the analysis of the behavior of the atomic X-ray scattering factor for the alloy components [73]. In general, this factor for the atom of μ kind has the form

$$f_q^{(\mu)} = \int n^{(\mu)}(\mathbf{r})e^{i\mathbf{qr}}d\mathbf{r}, \tag{1.2.5}$$

where $\eta^{(\mu)}(\bar{r})$ is the electron density for the atom of μ kind. The atomic factor can be represented as the sum of the contributions from the inner occupied shells $\eta_\alpha^{(\mu)}(\bar{r})$, the outer quasi-localized states $\eta_{d,f}^{(\mu)}(\bar{r})$, the weakly bounded electrons $\eta_s^{(\mu)}(\bar{r})$ [74], and the additional term appearing due to the hybridization $\eta_{s,d}^{(\mu)}(\bar{r})$, overlap and interatomic transitions $\eta_{d,d'}^{(\mu)}(\bar{r})$:

$$f_q^{(\mu)} = f_\alpha^{(\mu)}(\mathbf{q}) + f_{\alpha,f}^{(\mu)}(\mathbf{q}) + f_{sd}^{(\mu)}(\mathbf{q}) + f_{dd'}^{(\mu)}(\mathbf{q}). \tag{1.2.6}$$

If the electron density distribution for the inner (core) states is spherically symmetric, we have

$$f_\alpha^{(\mu)}(\mathbf{q}) = 2 \sum_{n^{(\mu)}=0}^{N^{(\mu)}} \sum_{l^{(\mu)}=0}^{n^{(\mu)}-1} (2l^{(\mu)}+1)\int_0^\infty |R_{n,l}^{(\mu)}(\mathbf{r})|^2 r^2 j_0(\mathbf{qr})\,d\mathbf{r}, \tag{1.2.7}$$

where the core state of the atom of kind μ is characterized by the set of the quantum numbers $\eta^{(\mu)}$ and $l^{(\mu)}$; $R_{nl}^{\mu}(r)$ is the radial part of the wave function; $j_l(\bar{q}\bar{r})$ is the

Bessel function of the order k. The expression (1.2.7) is written assuming that the wave functions of the core states are not being altered in passing from a free atom to an alloy.

The one-electron functions of the outer quasi-localized states (l=2 or 3)

$$\Psi^{(\mu)}_{nl\gamma j}(\mathbf{r}) = R^{(\mu)}_{nl}(\mathbf{r}) \,|\, l_{\mu}\gamma_j \rangle \qquad (1.2.8)$$

are proportional to the crystal harmonics

$$|\, l_{\mu}\, \gamma_j \rangle = \sum_{m=-l}^{l} (l_{\mu m} \,|\, \gamma j) \varphi^{(\mu)}_{lm}(\upsilon, \varphi), \qquad (1.2.9)$$

representing the linear combinations of the spherical harmonics $\Phi_{\mathrm{lm}}(\vartheta, \varphi)$ transforming according to the line j of the irreducible representation γ for the point group of the site occupied by the ion of the μ kind. Then, according to [75],

$$f^{(\mu)}_{d,f}(\mathrm{q}) = 4\pi \sum_{l'=0}^{\infty} \sum_{m'=-l'}^{l'} i^{l'} \varphi^{*}_{l'm'}(\upsilon, \varphi)(2l_{\mu}+1)\sqrt{\frac{2l'+1}{4\pi}} \times$$

$$\times \sum_{m_1, m_2=-l_{\mu}}^{l_{\mu}} \begin{pmatrix} l_{\mu} l_{\mu} l' \\ 0\,0\,0 \end{pmatrix} \begin{pmatrix} l_{\mu} l_{\mu} l' \\ -m_1 m_2 m' \end{pmatrix} (\gamma j \,|\, l_{\mu} m_1)(l_{\mu} m_2 \,|\, \gamma j) \times$$

$$\times \int_0^{\infty} R^{(\mu)\,2}_{nl}(\mathbf{r})\, r^2 j_{l'}(\mathrm{qr})\, d\mathbf{r}, \qquad (1.2.10)$$

where $\begin{pmatrix} l_1 l_2 l_3 \\ m_1 m_2 m_3 \end{pmatrix} 3j$ – are Wigner's 3j symbols. By making use of the properties of the 3j symbols, one can show that (1.2.10) includes only a few terms (with l'=0,2,4 for the d states and l'=0,2,4 for the f states).

The contribution of the s states can be considered in several ways. If the wave function of the s state is centered at a given ion (the traditionally adopted approximation), its contribution to the atomic factor is similar to that of d states as expressed by (1.2.10). As the s states are strongly collectivized, it would be more correct to consider them in the reciprocal $\overline{\mathrm{k}}$ space. In this case, the corresponding contribution will be

$$f^{(\mu)}_{s}(\mathrm{q}) = w_{\mu}(\mathrm{q})\, \mathrm{X}(\mathrm{q}), \qquad (1.2.11)$$

where $\mathrm{X}(\overline{\mathrm{q}})$ is polarizability of electron liquid, $\mathrm{W}_{\nu}(\overline{\mathrm{q}})$ is the form factor of the ion of μ kind. In a more complicated band structure of transition metal alloys, the s and d states are hybridized. In this case, there appears an additional term $f_{s,d}^{(\mu)}(\overline{q})$ in the atomic scattering factor which depends on the hybridization parameter $\Delta_{kd}^{(\mu)}$. And finally, when the d orbitals are strongly extended, the d shells of the neighboring ions overlap. Therefore, strictly speaking, X-ray scattering pattern cannot be divided into one-centered contributions. If one uses, nevertheless, this approximation, another contribution, $f_{dd'}^{(\mu\nu)}(\overline{q})$, appears that is directly determined by the overlap integrals $S_{dd'}^{(\mu\nu)}$ and the effective valency Z_{μ}^{*}. The corresponding contributions $f_{s,d}^{(\mu)}(\overline{q})$ and $f_{dd'}^{(\mu\nu)}(\overline{q})$ may be written in an explicit form using the formulas given in Chapter 5. Here we shall restrict ourselves to the discussion of the results obtained.

First, the atomic factor will be anisotropic if the d shells are completely filled. Second, the transition to the crystalline state may substantially change the atomic scattering factor. As a result, the use of the atomic factors calculated for the isolated atoms would be a rather crude approximation. Third, $f^{(\mu)}(\overline{q})$ is closely related to the band structure of the alloy, and the X-ray diffraction intensity depends on the hybridization of the s and d states and specific features of the dielectric screening. For the special form of the Fermi surface (if it contains, for example, the plane sections), the scattering factor may be a nonmonotonous function of the $\overline{q}$ vector. Fourth, the interatomic redistribution of the electron density affects the value of $f^{(\mu)}(\overline{q})$. It should be noted that the hybridization and interatomic density redistribution may be connected with the necessity to ensure the crystal symmetry.

Such detailed consideration in the present work of the behavior of the scattering factor is not accidental. The point is that the ordering alloys represent a convenient object for experimental finding of some of the effects discussed above. The intensity of the superstructure reflections in these alloys depends on the difference $[f_A(q) - f_B(q)]$ and, therefore, is more sensitive to the change of $f_A(q)$ and $f_B(q)$ than the intensity of the fundamental X-ray lines. Besides the superstructure reflections, the diffuse scattering may be used because it is particularly sensitive to the singularities in the electron subsystem [76].

Thorough experimental measurements of the intensities of fundamental and superstructure reflections on a series of the ordered Ni-based and Cu-based phases with $L1_2$ superstructure provided the evidence of the availability of the hybridization and electron redistribution effects [35]. It is worth noting, however, that the sensitivity of such experiments is not much higher than the amplitude of the measured effects so that the results of these measurements must be considered as tentative [364].

Cu_3Au. The completely filled d shells provide the required symmetry at the sites of both types, D_{4h} for Cu and O_h for Au. Only a slight increase has been noticed in the concentration of the s electrons near the Au ions. Presumably, this may be explained by the relatively great value of the orthogonalization holes centered at Cu and Au atoms [56].

Cu_3Pt. Nine d electrons of Pt can ensure only the D_h symmetry but not O_{4h}. In such case, one would expect stability only for a long-period structure of the $L1_2(M)$ type in which the symmetry of the Pt atom positions is not higher than D_{4h}. However, at low temperatures the $L1_2$ superstructure is found to be stable. In this case, the concentration of the s states decreases near Pt ions. Here, the mixing of the orthogonalization hole and s–d hybridization is likely to occur.

Ni_3Al. An elevated concentration of the d electrons at the Ni ion ($\Delta n \sim 0.4$ e) is observed. Probably, the d states of Ni are hybridized with the p states of Al.

Ni_3Ge. There takes place an increase in the concentration of the p states of Ge ($\Delta n \sim 1.2$ e) and, probably, the hybridization of the s, p, and d bands of Ni and Ge.

Ni_3Ga. The effects are negligible. The s, p, and d states are slightly hybridized.

Ni_3Mn. A considerable s–d hybridization takes place. The alloy contains ~ 0.7 electrons per atom.

The observed collectivization of the electrons in alloys significantly enhances (compared to pure metals) direct electrostatic ion-ion interaction. This may result in a higher stability of superstructures and the reduction in their homogeneity range. Both effects are pronounced in the series Cu_3Au - Cu_3Pt - Ni_3Al - Ni_3Ge.

In this section we have considered the influence of the size factor, electronic concentration, electronegativity, overlap of electronic shells, hybridization and interatomic density redistribution, and the change in the Fermi energy on the stability of the ordered state stability, symmetry and packing of superstructures, etc. The above consideration (as well as discussion in the preceding section) has brought us to the conclusion that the type of the forming superstructure and its relative stability are determined by the multifactor influence of different contributions into the superstructure energy. These results represent the physical foundation for the development of the quantum theory of ordered solid solutions and simultaneously for testing its performance.

The present consideration shows that the causes for ordering may be connected with the atomic properties of the components as well as with the change in the spectrum of the electronic states when going from a pure metal to a solid solution. The driving forces of ordering may be sufficiently diverse and competing. Further consideration in Chapter 5 of the energetic factors and the causes for ordering will show that other mechanisms exist that promote the long-range order; most of them will be treated there qualitatively.

We want to draw attention to another specific feature of the results described in this section. The above analysis shows that the influences of different factors are interrelated. The underlying reasons for ordering are the atomic properties of the components and the dependence of the type of the forming superstructure on the position of elements in the Periodic table (section 1.1). Atomic radii, atomic volumes, electronic configurations, electronegativities are qualitatively determined by the positions of elements in the Periodic system. Quantitatively these characteristics are also affected by the specific features of the interaction in solids.

1.3 Antiphase boundaries and mutual transformations of superstructures

Account of symmetry changes in the study of the ordering is important for several reasons. First, it enables the number of orientational and translational domains emerging during the transformation to be determined. This can be made by comparing the space groups of the initial disordered phase and the superstructure. The crystallographic description of superstructures is given in Tables 1 and 2, and that of the related disordered phases in Table 7. The corresponding Brillouin zones for disordered phases can be found in [23, 26, 27]. The comparison of the data represented in these tables allows one to obtain all information required.

In particular, employing the technique proposed in [77] provides the determination of the number of domains, orientational and translational, arising in each of the superstructures. The necessary data and the results of such treatment are given in Tables 8 and 9. The tables contain the Bravais lattices and the maximum numbers of the orientational and translational transformation domains. By contrast with the antiphase boundaries (APB), the boundaries of transformation twins are coherent only for particular orientations. The experiment shows that the twin boundaries tend to follow these special orientations. Thus the twin boundaries coincide with those of the orientational domains whilst the APBs are simultaneously the boundaries of the translational domains. In the low-symmetry superstructures, the structure and properties are mainly affected by the twin boundaries whereas in the high-symmetry, especially cubic, superstructures -- by the antiphase boundaries.

When a long-range order appears, the identical sublattices in different parts of the crystal may happen to be incoincident; in this case, such regions will be divided by the boundary that is called antiphase [78–81]. More accurate term for this boundary is the term "out-of-phase" [82], but it has not been adopted in the literature and we shall use the first one.

The regions bounded by the APBs are called the translational antiphase domains or simply the antiphase domains. The term "translational" points to the symmetry operation leading to the superpositions of the crystal lattices in two domains. If the ordering does not change the crystallographic singony, the ordered phase will contain only the translational antiphase domains. Otherwise, the ordered phase will comprise both the translational and orientational antiphase domains. The latter are more frequently called the transformation twins. Along with the rotational orientational domains that form due to the loss of the symmetry axes during the transformation, the ordering also leads to the appearance of antiparallel twins due to the loss of the planes of specular reflection [82].

There may be three different kinds of the **antiphase boundaries (APB)**: (1) thermal APBs, (2) deformation APBs arising due to the slip of dislocations (which are not complete in a given superstructure), (3) periodic APBs that can be disclosed by comparing a complex superstructure with a simple one.

Superstructure	Space group	Number of orientational domains	Number of translational domains
$L1_2$	O_h^1	1	4
$L1_0$	D_{4h}^1	3	2
$L1_1$	D_{3h}^5	4	2
$L1_3$	D_{2h}^{10} или D_{2h}^{25}	6	8
DO_{22}	D_{4h}^{17}	3	4
DO_{23}	D_{4h}^{17}	3	12
$L1_2$(M)	D_{4h}^{17}	3	f(M)
$L1_2$(MM)	D_{2h}^{13}	6	$f(M_1, M_2)$
Тип $MoPt_2$	D_{2h}^{25}	6	6
$D1_a$	C_{4h}^5	6	5

Table 8. Number of orientational and translational domains for superstructures arising from an f.c.c. (O_h^5) solid solution.

Solid solution	Superstructure	Space group	Number of orientational domains
O_h^9	B2	O_h^1	1
	B32	O_h^7	1
	$C11_B$	D_{4h}^{17}	3
	$D0_3$	O_h^5	1
D_{6h}^4	B19	D_{2h}^5	3
	$D0_{19}$	D_{6h}^4	1
	$D0_a$	D_{2h}^{13}	3
$A3'-D_{6h}^4$	$D0_{24}$	D_{6h}^4	1

Table 9. Number of orientational domains arising from b.c.c (O_h^9), h.c.p. (D_{6h}^4) and A3' (D_{6h}^4) solid solutions.

The antiphase boundary as a planar defect may be specified by the shear vector defining the displacement of one part of a crystal with respect to another and by the normal to the plane where the APB is located. If the shear vector lies in this plane then the boundary is conservative, i.e. its presence does not change the relative concentrations of the sites of different kinds. If the opposite is the case, the boundary is referred to as nonconservative and its presence changes, as a rule, the site concentrations as compared with the matrix of the ordered solid solution.

The type of boundary may be specified by the product (gr) [78], where g is the vector of the reciprocal lattice, or the vector orthogonal to the APB plane, and r is the shear (displacement) vector. If (gr) = 0, the APB will be conservative; the non-zero positive value of the scalar product (gr) determines the number of deficient planes while the negative value gives the number of interstitial planes.

The crystal lattice of a superstructure can be converted into that of another by means of the periodic shift by the appropriate vector, i.e. by creating the periodic APBs, stacking faults or more complicated defects of the structure. Different superstructures (say, of the f.c.c. types) within a certain crystal structure are related in such a way that each superstructure can be represented as another one, with periodic APBs, which are conservative for the superstructure of one stoichiometry and nonconservative for another. The distance between the periodic APBs is the antiphase period. It can be found from the geometric representation of a superstructure with periodic antiphase domains. Considering the superstructure with periodic APBs in the $\overline{k}$ space, one can define the antiphase period by the expression $P = 2\pi/\Delta\overline{k}$, with $\Delta\overline{k} = \overline{k}_s - \overline{k}_{s'}$, where $\overline{k}_s$ is the superstructure vector, $\overline{k}_{s'}$ -- that of the periodic domain structure, both vectors being located in the first Brillouin zone of the disordered solid solution [19].

The different possibilities of mutual transformations of the superstructure based on an f.c.c. solid solution with the shear vector a/2 <110> are given in Fig. 12. For one-dimensional periodic superstructure, the only transformation shown in Fig. 12 corresponds to the conservative antiphase boundaries observed experimentally. The case of the two-dimensional periodic superstructures of the $L1_2(MM)$ type involving 6 types of the periodic APB is too cumbersome and, therefore, not given in Fig.12. For each superstructure we show the orientation of the APB plane, the antiphase period (p) and the boundary type (gr).

Historically, the superstructures with the superstructure vectors located in the high-symmetry points of the Brillouin zone have not obtained special names, whereas the superstructures with the vectors in the low-symmetry points have been called long-period ones. In our treatment, the letter M relates to the long-period superstructures. It should be noted that M may be fractional. This implies that the mixture of two long-period structures is present in the structure of the alloy. Similarly, superstructures based on b.c.c. and h.c.p. lattices and observed experimentally may be related to one another. The mutual transformations of superstructures based on a b.c.c. lattice merely require the occurrence of the antiphase boundaries.

Situation with mutual transformations of the superstructures based on an h.c.p. lattice is more complicated. Three superstructures (DO_a, DO_{19}, B19) are formed from the A3 structure (h.c.p. with the ABAB packing), the fourth (DO_{24}) is based on the binary h.c.p. A3' structure with the ABAC packing. Mutual transformations of the structures based on different crystal lattices (A1, A2, A3, A3') are possible if the

$L1_2$	$100 \rightarrow 10\frac{1}{2}$ 1	$100 \rightarrow 10\frac{1}{4}$ 2	$100 \rightarrow 10\frac{1}{2M}$ M	$100 \rightarrow \frac{1}{2}\frac{1}{2}\frac{1}{2}$ $2/\sqrt{3}$
(001) 0	DO_{22}	$10\frac{1}{2} \rightarrow 10\frac{1}{4}$ 2	$10\frac{1}{2} \rightarrow 10\frac{1}{2M}$ $\frac{2M}{M-1}$	$10\frac{1}{2} \rightarrow \frac{1}{2}\frac{1}{2}\frac{1}{2}$ $\sqrt{2}$
(001) 0	(001) 0	DO_{23}	$10\frac{1}{4} \rightarrow 10\frac{1}{2M}$ $\frac{4M}{M-2}$	$10\frac{1}{4} \rightarrow \frac{1}{2}\frac{1}{2}\frac{1}{2}$ 4/3
(001) 0	(001) 0	(001) 0	$L1_2(M)$	$10\frac{1}{2M} \rightarrow \frac{1}{2}\frac{1}{2}\frac{1}{2}$ $\frac{2M}{\sqrt{2M^2+(M-1)^2}}$
$(1\bar{1}1)$ 0	$(\bar{1}10)$ 0	(221) 0	$(M\bar{M}\,1-M)$ 0	$L1_1$

$L1_0$	$100 \rightarrow 10\frac{1}{2M}$ 1	$100 \rightarrow \frac{1}{2}\frac{1}{2}\frac{1}{2}$ $2/\sqrt{3}$	$100 \rightarrow \frac{2}{3}\frac{2}{3}0$ $3/\sqrt{5}$ $100 \rightarrow 0\frac{2}{3}\frac{2}{3}$ $3/\sqrt{8}$	$100 \rightarrow \frac{2}{5}\frac{4}{5}0$ 1 $100 \rightarrow \frac{4}{5}\frac{2}{5}0$ $\sqrt{5}$
(001) 0	$L1_0(M = 1)$	$10\frac{1}{2} \rightarrow \frac{1}{2}\frac{1}{2}\frac{1}{2}$ $\sqrt{2}$	$10\frac{1}{2} \rightarrow \frac{2}{3}0\frac{2}{3}$ $6/\sqrt{5}$	$10\frac{1}{2} \rightarrow \frac{4}{5}0\frac{2}{5}$ $10/\sqrt{5}$
$(1\bar{1}1)$ 0	$(\bar{1}10)$ 0	$L1_1$	$\frac{1}{2}\frac{1}{2}\frac{1}{2} \rightarrow \frac{2}{3}\frac{2}{3}0$ $6/\sqrt{11}$	$\frac{1}{2}\frac{1}{2}\frac{1}{2} \rightarrow \frac{2}{5}\frac{4}{5}0$ $10/\sqrt{35}$
$(2\bar{1}0)$ ±1 (202) ±2	$(20\bar{1})$ ±2	$\bar{1}33$ ±2	тип $MoPt_2$	$\frac{2}{3}\frac{2}{3}0 \rightarrow \frac{4}{5}\frac{2}{3}0$ $3\sqrt{5}/2$
(210) ±3	(402)	$(1\bar{3}5)$	$(4\bar{2}0)$ ±2	$D1_a$

Fig. 12. Parameters of mutual transformations of b.c.c. lattice superstructures. Indicated are the transformations of superstructure vector, frequences of the occurrence of antiphase boundaries (APBs), planes where APBs occur, and the number of deficit or interstitial planes.

Fig. 12. (continuation)

	$L1_0$	$L1_0(M=1)$	$L1_1$	тип $MoPt_2$	$D1_a$
$L1_2$	$100 \to 000$ 1	$100 \to 10\frac{1}{2}$ 2	$100 \to \frac{1}{2}\frac{1}{2}\frac{1}{2}$ $2/\sqrt{3}$	$001 \to \frac{2}{3}\frac{2}{3}0$ $3/\sqrt{17}$	$100 \to \frac{4}{5}\frac{2}{5}0$ $5/\sqrt{5}$ $100 \to 0\frac{2}{5}\frac{4}{5}$ $\sqrt{5}/3$
$D0_{22}$	$10\frac{1}{2} \to 000$ $2/\sqrt{5}$	$001 \to 10\frac{1}{2}$ $2/\sqrt{5}$	$100 \to \frac{1}{2}\frac{1}{2}\frac{1}{2}$ $2/\sqrt{3}$ $10\frac{1}{2} \to \frac{1}{2}\frac{1}{2}\frac{1}{2}$ $\sqrt{2}$	$10\frac{1}{2} \to \frac{2}{3}0\frac{2}{3}$ $6/\sqrt{5}$	$10\frac{1}{2} \to \frac{4}{5}0\frac{2}{5}$ $2/\sqrt{5}$
$D0_{23}$	$10\frac{1}{4} \to 000$ $4/\sqrt{17}$	$10\frac{1}{4} \to 10\frac{1}{2}$ 4	$10\frac{1}{4} \to 10\frac{1}{2}\frac{1}{2}\frac{1}{2}$ 4/3	$10\frac{1}{4} \to \frac{2}{3}0\frac{2}{3}$ $12/\sqrt{41}$	$10\frac{1}{4} \to \frac{4}{5}0\frac{2}{5}$ 4
$L1_2(M)$	$10\frac{1}{2M} \to 000$ $\frac{2M}{\sqrt{4M^2+1}}$		$10\frac{1}{2M} \to \frac{1}{2}\frac{1}{2}\frac{1}{2}$ $\frac{2M}{\sqrt{2M^2+(M-1)^2}}$	$10\frac{1}{2M} \to \frac{2}{3}0\frac{2}{3}$ $\frac{6M}{\sqrt{4M^2+(3-4M)^2}}$	$10\frac{1}{2M} \to \frac{4}{5}0\frac{2}{5}$ $\frac{10M}{\sqrt{4M^2+(5-4M)^2}}$
$L1_3$	$\frac{1}{2}\frac{1}{2}\frac{1}{2} \to 001$ $2/\sqrt{3}$	$\frac{1}{2}\frac{1}{2}\frac{1}{2} \to 10\frac{1}{2}$ $\sqrt{2}$	$100 \to 000$ 1	$\frac{1}{2}\frac{1}{2}\frac{1}{2} \to \frac{2}{3}\frac{2}{3}0$ $6/\sqrt{11}$	$\frac{1}{2}\frac{1}{2}\frac{1}{2} \to \frac{4}{5}\frac{2}{5}0$ $10/\sqrt{35}$
$L1_2$	(010) ±1	$(00\bar{1})$ ±1	$(1\bar{1}\bar{1})$ ±1	(223) ±3	(524) ±5
$D0_{22}$	(201) ±2	$(\bar{2}01)$ ±2	$(1\bar{1}0)$	$(20\bar{1})$ ±1	(201) ±1
$D0_{23}$	(401) ±4		$(2\bar{2}\bar{1})$ ±4	$(40\bar{5})$	$(40\bar{3})$
$L1_2(M)$	(2M, 01) ±2M		$(M\bar{M}, 1-M)$ ±2M	(2M, 0,3 – 4M)	(2M, 05 – 4M)
$L1_3$	$(\bar{1}11)$ ±2	$(\bar{1}\bar{1}0)$ ±2	(100)	$(\bar{1}\bar{1}3)$	$(\bar{3}\bar{1}5)$

structures contain, besides the APBs, the defects of other types. If the packing in the superstructures is the same (as in the case of the f.c.c. ⇆ h.c.p. transformation), the combination of the periodic antiphase boundaries and stacking faults is needed for transforming one superstructure to another.

If the packings of crystal lattices are different (as in the case of the f.c.c. ⇆ b.c.c. transformation), extension or contraction along definite crystallographic axes or the shifts of sublattices are necessary.

2. PHASE DIAGRAMS OF TI-, FE-, NI-, CU-, AND AL-BASE ALLOYS IN THE ORDERED PHASE REGIONS

This chapter is devoted to a brief description, analysis and generalization of the phase diagrams of binary systems involving the most common ordered phases of the AB, AB_2, AB_3, AB_4 stoichiometry having the crystal structures of the ordered solid solutions. Those are: a cubic B2(CsCl)-type lattice of the AB composition, a tetragonal $L1_0$(AuCu)-type lattice of the AB composition, a tetragonal $C11_b$($MoSi_2$)-type lattice of the AB_2 composition, a cubic $L1_2$($AuCu_3$)-type lattice of the AB_3 composition, and a tetragonal $D1_a$($MoNi_4$)-type lattice of the AB_4 composition. The other superstructures, as a rule, occur not so commonly in metallic alloys. (We indicate the mode of formation and the main features of the corresponding phase diagram if these superstructures are present in alloys of the systems under consideration).

It is not possible in this work to describe the phase diagrams of about 800 binary phases, so we shall consider only the diagrams involving metals Ti, Fe, Ni, Cu, and Al, i.e. the basic metals used in the most important engineering and special alloys.

A short description and generalization of the phase diagram near the ordered phases are expedient in this work since the phase diagram gives a geometric representation of the ordered phase, showing the mode of its formation and the specific features of phase equilibria in the range of the phase existence. In addition, information regarding the main phase diagrams of alloys involving the ordered phases relieves one from a time-consuming search for the diagrams in numerous reference books on the phase diagrams of metallic alloys [39 - 47, 83 - 87]. In specifying the phase composition, we adhere the Kornilov's principle [83] based on the position of a given element in the electronegativity series: the more electropositive metal is taken to be the first element in the chemical formula (Ti in TiNi).

2.1 Ordered AB phases of the B2(CsCl) type

Ordered phases with a cubic structure of the CsCl (CuZn) type are most commonly found in metallic alloys. They are present in alloys of group VIII elements with elements of subgroups V to VIA, and I to IVB. As a rule, B2-type phases are formed congruently by crystallization of melt, in which case the liquid and solid phases have the same composition. The congruently melting phases have a constant melting temperature corresponding to an open maximum in the equilibrium phase diagram. For most B2 phases this maximum lies at the equiatomic composition.

Some of the B2-type phases, in particular those with rare-earth and low-melting metals, are formed by a peritectic reaction when the compositions of the liquid and solid phases do not coincide and the melting temperature for the alloy of stoichiometric composition corresponds to a hidden maximum in the phase diagram. More rarely, they may form in the solid state by ordering of a solid solution. Typical of the B2 phases is the formation of fairly extensive homogeneity regions, the β phases, which are stable to low temperatures or become unstable with respect to the martensitic and polymorphic transformations as the temperature decreases. We shall give further a brief description and the generalization of the phase diagrams in the regions close to the ordered B2 phases existing in Ti, Fe, Ni, Cu, and Al-base alloys.

Titanium-base B2-type phases

Titanium forms the ordered AB phases of the B2 type with all group VIII elements as well as with Au and Zn. All of them, except TiZn, crystallize from liquid with an open maximum corresponding, as a rule, to the stoichiometric composition. Ti-base B2 phases are homogeneous in the range 5 - 10 at.%; TiRh and TiIr show a homogeneity of 20% [84]. On cooling, the crystal lattices of B2 phases formed by Ti with Ni, Rh, Pd, Ir, Pt, and Au become unstable with respect to the martensitic B2 $\leftrightarrows$ B19 transformation. The martensitic transformation (MT) in ordered B2 phases is of a thermoelastic nature thus providing high parameters of martensitic inelasticity.

TiFe is formed congruently at 1500°C, and has a constant composition. The B2 phase is stable to the liquid helium temperatures [88].

TiCo crystallizes congruently at 1450°C. The congruent melting point corresponds to 53 at.% Co [84]. The ordering of the B2 type can be achieved only after prolonged annealing. The homogeneity range for TiCo does not exceed ~ 6% at 400°C. This phase is stable to the liquid helium temperatures. The homogeneity region is asymmetric; the TiCo/Ti_2Co boundary is shifted toward the TiCo stoichiometric composition, differing from the B2 phases formed by Ti with other group VIII elements.

TiNi crystallizes from liquid at 1250°C (Fig. 13a). The maximum of melting is flat and not coincident with the β phase stoichiometric composition. The homogeneity region is quite unusual: TiNi is almost a single B2-type phase whose homogeneity range has been studied in detail down to -200°C [89 - 91]. At 1100°C the β phase region extends from 48.5 to 55 at.% Ni. At temperatures below 400°C where this phase is absent, TiNi can be regarded as the phase of a constant composition. The phase boundary TiNi/TiNi + Ti_2Ni_3($TiNi_3$) in the phase diagram is inclined to the horizontal axis at smaller angle than the boundary TiNi/TiNi + Ti_2Ni. In the range where the TiNi phase exists, a horizontal line has been found at 670°C (on heating) or at 625°C (on cooling), but it is not clear as yet whether this line results from the formation of a new phase, Ti_2Ni_3, or from the solid-state transformation in the $TiNi_3$ compound [91, 92].

All the above features of the phase diagram are important for the optimum choice of the mode of heat treatment of TiNi-base alloys which are known to be the best engineering shape memory alloys. The TiNi alloy displays a high degree of shape recovery (95 - 100%), a high reactive stress developed upon shape recovering (up to 700 MPa), the maximum recovered strain among the known materials (up to 10%), and other beneficial technical characteristics.

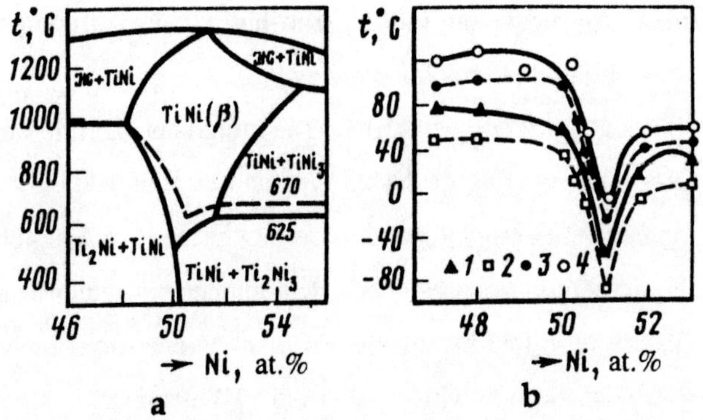

Fig. 13. (a) The phase diagram of the Ti-Ni system in the B2-phase region. (b) The critical points of the martensitic transformation in the TiNi phase region.
1 – M_H, 2 – M_K, 3 – A_H, 4 – A_K

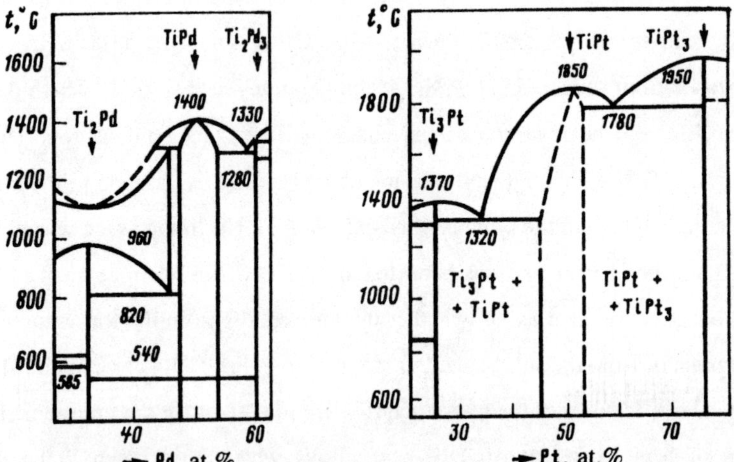

Fig. 14. The partial phase diagrams of the Ti-Pd and Ti-Pt systems in the B2-phase region.

The B2 phase lattice is unstable with respect to the MT. Upon quenching from high temperatures, alloys with 48 to 50 at.% Ni undergo the MT of the B2 ⇆ B19' type, where B19' signifies an orthorhombic martensite B19 (of the AuCd type) with a monoclinic distortion. A high-temperature B2 phase existing in alloys with more than 50.5 at.% Ni is stabilized by quenching. The MT in these alloys proceeds only at temperatures below room temperature (Fig.13b), the MT temperature being dependent strongly upon the alloy composition [90].

TiRu forms congruently at 2120°C; the B2 structure is stable down to the liquid nitrogen temperature range. According to [84, 93], the homogeneity range at temperatures below 1110°C does not exceed ~ 4%.

TiRh crystallizes congruently at 1940°C, exhibits a wide homogeneity range and is stabilized down to low temperatures by an excess, with respect to stoichiometric composition, amount of Ti [84, 93]. At the TiRh stoichiometric composition and at lower Ti content, the B2 structure is unstable with respect to the MT of the B2 ⇆ B19 type.

The **TiPd** phase forms congruently at 1400°C, has a narrow homogeneity range (of ~ 5%) and is unstable to the MT of the B2 ⇆ B19 type within the entire B2 phase homogeneity region (Fig. 14) [84, 93]. The temperatures of the beginning and the end of the direct and reverse MT have been established in [94]. Their concentration dependence is represented by the curve with a maximum at 550°C. The type of MT on alloying TiPd with Ni along the section of the system TiPd - TiNi has been studied in detail in several works [95–99].

The phase diagram of the **Ti - Os system** [84,93] is similar to that of the Ti - Ru. The B2 phase TiOs forming congruently at 2160°C is stable to liquid nitrogen temperature. The homogeneity range below 1100°C does not exceed 8%.

The phase diagram of the **Ti - Ir system** studied in [84] is similar to that of the Ti - Rh system. The B2 phase TiIr forming congruently at 2120°C has a wide homogeneity range. It is stabilized, down to low temperatures, by an excess, relative to stoichiometric, amount of Ti. Alloys with 41 to 57 at.% Ir undergo the martensitic B2 ⇆ B19 transformation in the range of homogeneous B2 phase.

The **Ti - Pt** phase diagram [84, 93] resembles that of Ti - Ir system. The TiPt phase crystallizing congruently at 1830°C has a narrow homogeneity range of ~ 5%. It is unstable with respect to the MT of the B2 ⇆ B19 type over the whole homogeneity range (Fig. 14). The MT temperatures have been determined in [94], their dependence on concentration is expressed by the curve showing a maximum at 1100°C. The authors of the work [100] have investigated the effect of Pt on the martensitic transformations in TiNi.

TiAu crystallizes congruently at 1490°C (Fig.15). The homogeneity range is sharply reduced with decreasing temperature, especially at the boundary TiAu/Ti_3Au. The crystal lattice of TiAu is unstable with regard to the MT of the
B2 $\leftrightarrows$ B19 type within the homogeneity range of the B2 phase [94]. As Au content decreases from 55 to 42%, the MT temperature is lowered. The dependence of the MT critical points on concentration shows no extremum. The evolution of the MT in the TiAu-base alloys with Ni has been investigated in [95 - 97]; the authors have presented the MT diagram for the TiAu - TiNi quasibinary system.

TiZn forms peritectically at 1070°C, has a constant composition and the boundary with another ordered phase, $TiZn_2$ (of the $MoSi_2$ type). The phase diagram of the Ti - Zn system is given in [41]. Titanium analogs Zr and Hf also form B2 phases with group VIII elements. The phase diagrams of these systems near the B2 phases are analogous to those of Ti-base systems [92]. According to these diagrams, B2 phases of Zr and Hf are formed chiefly by crystallization from liquid with an open maximum and have certain homogeneity ranges.

Iron-base B2-type phases

Ordered phases of the B2 type in Fe-base alloys have been found in the systems Fe - Ti, Fe - V, Fe - Co, Fe - Al, and Fe - Ga. The modes of formation are more diversified: B2 phases are typically formed by the ordering of b.c.c. solid solutions or by a peritectic reaction involving a liquid phase. In the last case, the homogeneity range of the B2 phase is markedly shifted away from the AB stoichiometry. Such phases are prone to polymorphic transformations and decomposition leading to the formation of new types of superstructures.

The Fe - Ti phase diagram near the B2 phase has been described earlier. Regarding to the Fe-V system, the works [101, 102] show that the formation of a stable σ phase near the equiatomic concentration due to the aging of the α (b.c.c.) solid solution at 600 °C is preceded by the precipitation of the metastable ordered B2-type phase, FeV. This fact has not been reflected, however, in the existing reference books on the phase diagrams.

FeCo forms by the ordering of the γ (f.c.c.) solid solution. The character of the interaction of this phase with other possible superstructures of the Fe_3Co and $FeCo_3$ compositions has not been established reliably as yet. The superstructure of the B2 type is observed in alloys containing 40.5 to 60.3 at.% Co. Kurnakov point corresponds to 730°C [42, 103].

FeRh is formed by the ordering of the γ (f.c.c.) solid solution at ~ 1450°C [86]. The cupola of the region where the ordered phase exists (Kurnakov point) at 1400°C corresponds to the stoichiometric composition; the homogeneity region for the α' phase is shifted to the Fe-rich end, ranging from 23 to 55 at.% Rh at 600°C.

FeAl is formed by the peritectic reaction L + αFe $\leftrightarrows$ α_2(FeAl) at 1310°C. The alloys of this system containing 0 - 50 at.% Al have been studied, after prolonged multiple-step annealing, in [104 - 107]. The polymorphic transformation of the α_2 phase, namely α_2(B) $\leftrightarrows$ α_2(H), has been shown to occur at 803°C and 51 at.% Al (Fig. 16). The secondary ordering, α_2(B) $\leftrightarrows$ α_2', associated with changing of the type of long-range order takes place in alloys containing 32 - 40 at.% Al in the range 1022 - 700°C. The peritectoid reaction α + α_2' $\leftrightarrows$ Fe_3Al occurs at 552°C and 26.8 at.% Al. Fe_3Al has an ordered atomic structure of the DO_3 type. The refined phase diagram of the Fe - Al system is given in the reference book [86]; the partial diagram involving the regions of the ordered phases is shown in Fig. 16 (K_1 and K_2 denote the regions where a specific decomposition of a solid solution takes place, i.e., the so-called K states exist).

FeGa forms peritectically at 1037°C (Fig.16). Its composition is strongly deviated from stoichiometry toward the Fe-rich side (to ~ 30 at.% Ga). The phase exhibits homogeneity of ~ 33 to 45 at.% Ga at 906°C. At 31.1 at.% Ga and 950°C the second

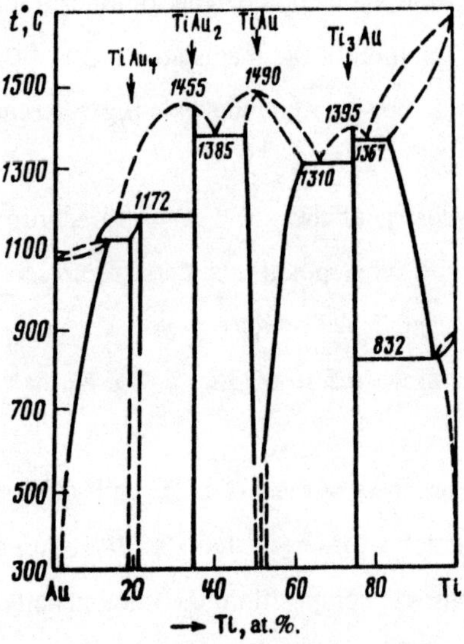
t,°C
TiAu4
TiAu2
TiAu
Ti3Au
1500
1300
1100
900
700
500
300
1455
1490
1395
1385
1367
1310
1172
832
Au
20
40
60
80
Ti
→ Ti, at.%.

Fig. 15

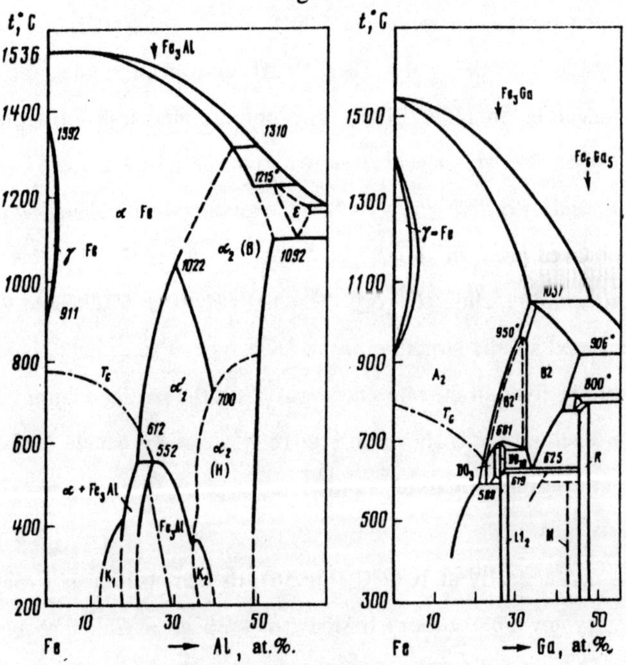
t,°C
1536
1400
1200
1000
800
600
400
200
Fe3Al
1392
1310
1215°
α Fe
γ Fe
ε
α2 (B)
1022
1092
911
TC
α'2
700
612
552
α2 (H)
α + Fe3Al
Fe3Al
K1
K2
10
30
50
Fe
→ Al, at.%.
t,°C
1500
1300
1100
900
700
500
300
Fe3Ga
Fe6Ga5
γ-Fe
1037
950°
906°
A2
B2
800°
B2'
TC
681
DO3
675
R
588
679
L12
M
10
30
50
Fe
→ Ga, at.%.

Fig. 16.

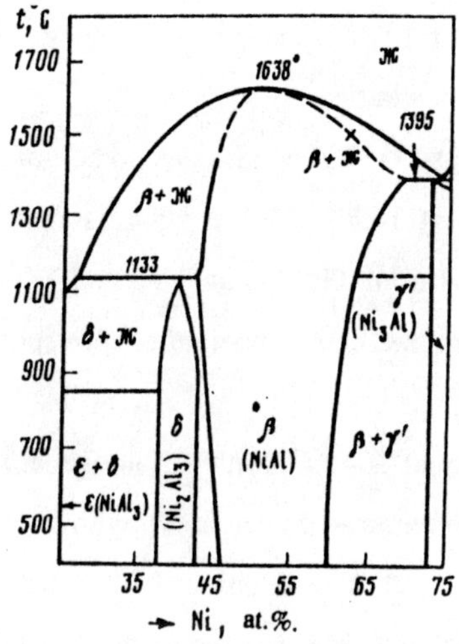

Fig. 17

ordered phase, B2', appears that in turn undergoes the first-order transition B2' $\leftrightarrows$ DO_{19} at 681°C. The ordered DO_3 phase forms at ~ 680°C through a degenerate peritectoid reaction. Another phase transition is worth mentioning, namely, the eutectoid decomposition of the B2 phase at 625°C into the DO_{19} phase and the intermetallic compound Fe_6Ga_5 (R phase). The ordered cubic phase Fe_3Ga of the $L1_2$ type with a narrow homogeneity range also forms through a degenerate peritectoid reaction at 619° C. The Fe-Ga system is characterized by the abundance of ordered phases and is in many respects similar to the Fe - Al system. The Fe - Ga equilibrium diagram in Fig. 16 is taken from the generalized data in [86].

Nickel-base B2-type phases

Ordered AB phases of the B2 type in Ni-base alloys typically form congruently (only in the case of NiGa the maximum melting temperature deviates from the stoichiometric composition) and have wide homogeneity ranges. The exceptions are YNi and HfNi, which are the phases of constant composition. Some Ni-base B2 phases display the lattice instability with respect to the martensitic and polymorphic transformations which is the characteristic property of the B2 phases. The Ti - Ni equilibrium diagram near the B2 phase has been described above.

YNi forms congruently at 1070°C and has no homogeneity region. According to [108], YNi has an orthorhombic lattice.

HfNi forms congruently at 1530°C and also has no homogeneity range [107].

NiBe forms congruently at 1600°C, has a wide asymmetric homogeneity region, called the β phase. It extends from 25 to 52 at.% Ni at 1000°C and reduces to ~ 8% at 600°C [110]. The maximum melting temperature corresponds to the stoichiometric composition.

NiZn forms peritectically at 1040°C, with a homogeneity range (β phase) of 10 - 15%. At 615°C this phase decomposes eutectoidally with the formation of the ordered β 1 phase of the AuCu type [111]. The β phase (of the B2 type) cannot be fixed by quenching. At room temperature there exists the β_1 phase of the AuCu type.

NiAl forms congruently at 1638°C (Fig.17); the β' phase extends from 45 to 60 at.% Ni at 500°C. The maximum melting temperature corresponds to the stoichiometric composition [112]. The B2 lattice is unstable with regard to the MT, in particular, at higher Ni content; at high cooling rate the shear transformation in the β' phase region leads to the formation of a tetragonal martensite. The MT occurs in all the alloys in the intermediate region between the high-temperature phases β' and γ' (NiAl - Ni_3Al). The martensite stability enhances with an increase in the Ni content, while the degree of tetragonality of the martensite lattice decreases [113].

The phase Ni-Al diagram in the region of the ordered B2 and $L1_2$ phases has been constructed in [114] with the use of the phenomenological model to describe equilibria between the solid and liquid phases and using the data on the thermodynamical properties of the components. The configurational entropy of disordered solid solutions and ordered phases has been calculated by using the cluster variation method in the tetrahedron approximation [114]. The alloys of the Ni - Al system in the region of the ordered phases constitute the basis for producing heat-resistant alloys.

NiGa forms congruently at 1220°C; the maximum melting temperature does not coincide with the equiatomic composition [115]. The homogeneity region is strongly displaced to the Ni-rich side of the diagram (36 at.% Ga at 1100°C) and significantly reduces with increasing temperature to the composition range 47 - 55 at.% Ga at 500°C.

NiIn forms in the solid state by the peritectoid β + ε' ⇆ NiIn reaction at 860°C. The homogeneity region is narrow [116].

Copper-base B2-type phases

The B2-type phases in copper alloys have been detected in the systems Pd - Cu, Be - Cu, Cu - Zn, Cu - Al. The mode of formation varies from crystallization from the liquid phase (ScCu) to the ordering of a solid solution (PdCu, BeCu). For some B2 phases, the homogeneity regions do not coincide with the AB stoichiometry. In addition, copper forms B2-type phases with most rare-earth and transuranium elements. The lack of reliable data on the equilibrium diagrams for the alloys of Cu with these elements does not allow one to reach general understanding of the formation mechanism for all Cu-base B2-type phases. Nevertheless, the analysis of the available phase diagrams shows that these phases have constant composition and form largely by a peritectic reaction [85].

ScCu forms congruently at 1125°C and has no homogeneity range [117].

PdCu forms by the ordering of the f.c.c. solid solution at 600°C, has a relatively wide homogeneity region (β phase) extending from 38 to 48 at.% Pd [118]. The ordered phase PdCu is likely to exist in the eutectoid equilibrium with the ordered Pd_3Cu phase at ~ 380 °C [85]. It should be noted that the tops of the ordered phase region cupola do not coincide with the stoichiometric composition (Fig.18).

BeCu (χ phase) forms, according to the calculated phase diagram [119], in the solid state by a peritectoid reaction at 880°C in the region between the intermediate phase $Be_{0.75}Cu_{0.25}$ and the intermediate b.c.c. phase β. Dashed line in Fig.18 shows the calculated concentration dependence of the temperature of the beginning of the ordering. According to the data of [42], a continuous β ⇆ β' transformation takes place in the region close to the equiatomic composition. The boundary of the β + β' region,

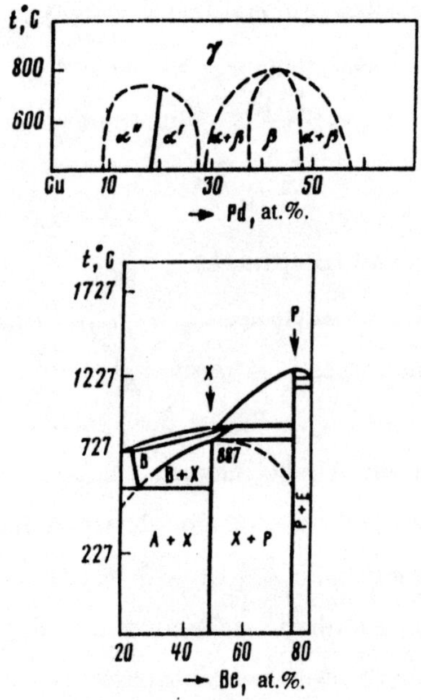

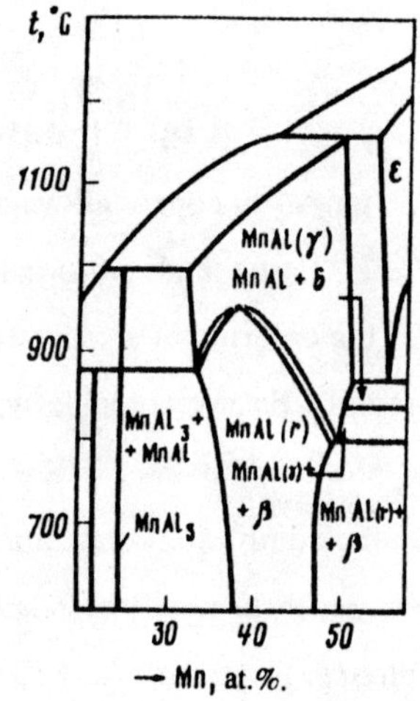

Fig. 18. The partial phase diagrams of the Cu -- Pd and Cu -- Be systems in the ordered phase regions.

Fig. 19. The partial phase diagram of the Mn - Al system in the B2-phase region.

with the critical point at 885°C, is also indicated in this work. The β' phase exhibits homogeneity of 47.2 to 48.8 at.% Be. This phase is stable and does not undergo a martensitic transformation, according to [120].

βCuZn forms peritectically at 900°C with a wide (39 - 55 at.% Zn at 800°C) homogeneity range, which, however, sharply reduces to the range 46 - 52 at.% Zn at 200°C in the region of the ordered β' phase [121]. The ordering transition β ⇆ β' of the A2 → B2 type occurs in the range from 468 to 450°C. On cooling, the B2 phase lattice becomes unstable with respect to the martensitic transformations. These transformations in Cu-, Ag-, and Au-base alloys are considered in detail in the monograph [120] containing the data on the crystal structures of the high-temperature phases, matrix phases and martensites forming under different conditions of quenching, deformation or preparation of thin foils. The high-temperature ordered B2-type phase is identified in [120] as the β_2 phase.

According to [120], two kinds of martensite are formed in 40 at.% Zn alloys: the twin β_2' martensite appears after special quenching that excludes the massive transformation $\beta_2 \leftrightarrows \alpha_m$; the martensite of the second type, β_2'', forms after an ordinary quenching owing to the thermal strains and has a superstructure of the $L1_0$ (AuCu) type. When thin foils of the disordered β phase are cooled, the as-quenched martensite has a distorted basic f.c.c. structure (also disordered). A greater deformation of the β phase gives rise to the formation of a disordered f.c.c. (α') or an h.c.p. (ζ') structure.

CuAl forms by peritectic reaction at 624°C, according to [42], has an insignificant homogeneity region. The calculated phase diagram of the Cu-Al system is given in [122]. The CuAl phase forms peritectically at 700°C with no homogeneity range.

Aluminum-base B2-type phases

B2 phases of Al in the Sc - Al, Co - Al, and Ru - Al systems are formed by crystallization from liquid, while in the other systems, including those with rare-earth metals La, Ce, Pr, Nd, Gd, Dy, through a peritectic reaction, with no homogeneity range.

ScAl crystallizes from liquid congruently at 1300°C, has no homogeneity range and is stable down to low temperatures [123].

MnAl forms peritectically at 1160°C, with a wide homogeneity range (γ phase) shifted away from the AB composition to the Al-side of the diagram. This phase covers a range of 32.5 to 50 at.% Mn at 1000°C (Fig.19) [42,43,124]. At lower temperatures, the B2 lattice undergoes distortions. The concentration dependence of the B2 transformation temperatures has a maximum at 950°C and 40 at.% Mn. The phase diagram of the Mn-Al system calculated by Kaufman [122] does not include the B2 phase near the equiatomic composition. Instead of this phase, the author suggests the phases Mn_5Al_4 and Mn_9Al_8 close to the MnAl stoichiometry.

The **CoAl** phase forms congruently at 1645°C [125], with an extensive range of homogeneity (β' phase), strongly toward the Co-rich side. At 1200°C it extends to 75 at.% Co, whereas at 800°C only to 48 at.% Co. Kaufman's calculations [126] confirm the type of the phase diagram observed near the CoAl phase.

RuAl also forms congruently at 2010°C, has a homogeneity region sharply reducing with decreasing temperature; at room temperature this region narrows to ~ 3%, at 1600° C it increases to ~ 10% [127].

ReAl forms peritectically at 1590°C. The homogeneity range has not been established with certainty [128].

IrAl has a structure of the B2 type, the mode of its formation has not been examined [41].

2.2 Ordered AB phases of the $L1_0$ type

Tetragonal phases of the AuCu type occur not so frequently in alloys compared to B2-type phases. Certain regularities in the mode of their formation with elements of Periodic table are not observed. It can be said that they occur more often in alloys of the subgroup IVA elements, mainly in titanium alloys as well as in alloys of Ni, Pd, and Pt with nontransition metals. Neither of the $L1_0$-type phases, in contrast to the B2-type phases, forms congruently by crystallization from liquid. More typically, they form in the solid state. Most $L1_0$ phases represent superstructures of the AuCu type formed by atomic ordering of solid solutions; they also have wide homogeneity ranges. Of the $L1_0$-type phases considered above, only titanium phases form peritectically involving the liquid phase and have no homogeneity region (except for TiAl).

Titanium-base $L1_0$-type phases

Ordered AB phases of the $L1_0$ type in titanium alloys occur in the systems Ti - Cu, Ti - Ag, Ti - Al, Ti - Ga, and Ti - In. Typically, they form through a peritectic reaction. Only TiAl has an extensive homogeneity region that is of great importance for developing heat-resistant alloys based on this compound [84]. It is typical of the above systems to form the ordered AB_3 phases of the DO_{19} or DO_{22} types.

According to [129], **TiCu** crystallizes congruently at 984°C (Fig. 20) and has a narrow region of homogeneity (from 48 to 52 at.% Cu), in which, depending on the Cu content, the morphotropic transition B11 $\leftrightarrows$ $L1_0$ takes place, according to the data of [130]. Rapid quenching from the melt gives rise to amorphization of the TiCu phase; crystallization in the amorphous state leads to the formation of the B11 structure which is stable down to room temperature [131]. In the Ti - Cu system, the Ti_2Cu phase forms congruently at 1051°C. This phase has an ordered structure of the $C11_8$ ($MoSi_2$) type in contrast to the calculation of Kaufman [132] who predicted the formation of this phase through a peritectic reaction.

TiAg forms peritectically at 1020°C; its homogeneity region at 900°C and below lies within a range of 2% [133]. The phase Ti_2Ag with a tetragonal ordered structure of the $MoSi_2$ type forms through a peritectoid reaction β_{Ti} + TiAg $\rightarrow$ Ti_2Ag at 945°C. It is interesting that TiAg has a hardness of 1800 MPa and can be easily cold-worked. Its ductility is not worse than that of the TiNi phase. It is supposed that TiAg exhibits the shape memory effect in the homogeneous state [2].

TiAl (γ phase) forms peritectically at 1460°C, has an extensive homogeneity region (from 48 to 64 at.% Al at 1200°C and from 50 to 60 at.% at 800°C). In alloys of Ti with Al, in addition to the $L1_0$-type phase, the ordered tetragonal phase $TiAl_3$ of the DO_{22} type and the hexagonal phase Ti_3Al (α_2 phase) of the DO_{19} type form. The nature and mode of formation of the latter have been a subject of discussions for many years and, correspondingly, different versions of the Ti - Al phase diagram have been suggested. This system makes up the basis for developing commercial titanium alloys.

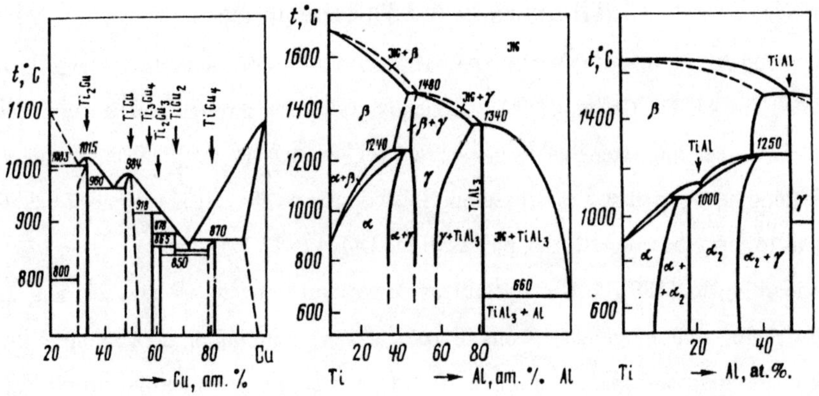

Fig. 20. The partial phase diagrams of the Ti - Cu and Ti - Al systems in the ordered phase regions.

A detailed analysis of the issue has been made by Kornilov in his monograph [84]. The author suggests the most plausible and repeatedly verified version of the phase diagram near the Ti_3Al composition which implies the formation of Ti_3Al from the β (f.c.c.) solution of Al in Ti at 1120°C via the atomic ordering of the DO_{19} type (Fig.20). The identity of the Ti_3Al phase is confirmed by the presence of singular points on composition vs. property curves [84] and by the general appearance of the diagram near the Ti_3Al composition. The α_2 phase has a wide homogeneity range of 20.7 - 33 at.% Al at 1100°C.

The phase diagram of the Ti - Al system in the range 0 - 50 at.% Al has been constructed in [134] by the thermodynamic calculation. The Ti_3Al phase of the DO_{19} type was shown to form by the atomic ordering of the Ti-base solid solution. These results confirmed the previous experimental data of [135] concerning this mode of formation of the Ti_3Al compound.

TiGa is the phase of a permanent composition which forms peritectically at 1170°C. In addition to this phase, two other phases form in the Ti-Ga system (as in the Ti - Al system) by the ordering of a solid solution: Ti_3Ga of the DO_{19} type forms through a peritectoid reaction between the β solid solution and intermetallic compound Ti_2Ga at

1030°C, and $TiGa_3$ of the DO_{22} type forms peritectically at 925°C. Both phases have no homogeneity regions [136]. According to the generalized data on the phase diagram of the Ti - In system [45], the ordered phase with the $L1_0$ structure corresponds to the composition Ti_3In_2. The mode of formation of this phase has not been established. The ordered phase Ti_3In of the DO_{19} type, by analogy with Ti_3Ga, forms in the solid state, supposedly by a peritectoid reaction. The existence of the $TiIn_3$ phase has not been confirmed.

Iron-based $L1_0$-type phases

The ordered AB phases of the $L1_0$ type (FeNi, FePd, FePt) represent classical superlattices. They were called Kurnakov's compounds by Kornilov. All these phases are formed by the ordering of f.c.c. (γ) solid solutions, by analogy with the formation of the AuCu compound. Their formation obeys the phase rule. In the phase diagrams these phases are considered as individual ones, with distinct interfaces between ordered and disordered phases. As shown in the classical works of Kurnakov, Ageev, Stepanov, Kornilov and others, stoichiometric compositions of ordered phases correspond to the singular points on property versus composition curves.

Since the ordering processes proceed slowly at low temperatures, phase equilibria in the respective regions of the phase diagrams are shown tentatively. It should be noted that reliable equilibrium diagrams for the systems Fe - Ni, Fe - Pt in the regions of the interaction between the ordered $L1_0$ and $L1_2$ phases are lacking, although these compositions constitute the basis for developing special magnetic materials, shape memory alloys and other practical alloys of special application. Below we present several variants of the Fe-Ni phase diagram.

FeNi. The existence of this phase has long been doubtful. It is absent, for example, in the equilibrium diagrams of Fe - Ni system presented in the reference books [41, 43, 45, 86]. In later works, it has been shown that alloys in the region close to the equiatomic composition are being ordered at 320°C in accordance with the reaction A1 → $L1_0$ [137]. Based on the results of experimental investigations of the alloys by methods of Mössbauer spectroscopy, X-ray and neutron diffraction and on the measurements of specific heat capacity and magnetic and mechanical properties, the authors of the work [138] have proposed their own version of the phase diagram for the region of the atomic ordering of FeNi and $FeNi_3$ (Fig.21). More detailed information on the nature and specific features of the phase equilibria in the Fe - Ni system can be found in the section devoted to the $L1_2$-type phases.

PdFe. According to the data on the Pd - Fe phase diagrams generalized in [86], this phase forms by the atomic ordering of the f.c.c. (γ) solid solution at 790°C and has a wide homogeneity region (γ_1 phase) extending form 48 to 60 at.% Pd. The maximum ordering temperature (Kurnakov point) does not correspond to the stoichiometric composition but to 57 at.% Pd. The PdFe phase is likely to exist in a eutectoid equilibrium with the second, $L1_2$-type (γ_2) phase at 760°C (Fig. 21).

According to [86], PtFe forms owing to the atomic ordering of the f.c.c. (γ) solid solution, has a wide homogeneity range (γ_2 phase) of 34 to 54 at.% Pt at 800°C. Kurnakov point at 1300°C is slightly deviated from stoichiometry. The phase is likely to exist in a eutectoid equilibrium with the ordered $L1_2$-type phase Pt_3Fe at 1150°C (Fig. 21).

Ni-base $L1_0$-type phases

AB phases of the $L1_0$ type in nickel alloys are formed, typically, in the solid state either by a peritectoid reaction (MnNi, NiZn) or in a classical mode corresponding to the formation of the AuCu compound. They have homogeneity regions near the $L1_0$ phases.

MnNi with a disordered b.c.c. lattice forms, according to [43, 139], on the base of the γ solid solution. The maximum transformation temperature (910°C) corresponds to 51 at.% Mn. On cooling, MnNi undergoes a series of transformations, the first occurring at 775°C through a peritectoid reaction leading to the appearance of the Ni-rich B' phase. The MnNi (B') phase then decomposes eutectoidally. The low-

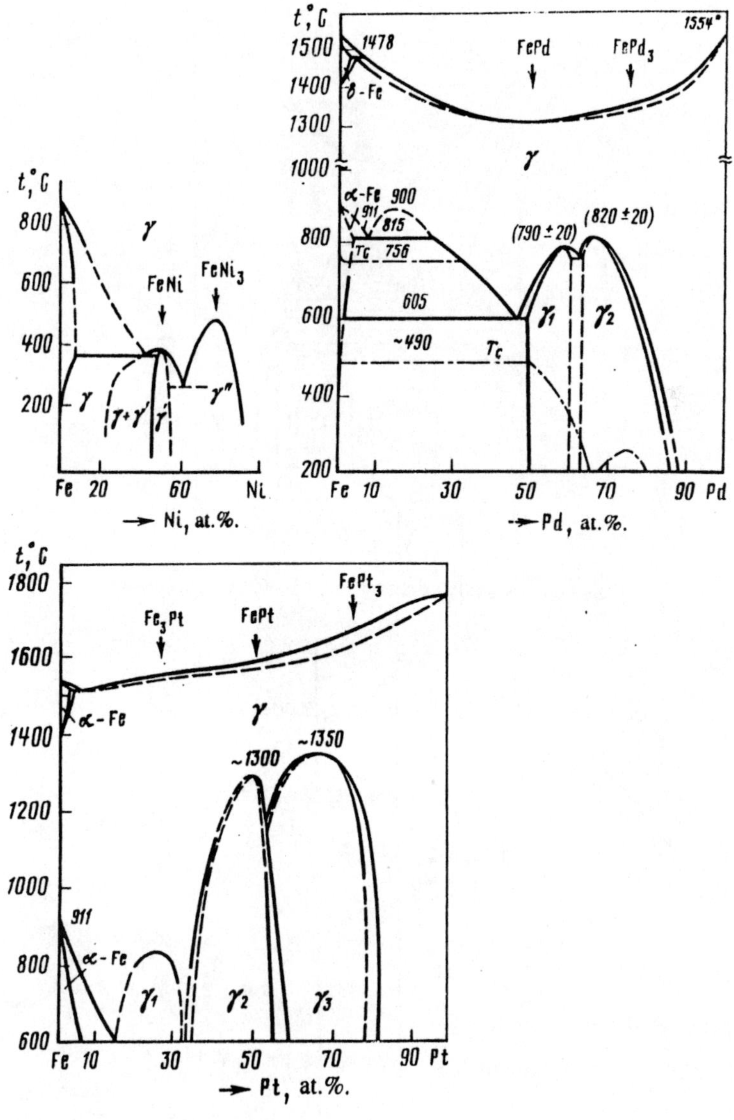

Fig. 21. The phase diagrams of the Fe - Ni, Fe - Pd and Fe - Pt systems.

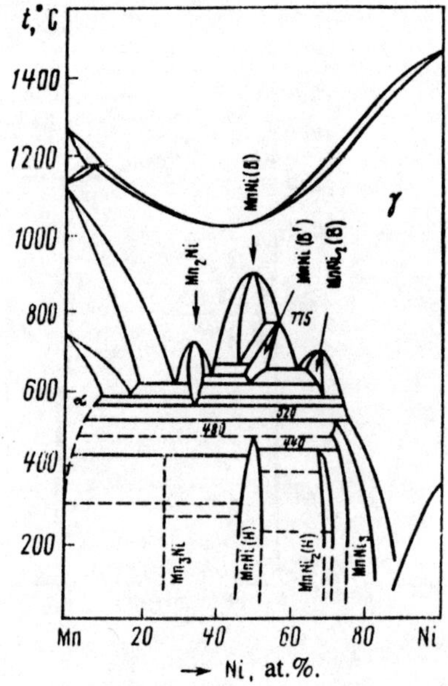

Fig. 22. The phase diagram of the Mn - Ni system.

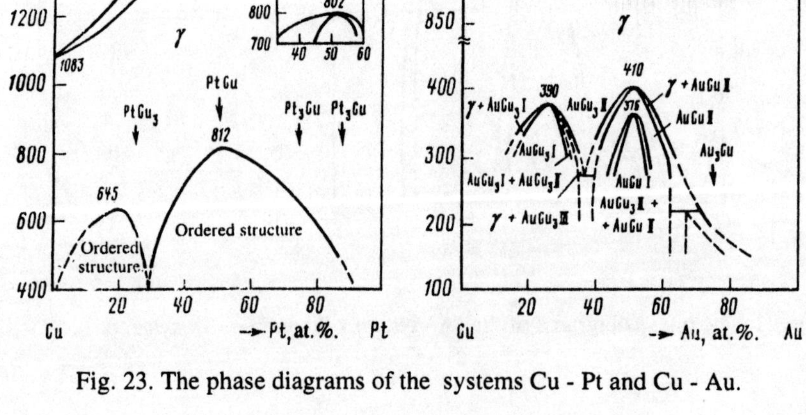

Fig. 23. The phase diagrams of the systems Cu - Pt and Cu - Au.

temperature modification of the AuCu type, MnNi(H), forms at 480°C through a peritectoid reaction between the Mn-base α solid solution and the $MnNi_3$ phase. The high- and low-temperature modifications of the MnNi phase have homogeneity regions extending over ~ 5% (Fig. 22).

PtNi forms through the atomic ordering of the f.c.c. (γ) solid solution of Ni and Pt. The homogeneity range of the $L1_0$ phase lies between 40 and 55 at.% Pt. Kurnakov point (645°C) coincides exactly with the stoichiometric composition. PtNi represents a classical example of the formation of the AuCu-type superstructure.

In the **Ni - Au** system, there exists a wide immiscibility gap in the solid state [45]. The investigation of the alloys from this region quenched from various temperatures has shown the formation of the metastable superstructures Ni_3Au, NiAu, and $NiAu_3$ at different stages of the decomposition of the solid solution at temperatures above 500°C. The study of thin films by electron microscopy has revealed a metastable modulated periodic structure in a range of 20 - 70 at.% Ni at 120°C. The ordered NiZn phase of the $L1_0$ type is formed in the Ni-Zn system as a result of the eutectoid decomposition of the B2 phase NiZn at 675°C [140]. At room temperature in the alloys of the NiZn composition, the structure of the AuCu ($L1_0$) type is always observed.

Copper-base $L1_0$-type phases

AuCu has two modifications: tetragonal AuCu I and orthorhombic AuCu II. The latter is a stable phase; the equilibrium position of the AuCu II ⇆ AuCu I transition at high temperatures is fairly well established. The temperature of the disorder-order transition T_c corresponding to the stoichiometric composition is accepted to be 410°C; the temperature of the AuCu II ⇆ AuCu I transition is equal to 385°C. In alloys deviated from the stoichiometric composition, the orthorhombic modification is more stable, even at low temperatures. The ordered phases AuCu and $AuCu_3$ are in a eutectoid equilibrium at 283°; the eutectoid point corresponds to 36 at.% Au [42]. The phase diagram of the system Au - Cu near the equiatomic region is shown in Fig. 23 [85]. Three phases are in equilibrium, namely: the disordered γ solid solution existing in the two modifications, AuCu(II) and AuCu(I), and their mixtures γ + AuCu(II) and AuCu(II) + AuCu(I). According to [141], all transformations in the solid state are of the second order (I - a tetragonal lattice, II - orthorhombic one). The phase diagram of the Au - Cu system, with extensive homogeneity ranges of superstructures $L1_2$ and $L1_0$, calculated by the cluster variation method is given in [142].

PtCu forms by the atomic ordering of the γ solid solution of Cu and Pt at $T_C = 812°$ C corresponding to the stoichiometric composition. The inset in Fig. 23 shows the position of two-phase regions which confirm a classical heterogeneous type of the ordering of PtCu. There are indications [85] that the ordered phases Pt_3Cu and Pt_7Cu exist, but their lattices have not been determined. The type of equilibrium between the ordered phases in the Pt - Cu system has not been investigated. The crystal structure of PtCu corresponds to the ordered phase [85] but the type of the ordering has not been determined as well as for Pt_3Cu and Pt_7Cu. Those two phases are supposed to continuously transform to each other.

2.3 Ordered AB phases of the $L1_2$ type

Phases with a cubic structure of the $AuCu_3$ type occur frequently in metallic alloys, especially in alloys of Mn, and group VIII elements such as Cu, Ag, and Au; they are formed between transition metals (Pt_3Mn, $PtMn_3$, Pt_3Co) as well as between nontransition ones ($MgAg_3$). More typical is their formation in alloys of Fe, Ni, and Pt. These phases are formed, as a rule, by atomic ordering of solid solutions of the type analogous to the formation of the $AuCu_3$ phase. They represent typical superstructures with wide homogeneity ranges. In the following, we shall describe the specific features of the formation of the $L1_2$ phases in alloys of Ti, Fe, Ni, and Cu.

Titanium-base $L1_2$-type phases

The ordered AB_3 phases of the $L1_2$ type in titanium alloys have been found in the systems Ti - Rh, Ti - Ir, Ti - Pt, Ti - Zn, and Ti - In. The $L1_2$ structure in these alloys forms, as a rule, by crystallization from the melt either congruently or incongruently; the alloys are also characterized by the formation of homogeneity regions of the mean concentration.

$TiRh_3$ crystallizes congruently at 1800°C [143]; the homogeneity region at 1200°C extends from 72.5 to 78 at.% Rh. TiRh forms peritectically at 1750°C [144] exhibiting homogeneity of 8 at.%.

$TiIr_3$ crystallizes peritectically at 2315°C [84], has a homogeneity range (ε phase) of 8 at.% and is stable down to low temperatures.

$TiPt_3$ crystallizes congruently at 1950°C [84] being the phase of a constant composition (Fig. 14).

$TiZn_3$ crystallizes congruently at 920°C [145], and has a constant composition.

The $L1_2$ phase has been found in the Ti-In system in the alloy with 33.8 at.% In [146]. The phase Ti_3In is assigned to have a hexagonal lattice of the DO_{19} type at 21 at.% In.

Iron-base $L1_2$-type phases

The ordered $L1_2$-type phases of the AB_3 composition in iron alloys have been found in the systems Fe - Ni, Fe - Pd, Fe - Pt, Fe - Ga, and Fe - Ge. Typically, all these phases are formed by ordering of the γ solid solution and show extensive homogeneity ranges. The maximum ordering temperature (Kurnakov point) usually corresponds to the stoichiometric composition. On cooling, the ordered phases $FeNi_3$, Pd_3Fe, $PtFe_3$ undergo the martensitic transformation (MT) whose temperature decreases with an increase in the degree of atomic order and the mechanism of transformation acquires the features of a thermoelastic transition.

$FeNi_3$ forms by the atomic ordering of the γ solid solution of Fe in Ni at 503°C (the ordering of the $AuCu_3$ type), and has a wide (62 - 82 at.% Ni) homogeneity range at

400°C [43]. The region of the ordering alloys in the Fe - Ni system has been the subject of numerous investigations. Discussion has been concentrated on the following issues: (1) the existence of the second, $L1_2$-type ordered phase of the Fe_3Ni composition, (2) the existence of the equiatomic phase FeNi with a structure of the AuCu type, and (3) the type of equilibrium between these two phases. Also important was the question about the relation between the martensitic $\gamma \leftrightarrow \alpha$ transformation in iron-rich alloys with other concurrent phase transformations in this region and with the atomic ordering. The part of the Fe - Ni phase diagram in a range of 69 - 78 at.% Ni based on the results of the Mössbauer spectroscopic studies [147] is shown in Fig. 24. The alloys were long annealed at 500°C upon deformation. The coordinates of Kurnakov point on the curves corresponding to the onset and the end of the order-disorder transformation are 515°C (788 K) and 73 at.% Ni. The two-phase region of the existence of the ordered and disordered phases corresponds to the hatched area in Fig. 24.

Mössbauer spectroscopy and X-ray diffraction analysis have been employed in [148] to investigate the alloys containing 25 - 50 at.% Ni after quenching from 1000°C and electron bombardment (E = 3 - 5 MeV). The authors present the phase diagram in which the decomposition of the γ solid solution is observed below 400°C. The Ni concentration in the Ni-rich regions amounts to ~ 50%, in the depleted regions ~ 25–27%. The processes of atomic ordering are likely to occur in the submicrovolumes with such content of Ni after long annealing at temperatures below 400°C. At higher temperatures, the equilibrium phase is the γ solid solution, free of the concentration inhomogeneities. The atomic ordering of the $L1_2$ type in the Fe - Ni system near the Fe_3Ni composition is still disputable [149–151]; its temperature and concentration limits as well as the superstructure type have not yet been precisely determined.

Usually, the problem of ordering near the Fe_3Ni composition is being discussed in connection with the nature of the invar properties of iron-nickel alloys. High microinheterogeneity and anomalies in the physical properties some investigators explain by the presence of the fine-dispersed ordered particles of Fe_3Ni emerging at the

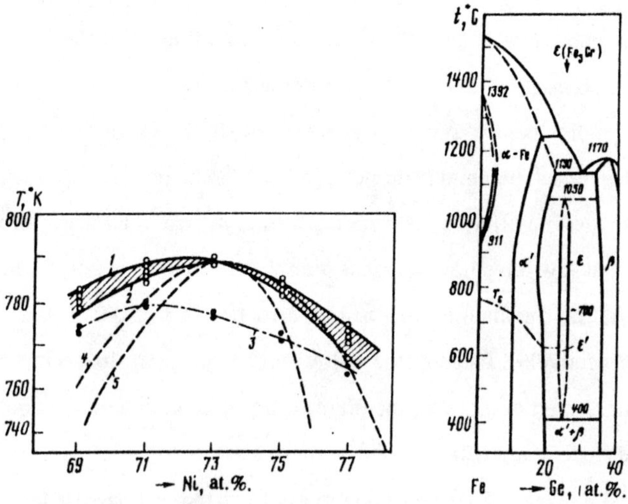

Fig. 24. The partial phase diagrams of the Fe - Ni and Fe - Ge systems in the $L1_2$ phase regions. The beginning of ordering at slow 1° step cooling (curve 1); The beginning of disordering (2); Ordering curve at ordinary cooling (3); Theoretical curves of the beginning and the end of ordering for the $L1_2$ superstructure (curves 4, 5).

earliest stages of precipitation. Low diffusive mobility at moderately high temperatures strongly impedes the processes of atomic redistribution and equilibration of the alloys (especially in a bulk form) and requires very long annealing at temperatures lower than 400°C [152]. The alloys within the concentration range 0 - 40 at.% Ni have been studied in [45]. As a result of the peritectoid reaction between the α and γ solid solution at 350°C, the α_1 phase of the Fe_3Ni composition is formed. Such type of a diagram has not been, however, confirmed later, in spite of attempts of many authors [153 - 155] to verify the existence of the phase Fe_3Ni by using different methods for accelerating the diffusion processes and bringing alloys into the equilibrium state.

Inglesfield [54] has investigated the structural transformations in iron alloys with 25 and 30 at.% Ni using Mössbauer technique and differential thermal analysis (DTA). The alloys were subjected to the 90% reduction in order to prepare thin foils (20 μm thick) and long annealed at 200 and 400°C. This preparation resulted in the decomposition of the f.c.c. (γ) phase into the regions containing the ferromagnetic Ni-rich γ_2 phase and the paramagnetic Ni-depleted γ_1 phase. These microvolumes undergo, in cooling, the martensitic f.c.c. ⇆ b.c.c. transformation at different temperatures depending on the Ni content. The Ni-depleted paramagnetic γ_1 phase in the 30 at.% Ni alloy has the A_s point below room temperature (-110°C) which is evidenced by the occurrence of the central paramagnetic peak in the Mössbauer resonance spectrum. The magnitude of the thermal effect corresponding to the inverse MT for the alloy subjected to heat-treatment at low temperature is less than that for the alloy homogenized at 1000 °C. The heats of transformation are estimated to be 0.68 ± 0.015 and 0.81 ± 0.05 kkal/g-atom, respectively. Indeed, the alloy subjected to a low-temperature heat-treatment contains, even at room temperature, the γ_1 phase, which does not participate in the inverse MT occurring within the basic matrix.

Matveeva and Kolyada [154] are prone to consider the γ_1 phase as the ordered one, by analogy with alloys $PtFe_3$ and Fe_3AlC in which the decrease of the A_s point takes place due to the atomic ordering in the lattice of a solid solution. In [155], the phase diagram has been constructed using the regular solution model and Weiss' assumption that the Fe atoms in the f.c.c. γ solid solution exist in two electronic states, the relative stability of which depends on the composition of the solid solution. The results of this calculation were checked experimentally. The specimens in the form of diffusion pairs were long annealed (during 800 hours) at 1100 - 800°C and lower. The presence of the two f.c.c. γ phases of different composition and magnetic structure in the microregions of Fe-rich alloys allows one to assume the different mode of their MT on cooling. In this respect, the work [155] maintains the correctness of the conclusions made in [154].

In accordance with the generalized data on the investigation of Fe-Pd alloys [86], the phase **Pd_3Fe** forms by atomic ordering of the γ solid solution at 820°C. The maximum ordering temperature corresponds to 70 at.% Pd. There exists a wide range of homogeneity of the ordered γ_2 phase extending from 62 to 84 at.% Pd at 300°C. This phase is in a eutectoid equilibrium with the second ordered $L1_0$-type (γ_1) phase at 750° C. It is known that in Fe-rich alloys of the Fe - Pd system quenching is accompanied by the martensitic $\gamma \leftrightarrow \alpha$ transformation similar to that observed in the Fe - Pt and Fe - Ni systems.

It was shown in [114] that in the alloys containing from 25 to 32 at.% Pd and quenched from 850°C the disordered f.c.c. austenite is converted into a b.c.c. or b.c.t. martensite, depending on whether the M_s point lies above or below the Curie point. In alloys with a b.c.t. martensite, a tetragonal austenite was detected after the inverse MT. These phenomena are completely analogous to those observed in the case of $PtFe_3$, the ordered $L1_2$ phase. According to [114], the MT of the f.c.c. ↔ b.c.t. type is supposed to be of a thermoelastic nature. It may be assumed that long-term annealing at temperatures above the MT temperature would result in the ordering of the atomic structure of the γ solid solution (of the $PdFe_3$ type) and in the decrease in the temperature of the subsequent MT, by analogy with the alloy $PtFe_3$ [149].

According to the data on the Pt-Fe phase diagram generalized in [86], the phases **Pt_3Fe** and **$PtFe_3$** form by atomic ordering of the γ solid solution at 1300 and 1350°C, respectively. They have extensive homogeneity regions of the existence of the γ_3 and γ_1 phases (Fig. 21). The state of atomic order in the γ_1 region substantially affects the MT in the alloys containing 24 to 27 at.% Pt.

During the 70s, a series of important investigations appeared concerning the influence of atomic ordering in solid solutions upon the nature of the martensitic transformation and the related inelasticity phenomenon. A great attention was paid to alloys near the $PtFe_3$ composition at which several competing reactions proceed in the solid state. There have been studied, particularly, kinetics of the $\gamma \leftrightarrow \alpha$ MT [149, 156, 157], the crystal structure of austenite and martensite [158–161], the dependence of the shape memory effect [149, 157] on temperature and annealing time for alloys close to the $PtFe_3$ composition where atomic ordering of the $A1 \rightarrow L1_2$ type takes place.

According to [156], the main factor determining the thermoelastic nature of the MT is the position of the M_s point depending on the degree of long-range order; this point must lie below the Curie point. In [162], the thermal effects and the temperatures of the $\gamma \leftrightarrow \alpha$ transformations for the alloys $Fe_{75}Pt_{25}$ and $Fe_{76}Pt_{24}$ undergoing atomic ordering at 550°C have been measured by calorimetric technique. Specimens were held

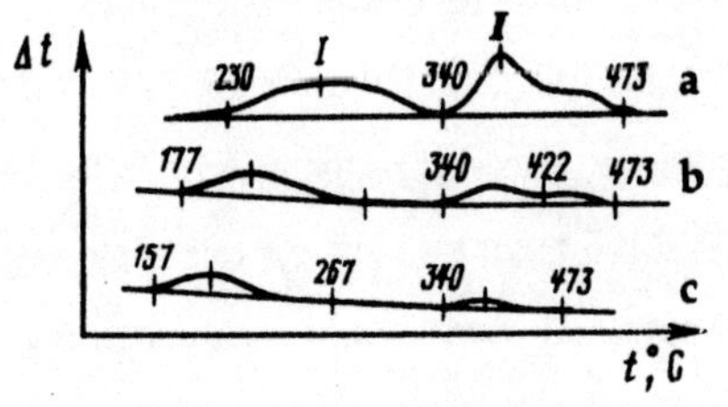

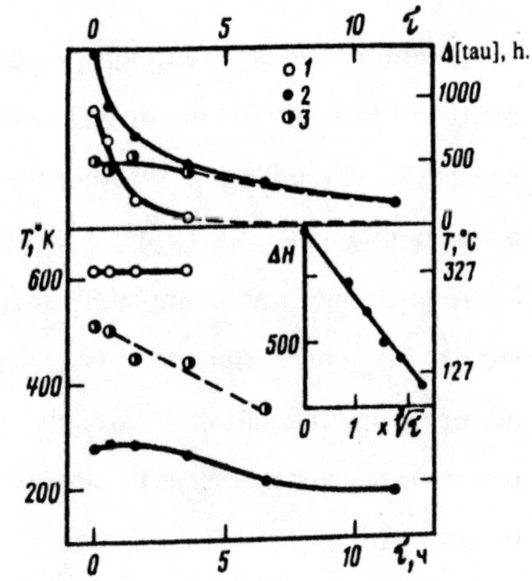

Fig. 25. Heating curves for the $Fe_{75}Pt_{25}$ alloy, three cycles. Annealing: 0.5 h (a); 1.5 h (b); 3.5 h (c).

Fig. 26. Variation of the temperatures and heat of the $\gamma \leftrightarrow \alpha$ transformation in the $Fe_{75}Pt_{25}$ alloy with annealing time.

for some time in a DSC-111 Setaram calorimeter at 550°C, cooled to -125°C and heated repeatedly to 500°C. Each cycle consisted of annealing followed by cooling and subsequent heating.

The heating curves for three such cycles for the alloy $Fe_{75}Pt_{25}$ are presented in Fig. 25; they involve two thermal effects related to the inverse $\alpha \rightarrow \gamma$ transformation; the values of these effects, like those in the direct $\gamma \rightarrow \alpha$ transformation, successively decrease. The point A_s and the enthalpy $\Delta H_{\alpha \rightarrow \gamma}$ of the first effect (I) decrease with annealing time; the second effect (II) point does not change up to the complete disappearance of the effect after 6 h annealing which is accompanied by a drastic decrease of the $\Delta H_{\alpha \rightarrow \gamma}$ value. This gives evidence that the process of the A1 - $L1_2$ ordering proceeds in a heterogeneous way; the ordered and disordered austenites are present simultaneously in the alloy, their relative amount changing during the annealing process.

The temperature of the first ($\alpha \rightarrow \gamma$) effect drops which may imply, in accordance with the results of [149], an inelastic nature of the MT in the case of the ordered high-temperature γ phase. This conclusion is also confirmed by a high degree of shape recovery in the process of the inverse $\alpha \rightarrow \gamma$ MT upon heating the alloy $Fe_{76}Pt_{24}$ subjected to preliminary 20 h annealing at 550°C. The temperatures and the heats of the $\gamma \leftrightarrows \alpha$ transformation are plotted against annealing time in Fig. 26.

Analysis of these data shows that the decrease of $\Delta H_{\gamma \leftrightarrows \alpha}$ with annealing time should be connected at least with two different processes. At the first stages of annealing when ΔH rapidly declines but the temperature of the beginning of the transformation remains unchanged, it may be assumed that merely the amount of transformed phase decreases. Further, at longer annealing time, the thermodynamical parameters are likely to change. This is evidenced by changing of the transformation temperature. The dependence of ΔH upon annealing time is well approximated by the equation $\Delta H_{\gamma \rightarrow \alpha} = k\sqrt{\tau} + \Delta H_{\tau=0}$ (Fig. 26c). Such dependence may appear if the decrease of the amount of the transformed phase is due to diffusion processes. The dependences of $\Delta H_{\alpha \rightarrow \gamma}(II)$ and $A_s(II)$ on time do not differ from those observed for $\Delta H_{\gamma \rightarrow \alpha}$ and M_t at short annealing time.

The processes of $\alpha \rightarrow \gamma(I)$ transition proceed in a different way. The value of $\Delta H_{\alpha \rightarrow \gamma}$ (I) slightly changes with time, while the temperature of the beginning of the transition drops sharply. This temperature seems to be a more sensitive characteristic than ΔH and, if so, the absence of the pronounced dependence of ΔH on time (Fig. 26) gives evidence of a small change in the amount of the transformed phase. X-ray diffraction data confirm these assumptions. The phase composition of the alloys subjected to 10 h annealing at 550°C and 600°C is different. Upon annealing at 550°C, the alloys contain, along with the partially ordered austenite, also the α phase which is absent in alloys annealed at 600°C. The degree of order in the γ phase is higher but does not exceed 0.5. This gives explicit evidence that at temperatures just below 600°C, the eutectoid decomposition of the γ phase ($\gamma \rightarrow \alpha + PtFe_3$) occurs and the amount of this phase decreases.

In annealing of the quenched alloy containing a high concentration of thermal vacancies, the initial rate of the eutectoid decomposition of the γ phase is rather high. In cooling, the decomposition is broken in the martensitic transformation and "freezes down". Then, in the course of annealing, the atomic ordering proceeds predominantly, affecting the subsequent martensitic transformation. The drop in $\Delta H_{\gamma \leftrightarrows \alpha}$ is slowing down. After 10 h annealing of the $Fe_{76}Pt_{24}$ alloy, $\Delta H_{\gamma \rightarrow \alpha}$ falls down to the value of 522 J/mole.

Hence, isothermal annealing of alloys close to $PtFe_3$ at 550°C followed by cooling and heating within the range of the $\gamma \leftrightarrows \alpha$ transformation causes, at the first cycles, both the atomic ordering of the A1 $\rightarrow$ $L1_2$ type and the eutectoid decomposition $\gamma \rightarrow \alpha +$ $PtFe_3$. The decrease in the amount of the γ phase leads to a sharp drop in the $\gamma \rightarrow \alpha$ transformation heat. At subsequent cycles, the MT is mainly influenced by the heterogeneous atomic ordering, accompanied by the decrease of the critical points of the $\gamma \leftrightarrows \alpha$ transition. The MT in ordered alloys is of a thermoelastic nature.

Fe_3Ga forms peritectoidally in the solid state at 619°C [163]. The phase contains 21.6 at.% Ga at 605°C and 29.5 at.% Ga at 619°C. At 300°C it extends from 26.29 to 28.6 at.% Ga (Fig. 22).

Figure 24 shows the Fe-rich part of the Fe - Ga phase diagram. The ordered phase Fe_3Ge of the DO_{19} type (ε phase) forms by a peritectoid reaction at 1050°C and undergoes the homogeneous transition, $DO_{19} \leftrightarrows L1_2$, at 700°C in a range of 23.8 - 26 at.% Ge. The L12-type (ε') phase decomposes eutectoidally into the ordered α' phase and the β phase of the NiAs type at 400°C [164].

Nickel-base $L1_2$-type phases

Ordered AB_3 phases of the $L1_2$ type are distinctly revealed in the systems Ni - Mn, Ni - Fe, Ni - Al, Ni - Ga, Ni - In, and Ni - Si. They form either peritectically or in the solid state through a peritectoid reaction. In both cases, the compositions of the liquid and solid phases from which the ordered phases have formed do not correspond to the composition of the latter. The composition AB_3 in the phase diagram corresponds to a closed maximum. The $L1_2$ phases in the Ni - Fe system are formed by the atomic ordering of the γ solid solution (see Section 2.1; Fe-base alloys).

$MnNi_3$ with a structure of the $L1_2$ type [139] forms through a peritectoid reaction between αMn and γ solid solution at 520°C. The ordering in the $MnNi_3$ region proceeds by the nucleation and growth mechanism, with the coexisting ordered and disordered phases. The phase diagram of the Ni-Mn system in the range of the ordered phases is presented in Fig. 22. The calculated diagram [166] confirms a complex mechanism of the formation of the ordered phases in the MnNi and $MnNi_3$ alloys in the solid state.

The reference book [43] contains just an indication on the formation of the ordered Ni_3Co phase in the Ni - Co system.

Numerous investigations showed that an unlimited miscibility takes place in the system Ni - Au in the liquid state, whereas at high temperatures it occurs both in the liquid and solid states. It has been established that below 820°C a homogeneous solid solution decomposes into the mixture of Au- and Ni-base solid solutions in accordance with the two-phase mechanism. At certain stages of decomposition above 500°C, the ordered structures (stable in a narrow temperature range) $NiAu_3$, NiAu (of the AuCu type), and Ni_3Au (of the $AuCu_3$ type) are formed . The Ni_3Au composition corresponds to a sharp maximum on the specific resistivity vs. concentration curve [45].

Ni_3Al forms peritectically at 1395°C, with a narrow homogeneity region (γ_1 phase) extending from 72 to 77 at.% Al at 800°C. According to [42], this phase is stable down to low temperature. The boundaries of the phase region $\beta + \gamma'$ (γ' is the ordered phase) are shown in Fig. 17. The phase diagram of the Ni - Al system has been constructed by many workers using the thermodynamical data for the constituent elements. The type of the calculated diagram coincides with the experimental one.

Ni_3Ga forms peritectically at 1207°C, showing a homogeneity range within 24 - 28 at.% Ga at 600°C [168].

Ni_3In forms by a peritectoid reaction at 848°C. The homogeneity of Ni_3In is very narrow [169].

The phase diagram of the Ni - Si system has been studied in detail experimentally [43, 170] and by the calculation methods [171]. Part of the experimental diagram is given in [170]; the β phase (Ni_3Si) exists in three modifications, β_1, β_2, and β_3. The β

3 phase forms peritectically at 1180°C. The phase transition $\beta_3 \leftrightarrows \beta_2$ occurs at 1125°C. The phase β_1 forms through the peritectoid reaction $\alpha + \beta_2 \rightarrow \beta_1$ at 1010°C. All the β-type phases have narrow homogeneity regions. The β_1 phase has the ordered structure of the $L1_2$ ($AuCu_3$) type, its homogeneity region is shifted to the Ni-rich side and lies between 21.8 and 24.2 at.% Ni at 900°C [170]. The β_3 phase has an orthorhombic lattice of the Fe_3C type.

Copper-base $L1_2$-type phases

The ordered $L1_2$-type phases in copper alloys based on a cubic lattice are reliably identified in the systems Pd - Cu, Pt - Cu, and Au - Cu. Typically they form in a classical fashion as a result of the atomic ordering of the γ solid solutions, show wide homogeneity ranges and are stable down to low temperatures.

In the reference book [85], we can find but a mention of the possibility of forming the ordered $L1_2$-type phases at 50, 75 and 25 at.% Rh in the system Rh - Cu. However, the phase diagram of this system represents the diagram with a binodal curve within a range of 20 to 90 at.% Rh which bounds the decomposition region with the cupola at 1100°C.

$PdCu_3$ forms by the atomic ordering of the γ solid solution. The concentration dependence of the ordering temperatures shows a flat maximum within the range 15–20 at.% Pd at 500°C; the region of the homogeneous phase extends from ~ 10 to just above 25 at.% Pd (Fig. 18). The distinction is made in the homogeneity region between two types of ordering. The α'' phase has an f.c.c. lattice of the $L1_2$ type, the α' phase is tetragonal, with c/a=0.986 [85].

$PtCu_3$. forms by the atomic ordering of the γ solid solution of Cu and Pt [85]. Kurnakov point (645°C) does not coincide with the stoichiometric composition. There exists an extensive region of homogeneity based on the $L1_2$ structure (Fig. 23).

As Hansen still noticed, the Au - Cu system became a favorable object for investigating the ordering phenomenon by experimental and theoretical methods. A great number of works has been devoted to various issues of the order-disorder process.

The investigations in the 50s have come to the important conclusion that the ordering in alloys of this system is of a classical heterogeneous type. In the present chapter we consider only the questions concerning the Pt - Cu equilibrium diagram in the range of ordered phases.

$AuCu_3$. The most reliable position of Kurnakov point is estimated to be 388 ± 3°C. According to [43, 85], the homogeneity region ranges from 20 to 34 at.% Au at 300°C. The $AuCu_3$ phase is in a eutectoid equilibrium at 285°C with the orthorhombic ordered phase AuCu (II) (Fig. 23). Like the ordered phase AuCu, the phase $AuCu_3$ has thestructure $AuCu_3$ (II), which is described as consisting of 18 unit cells $AuCu_3$ (I) with the antiphase domain boundary occurring after each 9 cells. The $AuCu_3$ (II) phase is assumed to exist between 347 and 357°C at 31 at.% Au. The $AuCu_3$(I) phase has a cubic structure of the $L1_2$ type.

2.4 Ordered AB_4 phases of the D1a type

There exist a number of phases with a tetragonal structure of the $MoNi_4$ type. They form in alloys of Au with transition IV to VII group metals as well as in alloys of Ni with Mo and W. As a rule, these phase form in the solid state either through a peritectoid reaction or by the ordering of a solid solution.

A typical representative of this class is the $MoNi_4$ phase which forms peritectoidally at 865°C after a long-term annealing, showing an insignificant homogeneity range (β phase, Fig. 27). Later investigations confirmed the type of the diagram near $MoNi_4$ suggested in [172]. According to the calculated equilibrium diagram [173], the phase $MoNi_4$ forms by the atomic ordering of the γ solid solution of Mo in Ni at ~ 900°C.

WNi_4 forms by a peritectoid reaction at 1000°C and has a constant composition. The phase diagram of the system W - Ni constructed by calculation and experimental methods is given in [174]. The alloys were prepared by pressing powders and subsequent sintering them in vacuo during 6 hours at 1130°C.

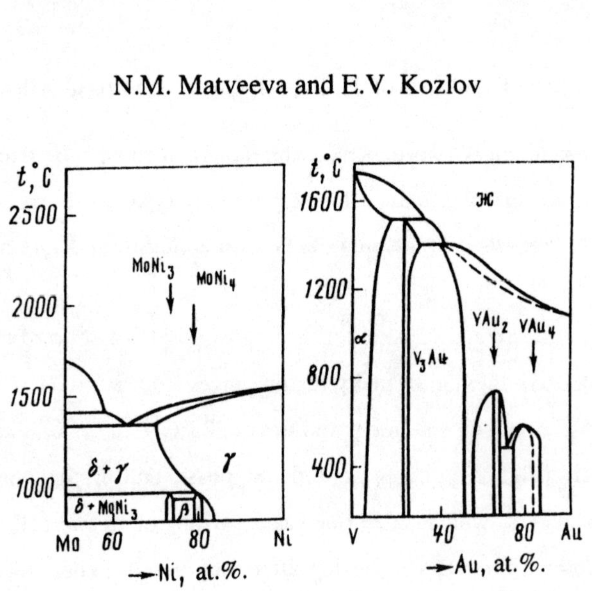

Fig. 27. The partial phase diagrams of the Mo - Ni and V - Au systems in the region of the $D1_a$-type ordered phases.

The **$TiAu_4$** phase with a structure of the $MoNi_4$ type forms in the Ti - Au system by a peritectoid reaction at 1172°C and has a homogeneity of ~ 5 at.% (Fig.15). This is the single phase of $MoNi_4$ type that crystallizes with the participation of a liquid phase [84, 85]. The system Au - Hf was investigated in [175]. It has been found that $HfAu_5$ (85 at.% Au) has a structure of the $MoNi_4$ type, while the $HfAu_4$ compound has an orthorhombic structure of the $ZrAu_4$ type that is not inherent in superstructures. The mode of phase formation in this system is not established.

VAu_4 forms by the atomic ordering of the γ solid solution of V in Au [176, 177]. Ordering temperature versus concentration curve has a maximum (Kurnakov point) at 40°C corresponding to the stoichiometric composition. The phase has a wide homogeneity range (Fig. 27).

$CrAu_4$ forms by the atomic ordering of the γ solid solution of Cr in Au at temperatures from 200 to 400°C and concentrations in a range of 10 to 20 at.% Cr. A 40-day annealing time at 315°C was required to detect the superstructure of the $MoNi_4$ type in the 20 at.% Cr alloy [45].

$MnAu_4$ is the phase of a variable composition formed by a peritectoid reaction at 450°C [178]. The phase diagram of the Mn - Au system comprising several ordered phases of different types was constructed by generalizing the data on phase equilibria and crystal structures of intermediate phases [179].

2.5 Martensitic transformations in B2 phases

Systematization of the ordered phases showed that B2-type phases are the most abundant class of superstructures. A specific feature of these phases is the instability of the B2 lattice with respect to the martensitic transformations [120]. As shown in [156 - 159], the degree of long-range order determines a thermoelastic nature of the martensitic transformation in these phases and the coherence of the interphase boundaries due to the defectless joining of the lattices that transform to one another. The thermoelastic MT explains an inelastic deformation behavior of the alloys; this transformation is accompanied by such unique phenomena as superelasticity, shape recovery and a high generation of internal stresses in the shape memory effect, a cyclic resistance observed in the temperature range of the MT, etc. An extraordinary scientific and practical interest in these alloys is quite natural.

The thermoelastic martensitic transformations occur in B2 phases formed by nontransition metals (especially in Cu-, Ag- and Au-base phases) as well as by transition metals, in particular, Ti. The first group of alloys is considered in detail in the book [120]. The most detailed investigations of the second group have been carried out for the B2 phase of the intermetallic compound TiNi [43, 88, 90 - 95, 98]. With regard to other representatives of the Ti-base B2 phases, most of the papers have been published in the last decade because some international and especially all-union conferences on the problem of the shape memory effect were held that time.

The study of the MT in the quasibinary systems based on the B2 phases is a new direction of investigations developed primarily in our country [95 - 100]. The major

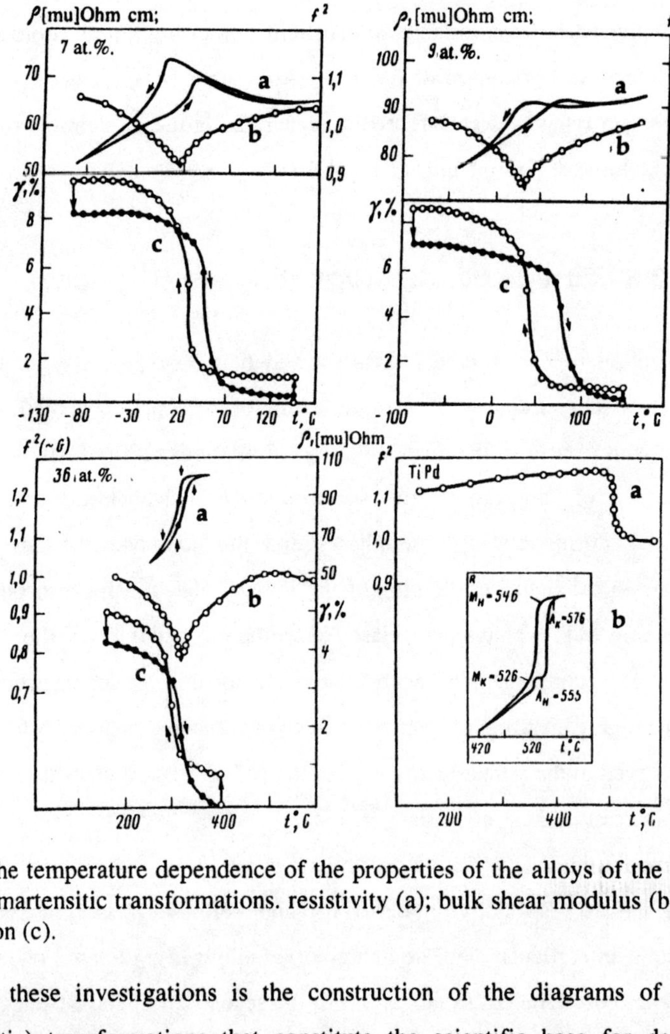

Fig. 28. The temperature dependence of the properties of the alloys of the TiNi - TiPd system at martensitic transformations. resistivity (a); bulk shear modulus (b); reversible deformation (c).

result of these investigations is the construction of the diagrams of diffusionless (martensitic) transformations that constitute the scientific base for devising shape memory alloys. Such diagrams give information on temperatures, the type of MT, and variations in the crystal structure. Using these diagrams, in parallel with the investigation of a variety of physicochemical and thermomechanical properties as a function of the concentration, enables one to optimize shape memory characteristics. The construction of the MT diagrams for a series of quasibinary systems provides the possibility of classifying them by the type of MT, the stabilization of transforming phases, the sequence and the evolution of MT, and, as a result, allows one to conduct a directed search for the alloys with the required properties.

Analysis of the literature data showed that the B2 phases TiNi, TiCo, TiFe, TiPd, TiPt, and TiAu are isomorphous at high temperatures and form continuous solid solutions of the B2 (CsCl) type in their mutual systems. The initial B2 structure is unstable with respect to the two martensitic transformations, B2 → B19 and B2 → R(ω), where B19 is an orthorhombic martensite of the AuCu type and R(ω) is a rhombohedrically distorted B2 phase. The temperatures of the B2 lattice stability loss with respect to these two types of atomic displacements (M_H and T_R) depend in different ways upon the composition. For example, alloying TiNi with Pd, Pt and Au gives rise to the stabilization of the martensite structure; on alloying TiNi with Co, Fe, Ni, the temperature T_R of the B2 → R(ω) MT decreases; the martensite phase region remains limited down to the liquid nitrogen temperature range. Similar dependence of the MT temperatures is observed upon alloying the other Ti B2 phases. This fact allows one to single out two types of the MT diagrams of the quasibinary systems with B2 phases in accordance with their geometric consruction and final crystal structure of the martensite phase. Below we give the experimental results of the investigation of the MT evolution in the Ti-base B2-phase systems for the two types of the MT diagrams using, as examples, the systems TiNi - TiPd and TiNi - TiCo.

The investigations were performed by X-ray diffraction methods (heating and cooling), DTA and the measurement of the bulk shear modulus and electric resistivity as a function of temperature. The shape memory effect (the value of the reversible deformation), the temperature range where this effect occurs, the degree of shape recovery, and the MT temperature hysteresis have been studied in torsion strain experiments.

It was shown [180] that in the stoichiometric TiNi alloy the phase transition takes place on cooling, with the formation of the monoclinic martensite B19'. But after thermal cycling or additional deformation, i.e., under the ordinary conditions in which the shape memory effect is observed, the MT scheme changes for the B2 → R(ω) →

B19' transformation. The substitution of Ni for atoms of greater size (Pd, Pt, Au) gives rise to the same effect causing significant internal stresses in the lattice of the initial B2 phase.

For this reason, when constructing the MT diagram of the TiNi - TiMe (Me = Pd,Pt,Au) systems, the actual phase composition of the TiNi alloy at the moment when, on cooling, the second martensitic transformation begins to proceed in the two-phase region B2 + B19' has been taken into account. It is known that this transformation (T_R) is accompanied by an unusually sharp rise in resistivity and an abrupt drop in the shear modulus. The temperature dependences of various properties of the TiNi - TiPd system alloys for different regions of the phase diagrams are shown in Fig. 28. The MT diagram obtained by cooling of the alloys is given in Fig. 29.

The diagram includes the region where the B2(ω) state exists. This state has a structure of the B2 phase with uncorrelated static displacements of atoms similar to those in the ω phase, i.e., it is characterized by a longitudinal atomic displacement wave 2/3<111><111> in the static plane-wave model. The temperature T_ω corresponding to the B2(ω) state in alloys of different composition was deduced from resistivity vs. temperature curves; it is associated with the onset of a slow rise in elastic resistivity on cooling (Fig. 28).

The most typical for the alloys of the quasibinary systems with the stabilized B19 structure is the phase transition B19 → B19', close to the second-order transitions. It proceeds in all the first-type systems studied (including TiNi - TiPd system) in alloys of a definite composition (~ 10 at.% V at room temperature). These alloys were called the boundary ones [96]. At the boundary concentrations of the alloying elements, the sequence of the martensitic transformations changes from a simple transition B2 → B19 (in alloys with a high content of Pd, Pt or Au) through the complex chains B2 → B2 + B19' → R(ω) + B19' → B19' and B2 → R(ω) → R(ω) + B19' → B19' to another simple transition B2 → B19' (corresponding to small concentrations of Pd, Pt, Au, Cu). In alloys of the TiNi - TiPd(Pt,Au) systems containing more than 15% Pd(Pt,Au), the final martensitic phase will be of the B19 type. A more detailed investigation of the crystal structures in the martensitic transformations occurring in the above systems is described in [95–100].

The MT diagrams in the systems TiNi(Pd,Pt,Au) - TiCo(Fe) refer to the diagrams of the second type; their characteristic features are a wide region of the high-temperature B2 phase, the limited regions of the martensitic phases B19' and R(ω), and the absence of the B19 structure in the systems TiNi - TiCo(Fe). The specific features of the MT in alloys with the stabilized B2 structure are exemplified by the alloys of the TiNi - TiCo system in Fig. 30 [181, 182]. As the Co content in alloys increases, the MT scheme B2 → B2 + B19' → R(ω) + B19' → B19' changes for the sequence B2 → R(ω) → R(ω) + B19' → B19'. Simultaneously, the temperature range between T_R (B2 → R(ω)) and M_H (R(ω) → B19') increases. In alloys containing more than 10 at.% Co, a monoclinic martensite does not realize; only the M_T of the B2→R(ω) type occurs in the alloys with 10 to 15 at.% Co. A slight rhombohedric distortion of the B2 lattice is accompanied by the broadening of the Bragg reflection (110) while a strong distortion leads to the appearance of the $(110)_{B2}$ doublet. The MT diagram corresponding to the cooling of the alloys of the TiNi - TiFe system is shown in Fig. 31 [183].

In the investigation of the martensitic transformations in the quasibinary systems based on the equiatomic titanium compounds with the B2 structure, close attention was focused on the premartensitic states of the initial high-temperature B2-type lattice [182]. The signs of the instability of the B2 lattice manifest themselves in an anomalous behavior of the shear modulus (dG/dI<0) and electrical resistivity (dρ/dT<0) near the MT. In a wide concentration range of the B2 structure where different sequences and types of the MT are realized, one can trace the degree and the range of the premartensitic "softening" of the initial B2 lattice. It has been established that the degree and range of the softening are considerably diminished near the compositions TiPd, TiPt, TiAu where the transformation B2 → B19 takes place at high temperatures. As the Ni content increases in these alloys, with the simultaneous change of the

Fig. 29.

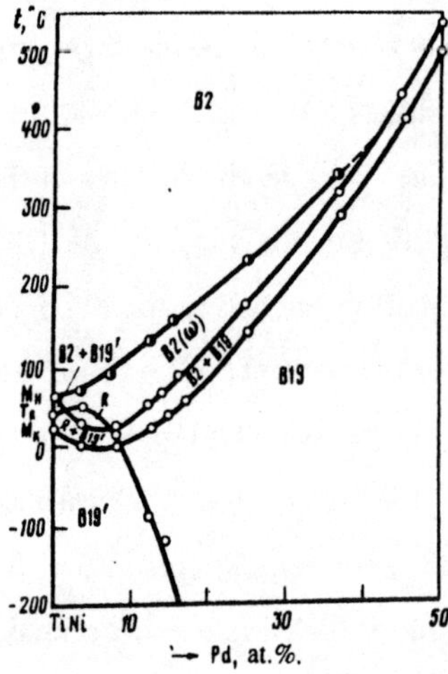

Fig. 30.

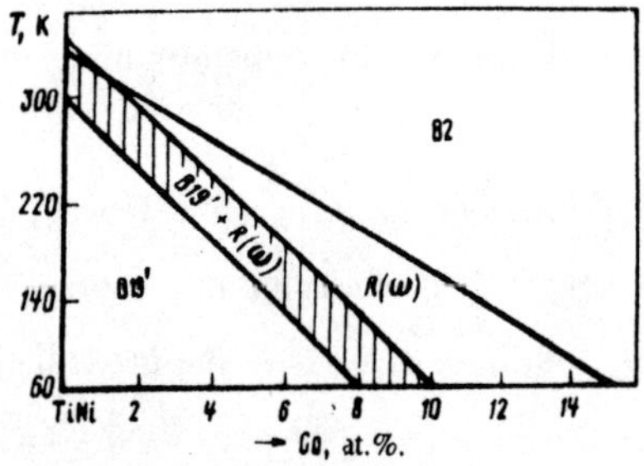

Fig. 31.

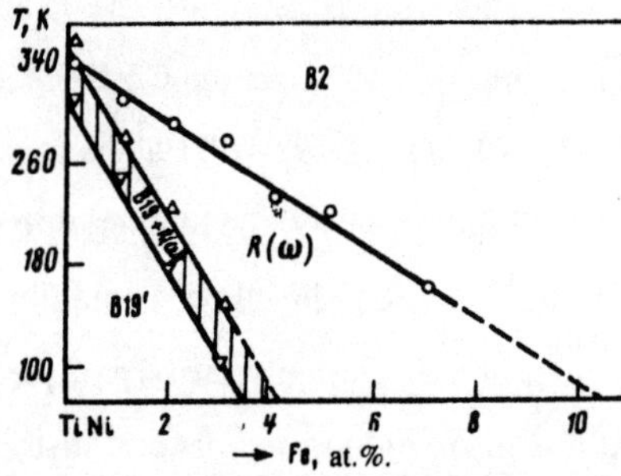

Fig. 29. The diagram of martensitic transformations (on cooling) in the TiNi - TiPd system.

Fig. 30. The diagram of martensitic transformations (on cooling) in the TiNi - TiCo system.

Fig. 31. The diagram of martensitic transformations (on cooling) in the TiNi - TiFe system.

sequence of the martensitic transformations from B2 → B19 through B2 → B19 → B19' and B2 → R(ω) → B19' to the B2 → B19' transition, the degree and the range of the premartensitic softening increase.

Specific features of the evolution of the premartensitic state have been clearly traced in the system TiFe - TiPd [184]. The B2 structure of TiFe is stable down to 0 K; TiPd undergoes the martensitic transition at 780 K. For this reason, on alloying with Pd, one can seen the first signs of the loss of the B2 lattice stability with respect to the B2 → B19 transformation. The exact elemental composition at the different stages of destabilization of the B2 structure and the temperatures of these stages can also be determined. The process of destabilization proceeds gradually as the lattice distortion increases and a greater volume of the material becomes involved in the MT. For example, the premartensitic "softening" of the shear modulus strongly increases in the alloy with 33 at.% Pd. The martensitic B2 → B19 transformation occurs at high temperatures and covers the whole volume of the material.

Consequently, the "softening" of the lattice prior to the MT is the necessary condition for the occurrence of the thermoelastic MT in the investigated alloys; the rate of the evolution of this process determines the magnitude of the lattice deformation and, consequently, the value of the reversible deformation in the shape memory effect.

The systematic investigation of the MT in more than six quasibinary systems based on Ti B2 compounds and analysis of the literature data on the MT in TiNi enable the total general scheme of martensitic transformations in the complex TiNi-base systems to be obtained:

$$\begin{array}{ccccccccccc}
B2 & \rightarrow & B2 & \rightarrow & B2(\omega) & \rightarrow & B2(\omega) & \rightarrow & B2(\omega) & \rightarrow & R(\omega) \\
\downarrow & & \downarrow & & \downarrow & & \downarrow & & \downarrow & & \\
B19 & & B19' & & B19' & & B19 & & R(\omega) & & \\
 & & & & \downarrow & & \downarrow & & \downarrow & & \\
 & & & & R(\omega) & & B19' & & B19' & & \\
 & & & & \downarrow & & & & & & \\
 & & & & B19' & & & & & &
\end{array}$$

A particular sequence of the martensitic transformations can be realized only in a definite concentration range of the alloys of quasibinary systems.

To predict the type of the interaction between B2 phases in ternary systems, the metallochemical approach [83, 84] can be adopted, involving the isomorphism of the phases, the mode of their formation in binary systems, the size and other factors as well as the data from Chapter 2. It may be assumed, for example, that the quasibinary type of equilibrium between the Ti-base B2 phases undergoing the MT would be realized in the mutual systems of the following phases: TiNi, TiCo, TiFe, TiRh, TiPd, TiIr, TiPt, TiAu, and TiZn. The type of the phase diagram would be determined by the presence of the MT in both (type I) or two (type II) of the B2 phases.

3. ATOMIC ORDER IN SOLID SOLUTIONS

3.1 Short-range and long-range order parameters. Correlation functions and fluctuations

In metallic solid solutions atoms are rarely distributed over lattice sites in a completely random manner. Usually, there exists a tendency for any particular kind of atom to have in the nearest "coordination shells" an excessive number of like or unlike atoms compared to a random distribution. A binary alloy is considered to exhibit short-range clustering if the atoms have, on the average, an excess of the same kind of atoms in the nearest neighborhood. In the opposite case, when the number of unlike atoms on the neighboring sites is greater than the random value, the alloy is said to have short-range order [186, 187]. Such classification is however quite arbitrary because our knowledge of the nearest atom pairs is insufficient to completely describe the state of atomic order in a solid solution; one has to know the dependence of the number of different pairs on the distance and the occurrence of different atom complexes in the alloy as well. Nevertheless, using this well-established classification we want to note that all what follows is related to ordering alloys, i.e., to ones showing a long-range or short-range order in a solid solution.

Usually, long-range order exists in a solid solution at temperatures below the so-called critical temperature T_C [188, 189]. The order-disorder transition may be of the first or second order or proceed through a two-phase equilibrium region. In this sense the concept of the critical temperature seems to be somewhat uncertain. However, this term will be used henceforth. When long-range order is formed in an alloy, the probabilities of finding different kinds of atoms at different sites become unequal and the lattice may be divided into several sublattices the number and mutual position of which defines a given superstructure. The formation of a superstructure is always accompanied with the lowering of the lattice symmetry [19, 24].

The state of atomic order in an alloy is uniquely determined by **the set of random numbers,** $\{C_i^\alpha\}$ where $\{i\}\equiv R_i$ denotes the set of lattice sites, and α defines the atom species:

$$C_i^\alpha = \begin{cases} 1 & \text{if the site i is occupied by atom } \alpha, \\ 0 & \text{if the site i is occupied by an atom of different kind} \end{cases} \quad (3.1.1)$$

Most suitable for specifying the state of order are **the *a priory* one-atom probabilities** P_i^α, the pair probabilities $P_{ij}^{\alpha\beta}$ or, in general, many-atom probabilities of finding different kind of atoms at different sites. These parameters or the correlation functions are defined as the quantities averaged over the ensemble of random numbers C_i^α or their products, i.e., they represent the initial moments of deviations from the average values

$$\begin{aligned} P_i^\alpha &= \langle C_i^\alpha \rangle \\ P_{ij}^{\alpha\beta} &= \langle C_i^\alpha C_j^\beta \rangle \\ &\vdots \\ P_{i_1 i_2 \ldots i_n}^{\alpha_1\alpha_2\ldots\alpha_n} &= \langle C_{i_1}^{\alpha_1} C_{i_2}^{\alpha_2} \ldots C_{i_n}^{\alpha_n} \rangle. \end{aligned} \quad (3.1.2)$$

The brackets <...> denote averaging over the statistical ensemble. For the n-component alloy, the following **normalization relations** between the *a priori* probabilities hold:

$$\sum_{\alpha=1}^{n} P_i^\alpha = 1;$$

$$\sum_{\alpha=1}^{n} P_{ij}^{\alpha\beta} = P_j^\beta; \quad \sum_{\beta=1}^{n} P_{ij}^{\alpha\beta} = P_i^\alpha; \quad \sum_{\alpha,\beta=1}^{n} P_{ij}^{\alpha\beta} = 1 \ldots \quad (3.1.3)$$

The expressions for the correlation functions may be written in a different form to underline the difference between the averaging over a given state and over the ensemble of states (see, for example, [190]).

$$P_i^{\alpha} = \sum_{\gamma} \langle \gamma | \rho(\gamma) \frac{1}{N} \sum_{k=0}^{N-1} C_{i+k}^{\alpha} | \gamma \rangle ,$$

$$P_{ij}^{\alpha\beta} = \sum_{\gamma} \langle \gamma | \rho(\gamma) \frac{1}{N} \sum_{k=0}^{N-1} C_{i+k}^{\alpha} C_{j+k}^{\beta} | \gamma \rangle . \tag{3.1.4}$$

The inner summation (N is the number of lattice sites) relates to the averaging over the state and the outer one -- over the ensemble of states. Instead of the *a priori* probabilities, one can introduce **the *a posteriori* probabilities** [82, 191] for atoms of different kind to occupy sites in the lattice provided the one-atom probabilities P_i^{α} are fixed:

$$P\left({}_{i}^{\alpha} \,|\, {}_{j}^{\beta}\right); \;\; P\left({}_{i}^{\alpha} \,|\, {}_{jk}^{\beta\gamma}\right) \ldots \tag{3.1.5}$$

In some cases it is more convenient to use a set of special order parameters. Among them the most commonly used are the Bragg-Williams long-range order parameters [192, 193]:

$$\eta = \frac{P_i^A - C_A}{1 - \nu} . \tag{3.1.6}$$

The index i denotes the site of kind α, ν is the concentration of i sites, c_A is concentration of A atoms. The parameter η is equal to unity for the perfectly ordered alloy of stoichiometric composition and to zero for the alloy without long-range order. This parameter is widely used in experimental and theoretical description of the extent of superstructure perfection. Another parameter, η^*, is often used for the alloys of nonstoichiometric composition [188]:

$$\eta^* = \frac{P_i^A - C_A}{\gamma} , \tag{3.1.7}$$

where

$$\gamma = \begin{cases} C_A \dfrac{1-\nu}{\nu}, & \text{when } C_A \leqslant \nu, \\ 1 - C_A, & \text{when } C_A > \nu. \end{cases} \tag{3.1.8}$$

Unlike η, the maximum degree of order in an alloy of nonstoichiometric composition corresponds to $\eta^*=1$. Several order parameters are required to specify the state of order in more complex superstructures. Their number will depend upon the number of nonequivalent sublattices Q and the number of components n [188]:

$$M = (Q-1)(n-1). \tag{3.1.9}$$

Some examples will be given later in connection with the superstructures studied.

Pair correlation in alloys is often described by the Cowley-Warren parameter [194, 195]:

$$\alpha_{ij} = 1 - \frac{P_{ij}^{AB}}{C_A C_B}. \tag{3.1.10}$$

which is most appropriate for processing the results of the measurement of local order in alloys by the X-ray diffraction method.

Another convenient types of the order parameters are the correlation parameters or central moments:

$$\begin{aligned} \epsilon_{ij}^{\alpha\beta} &= \langle (C_i^\alpha - \langle C_i^\alpha \rangle)(C_j^\beta - \langle C_j^\beta \rangle)\rangle, \\ \epsilon_{ijk}^{\alpha\beta\gamma} &= \langle (C_i^\alpha - \langle C_i^\alpha \rangle)(C_j^\beta - \langle C_j^\beta \rangle)(C_k^\gamma - \langle C_k^\gamma \rangle)\rangle, \\ \epsilon_{i_1 i_2 i_3 \ldots i_n}^{\alpha_1 \alpha_2 \alpha_3 \ldots \alpha_n} &= \langle (C_{i_1}^{\alpha_1} - \langle C_{i_1}^{\alpha_1} \rangle)(C_{i_2}^{\alpha_2} - \langle C_{i_2}^{\alpha_2} \rangle) \ldots (C_{i_n}^{\alpha_n} - \langle C_{i_n}^{\alpha_n} \rangle)\rangle. \end{aligned} \tag{3.1.11}$$

They are connected with the correlation functions as follows:

$$\begin{aligned} \epsilon_{ij}^{\alpha\beta} &= \langle C_i^\alpha C_j^\beta \rangle - \langle C_i^\alpha \rangle \langle C_j^\beta \rangle = P_{ij}^{\alpha\beta} - P_i^\alpha P_j^\beta), \\ \epsilon_{ijk}^{\alpha\beta\gamma} &= \langle C_i^\alpha C_j^\beta C_k^\gamma \rangle - \langle C_i^\alpha C_j^\beta \rangle \langle C_k^\gamma \rangle - \langle C_i^\alpha C_k^\gamma \rangle \langle C_j^\beta \rangle - \\ &- \langle C_j^\beta C_k^\gamma \rangle \langle C_i^\alpha \rangle + 2 \langle C_i^\alpha \rangle \langle C_j^\beta \rangle \langle C_k^\gamma \rangle = \\ &= P_{ijk}^{\alpha\beta\gamma} - P_{ij}^{\alpha\beta} P_k^\gamma - P_{ik}^{\alpha\gamma} P_j^\beta - P_{jk}^{\beta\gamma} P_i^\alpha + 2 P_i^\alpha P_j^\beta P_k^\gamma. \end{aligned} \tag{3.1.12}$$

The order-disorder phase transition, i.e., the appearance of a superstructure implies an abrupt increase in the correlation radius. In the state of short-range order, the correlation in atomic positions decreases with the distance. In alloys with long-range order the correlation radius tends to infinity or, at least, to the extent of the antiphase domains achieved in a given alloy. The difference between the two states of order can be characterized, for example, by the asymptotic behavior of the *a posteriori* pair correlation function $P^{\alpha\beta}(\mathbf{r})$ at the great $\mathbf{r}$ values [196] (α, β are the atom species A and B). For the completely disordered state, $\lim_{\mathbf{r}\to\infty} P^{\alpha\beta}(\mathbf{r}) = C_\beta$, i.e. the concentration of the respective component. For the ordered state of an alloy with single kind of domains, $\lim_{\mathbf{r}\to\infty} P^{\alpha\beta}(\mathbf{r}) = P_\beta$, the one-atom probability of finding β atom at the lattice site. In this case, the function $P^{\alpha\beta}(\mathbf{r})$ oscillates asymptotically; its value being dependent on the type of the site at the end of vector $\mathbf{r}$, and the amplitude of oscillation proportional to the long-range order parameter η.

The arrangement of atoms in a solid solution with a certain degree of order is governed by the statistical laws and can be described by the atomic density distribution functions. Knowledge of these functions, though hardly possible in principle, would permit to completely specify the arrangement of atoms in an alloy. The correlation functions and correlation parameters are, respectively, the initial and central moments of the density function (i.e., the differential distribution function). In general, this function is asymmetric and, probably, polymodal. Its variance is determined by the pair correlation parameter, while the asymmetry is characterized by the parameters of three-atom correlation. In the limiting case of a perfectly ordered alloy, the density function takes the form of δ-function.

Fluctuations in the occupation of lattice sites are related to the correlation parameters but not reduced to them altogether. They can be expressed through the moments of the density function; for example,

$$\begin{aligned} \Delta C_i^\alpha &= C_i^\alpha - \langle C_i^\alpha \rangle, \\ \langle (\Delta C_i^\alpha)^2 \rangle &= \epsilon_{ii}^{\alpha\alpha} - P_{ii}^{\alpha\alpha} - (P_i^\alpha)^2, \\ \langle (\Delta C_i^\alpha)^3 \rangle &= \epsilon_{iii}^{\alpha\alpha\alpha} - P_{iii}^{\alpha\alpha\alpha} - 3P_{ii}^{\alpha\alpha} P_i^\alpha + 2(P_i^\alpha)^3. \end{aligned} \quad (3.1.13)$$

The N-atom density function is obviously connected with the structure factor of the solid solution crystal:

$$S_N(\mathbf{q}_1, \mathbf{q}_2, \ldots, \mathbf{q}_N) = \frac{1}{N} \sum_{i_1 i_2 \ldots i_N} \langle C_{i_1} C_{i_2} \ldots C_{i_N} \rangle e^{i\mathbf{q}_1\mathbf{R}_1} e^{i\mathbf{q}_N\mathbf{R}_N}, \tag{3.1.14}$$

which represents the characteristic function [199, 200] for the density function. For the one-atom and pair correlation functions the characteristic functions are, respectively, the structure amplitude [201] and the structure factor as given by the second-order perturbation theory [32]. This representation corresponds to the Born approximation in the scattering theory [32] or to the kinematical structure factor in the X-ray diffraction [203]:

$$S_1(\mathbf{q}) = \frac{1}{N} \sum_{i=1}^{N} \langle C_i \rangle^{i\mathbf{q}\mathbf{R}_i},$$

$$S_2(\mathbf{q}) = \frac{1}{N} \sum_{ij} \langle C_i C_j \rangle e^{i\mathbf{q}(\mathbf{R}_i - \mathbf{R}_j)}. \tag{3.1.15}$$

The above relationships make up the basis for the experimental determination of the correlation function by the X-ray diffraction technique as well as for the theoretical calculation, by the pseudopotential method, of the ordering energy of a solid solution corresponding to different degree of order.

It should be noted that at present only one-atom and pair correlation functions are accessible to the direct measurement with the use of superlattice reflections and the diffuse scattering of X-rays [204, 205]. Nevertheless, the results of computer X-ray data processing [206 - 210] and information derived from Mössbauer spectra give an evidence of the important role of multiatomic correlations in ordering alloys. This role is also evidenced by the results of short-range order structure modeling with the use of X-ray diffraction data [211 - 214].

We have presented only one way in which the correlation functions, correlation parameters and the moments of the density distribution function (3.1.11), (3.1.12) may be interrelated. By combining (3.1.11) and (3.1.12) one can obtain the more convenient relations between the correlation parameters and correlation functions:

$$P_{12}^{\alpha\beta} = P_1^\alpha P_2^\beta + \epsilon_{12}^{\alpha\beta},$$
$$P_{123}^{\alpha\beta\gamma} = P_1^\alpha P_2^\beta P_3^\gamma + P_1^\alpha \epsilon_{23}^{\beta\gamma} + P_2^\beta \epsilon_{13}^{\alpha\gamma} + P_3^\gamma \epsilon_{12}^{\alpha\beta} + \epsilon_{123}^{\alpha\beta\gamma},$$
$$\ldots$$
$$\ldots$$
$$P_{i_1 \ldots i_N}^{\alpha_1 \ldots \alpha_N} = P_{i_1}^{\alpha_1} P_{i_2}^{\alpha_2} \ldots P_{i_N}^{\alpha_N} + P_{i_1}^{\alpha_1} \epsilon_{i_2 \ldots i_N}^{\alpha_2 \ldots \alpha_N} + \ldots +$$
$$+ P_{i_N}^{\alpha_N} \epsilon_{i_1 \ldots i_{N-1}}^{\alpha_1 \ldots \alpha_{N-1}} + \ldots + \epsilon_{i_1 \ldots i_N}^{\alpha_1 \ldots \alpha_N}. \qquad (3.1.16)$$

Only the first line in (3.1.16) remains unchanged after this transformation; all the other relations have not been used earlier in the theory of order-disorder transformations. Apart from formulas (3.1.11), (3.1.12) and (3.1.13), the quantities specifying the state of order are related to each other by means of the **equilibrium equations** and **the normalization and fluctuation relationships.** The former will be discussed in section (4.3). Let us consider now the normalization and fluctuation relationships between the correlation functions.

Normalization relations for the correlation functions are determined by the probability properties. We shall consider them for a binary alloy A - B. It can be shown for any particular superstructure of arbitrary stoichiometry [215] that

$$\sum_{\alpha = A, B} P_i^\alpha = 1$$
$$\sum_{\beta = A, B} P_{ij}^{\alpha\beta} = P_i^\alpha$$
$$------$$
$$------$$
$$\sum_{\alpha_N = A, B} P_{i_1, i_2 \ldots i_N}^{\alpha_1 \alpha_2 \ldots \alpha_N} = P_{i_1, i_2 \ldots i_{N-1}}^{\alpha_1 \alpha_2 \ldots \alpha_{N-1}}. \qquad (3.1.17)$$

Here indices i and j denote the lattice sites. The equations (3.1.17) are exact. They suggest the possibility to fully specify the state of order in an alloy by introducing the probabilities of finding atoms of one particular kind on different sites. We shall make use of this fact in the following consideration. From (3.1.17) one can get the following normalization relations between the correlation parameters

$$\sum_{\beta = A,B} \epsilon_{ij}^{\alpha\beta} = 0$$

$$\sum_{\gamma = A,B} \epsilon_{ijk}^{\alpha\beta\gamma} = 0$$

— — — — —

— — — — —

$$\sum_{\alpha_N = A,B} \epsilon_{i_1 i_2 \ldots i_N}^{\alpha_1 \alpha_2 \ldots \alpha_N} = 0. \tag{3.1.18}$$

It is worthwhile pointing out that the relations like (3.1.18) were written earlier only for one-atom and pair correlations.

Along with the exact equations, the approximate normalization relations are also used in statistical mechanics without specifying, as a rule, the degree of approximation. For a solid solution we have, for example,

$$\sum_i P_i^A = N_A$$

$$\sum_j{}' P_{ij}^{AA} = (N_A - 1)P_i^A$$

— — — — — — — — —

— — — — — — — — —

$$\sum_{i_N}{}' P_{i_1 i_2 \ldots i_N}^{AA \ldots A} = (N_A - 1)P_{i_1 i_2 \ldots i_{N-1}}^{AA \ldots A} \tag{3.1.19}$$

The prime over the summation sign means that $j \neq i$. The first equation in (3.1.19) is the exact one. It is used in the theory of order-disorder transition in the form

$$\sum_i \nu_i P_i^A = C_A, \tag{3.1.20}$$

where i denotes the site type. The remaining relations similar to those used in the theory of liquids (see, e.g., [216]) will be accurate only in the thermodynamical limit ($N_A \to \infty$). The sense of this approximation can be understood by considering, for example, the second relation (3.1.14):

$$\sum_j{}' P_{ij}^{AA} = (N_A - 1)P_i^A + \sum_j{}' \epsilon_{ij}^{AA}. \tag{3.1.21}$$

Since the second term on the right-hand side of formula (3.1.21) includes N terms, its value will depend on how rapidly correlation decreases with the distance. If correlations are non-zero for a limited number of the neighboring coordination shells, expression (3.1.19) can be used as the normalization condition. If correlations cover the macroscopic volume of the system, the above relations (3.1.21) are not rigorously valid. In fact, the normalization equations (3.1.21) are equivalent to the conditions for the correlation damping. In other words, their use corresponds to the "roughening" of the distribution function and leads to the increase in entropy [191].

Fluctuation relationships. There exists another important class of relations between the order parameters which has not been discussed earlier in the theories of order-disorder transition, namely, the fluctuation relations. They originate from the restrictions imposed on the statistical ensemble when developing the theory of order-disorder transition. Let the ensemble be chosen so as to exclude the fluctuations in the number of particles. In this case

$$\begin{aligned}
&\langle N_A^2 \rangle - \langle N_A \rangle^2 = 0 \\
&\langle N_A^3 \rangle - \langle N_A \rangle^3 = 0 \\
&- - - - - - - \\
&- - - - - - - \\
&\langle N_A^n \rangle - \langle N_A \rangle^n = 0.
\end{aligned} \tag{3.1.22}$$

Each of relations (3.1.22) entails the connections between the order parameters and correlations (i.e., the fluctuation relationships):

$$\begin{aligned}
&\sum_i P_i^A(1 - P_i^A) + \sum_{ij}{}' \epsilon_{ij}^{AA} = 0 \\
&\sum_i P_i^A((P_i^A)^2 - 1) - 6\sum_{ij}{}' P_i^A \epsilon_{ij}^{AA} + \sum_{ijk}{}' \epsilon_{ijk}^{AAA} = 0 \\
&- - - - - - - - - - - - - - - - - - \\
&- - - - - - - - - - - - - - - - - - .
\end{aligned} \tag{3.1.23}$$

Look at the first line of (3.1.23). If we take into account the number of terms in the first and second sums, it will become apparent that the neglect of the fluctuations in the number of particles is equivalent to the assumption that correlation decreases with the distance. Correlation on the distant atomic shells will be of the order of 1/N. The substitution of the first equation (3.1.23) into (3.1.21) leads to (3.1.19), which means that the neglect of the fluctuations in the particle number substantiates the normalization relations (3.1.19).

In the case of the B2 superstructure (in the nearest neighbors approximation), equations (3.1.23) take the form

$$\epsilon_{12}^{AA} = -\frac{1 - \eta^2}{32}, \tag{3.1.24}$$

For the L1$_2$ superstructure we have

$$\epsilon_{12}^{AA} + \epsilon_{22}^{AA} = -\frac{1-\eta^2}{32}. \tag{3.1.25}$$

These expressions give the temperature-independent relation between the correlation and long-range order parameters. They imply the existence of non-zero correlations at η =1. The theories neglecting the correlations prove to be inconsistent with the underlying assumptions. On the other hand, these theories can be artificially improved if one uses (3.1.24) and (3.1.25) together, for example, with the expressions of the Gorskii--Bragg--Williams method. The theories developed in the relevant approximations must be subject to these relations.

If the theory is being constructed on the basis of microcanonical ensemble, the energy fluctuations are eliminated from the treatment

$$\begin{aligned}
&\langle E^2 \rangle - \langle E \rangle^2 = 0 \\
&\langle E^3 \rangle - \langle E \rangle^3 = 0 \\
&-\,-\,-\,-\,-\,-\,-\,- \\
&-\,-\,-\,-\,-\,-\,-\,- \\
&\langle E^n \rangle - \langle E \rangle^n = 0.
\end{aligned} \tag{3.1.26}$$

Such a restriction of the ensemble yields additional fluctuation relationships. Taking into account only the interaction between the nearest neighbors, the first equation in (3.1.26) may be written as

$$\sum_{ijkl} (P_{ijkl} - P_{ij} P_{kl}) + 4 \tilde{\sum_{ijk}}{}' (P_{ijk} - P_{ij} P_{jk}) + 2 \tilde{\sum_{ij}} (P_{ij} - P_{ij}^2) = 0. \tag{3.1.27}$$

The sign "~" means that the spacing between the sites involved in the pair probability function is equal to the radius of the first coordination shell. For the B2 superstructure it follows from (3.1.27)

$$17 - 2\eta^2 - 15\eta^4 - 12\epsilon_{12}^{AA} - 4\epsilon_{12}^{AA}\eta^2 - 368(\epsilon_{12}^{AA})^2 + \ldots = 0. \tag{3.1.28}$$

Analogous expression for the L1$_2$ superstructure comprises the four-atom correlation parameters:

$$\frac{1}{16}(1-\eta)^2+\epsilon_{22}^{AA}+\frac{1}{16}(1-\eta)(1+3\eta)+\epsilon_{22}^{AA}-\frac{1}{256}(1-\eta)^4-$$
$$-\frac{1}{8}(1-\eta)^2+\epsilon_{22}^{AA}-[\epsilon_{22}^{AA}]^2-\frac{1}{256}(1-\eta)^2(1+3\eta)^2$$
$$-\frac{1}{8}(1-\eta)(1+3\eta)-[\epsilon_{12}^{AA}]^2+\ldots=0. \tag{3.1.29}$$

In most of the theories of order-disorder transition imposing restrictions on the ensemble, at least a long-range order parameter is fixed. For the solid solution specified by a single long-range order parameter, this condition leads to the following fluctuation relations:

$$\langle\eta^2\rangle-\langle\eta\rangle^2=0$$
$$\langle\eta^3\rangle-\langle\eta\rangle^3=0$$
$$-\,-\,-\,-\,-\,-\,-\,-\,-$$
$$-\,-\,-\,-\,-\,-\,-\,-\,-$$
$$\langle\eta^n\rangle-\langle\eta\rangle^n=0. \tag{3.1.30}$$

In contrast to the derivation of equations (3.1.23) and (3.1.27), the derivations given below involve some specific details. The definition of a long-range order parameter itself implies the averaging procedure. From (3.1.30) we obtain

$$\eta=\frac{1}{\nu N}\sum_{i\in L_1}C_i^A-\frac{1}{(1-\nu)N}\sum_{i\in L_2}C_i^A, \tag{3.1.31}$$

where ν is the concentration of the first type of sites, L_1 and L_2 are the sets of the sites of the first and second type, respectively. Hence, (3.1.30) takes the form

$$\frac{1}{\nu^2}\sum_{i\in L_1}P_i^A(1-P_i^A)+\frac{1}{(1-\nu)^2}\sum_{i\in L_2}P_i^A(1-P_i^A)+$$
$$+\frac{1}{\nu^2}\sum_{ij\in L_1}{}'\,\epsilon_{ij}^{AA}+\frac{1}{(1-\nu)^2}\sum_{ij\in L_2}{}'\,\epsilon_{ij}-\frac{2}{\nu(1-\nu)}\sum_{\substack{i\in L_1\\ j\in L_2}}\epsilon_{ij}^{AA}=0. \tag{3.1.32}$$

Expression (3.1.32) resembles the fluctuation relations resulting from the condition of fixed concentration (see (3.1.23)). It is easily seen if we combine the sums by pairs.

After such operation, one can see that the fixation of the long-range order parameter defines new relationships between correlations. For the B2 superstructure in the nearest-neighbor approximation, we obtain

$$\epsilon_{12}^{AA} = \frac{1-\eta^2}{32}. \tag{3.1.33}$$

For the $L1_2$ superstructure,

$$3\epsilon_{12}^{AA} - \epsilon_{22}^{AA} = \frac{1}{64}(1-\eta)(6-13\eta). \tag{3.1.34}$$

These expressions look like the fluctuation relationships (3.1.24) and (3.1.25) but do not coincide with them. Thus the restriction of the partition function with respect to η (Kirkwood's ensemble) provides new relations between the order parameters and correlations.

In some theories the probabilities of finding atoms at different lattice sites, one-particle (P_1), pair ($P_{1,2}$), and others, are fixed. Obviously, such restriction leads to the new fluctuation relations:

$$\begin{aligned} &\langle (P_i^A)^2 \rangle - \langle P_i^A \rangle^2 = 0 \\ &\langle (P_i^A)^3 \rangle - \langle P_i^A \rangle^3 = 0 \\ &\quad \ldots \end{aligned} \tag{3.1.35}$$

$$\begin{aligned} &\langle (P_{12}^{AA})^2 \rangle - \langle P_{12}^{AA} \rangle^2 = 0 \\ &\langle (P_{12}^{AA})^3 \rangle - \langle P_{12}^{AA} \rangle^3 = 0 \\ &\quad \ldots \end{aligned} \tag{3.1.36}$$

and so on.

It should be pointed out that the pair probabilities depend on the kinds of the lattice sites and the distance between them, while the many-particle probabilities -- on cluster geometry as well. It would be unpracticable to describe all cases because the corresponding formulas are too cumbersome. Further, we shall restrict ourselves to the relations resulting from the neglect of the fluctuations of one-particle and pair probabilities.The latter are considered only for the first coordination shell. In this case, (3.1.35) and (3.1.36) take the form, respectively,

$$\begin{aligned} &P_1^A(1-P_1^A) + \sum_{\substack{R \in L_1 \\ R \neq 0}} \epsilon^{AA}(R) = 0 \\ &\qquad\qquad \ldots \end{aligned} \tag{3.1.37}$$

and

$$\tilde{\sum_{i\in L_1}} P_{ij}^{AA}(1-P_{ij}^{AA}) + \tilde{\sum_{i\in L_2}} P_{ijj}^{AAA} +$$

$$+ \tilde{\sum_{\substack{i,i'\in L_1\\ j\in L_2}}} P_{ii'j}^{AAA} + \tilde{\sum_{\substack{i,i'\in L_1\\ j,j'\in L_2}}} (P_{iji'j'}^{AAAA} - P_{ij}^{AA}P_{i'j'}^{AA}) = 0. \qquad (3.1.38)$$

For the B2 superstructure, the expression (3.1.37) in the nearest-neighbor approximation will be written as

$$P_1^A(1-P_1^A) = 0. \qquad (3.1.39)$$

For the $L1_2$ superstructure, if ν=1/4, we have

$$P_1^A(1-P_1^A) = 0 \qquad (3.1.40)$$

and if ν=3/4,

$$\epsilon_{22}^{AA} = -\frac{1}{8}P_2^A(1-P_2^A). \qquad (3.1.41)$$

To the above-written fluctuation relationships, we can add several others. In particular,

$$\begin{aligned} &\langle(\epsilon_{ij}^{AA})^2\rangle - \langle\epsilon_{ij}^{AA}\rangle^2 = 0,\\ &\langle(\epsilon_{ij}^{AA})^3\rangle - \langle\epsilon_{ij}^{AA}\rangle^3 = 0,\\ &\qquad\dots \end{aligned} \qquad (3.1.42)$$

$$\begin{aligned} &\langle(\epsilon_{ijk}^{AAA})^2\rangle - \langle\epsilon_{ijk}^{AAA}\rangle^2 = 0,\\ &\langle(\epsilon_{ijk}^{AAA})^3\rangle - \langle\epsilon_{ijk}^{AAA}\rangle^3 = 0,\\ &\qquad\dots \end{aligned} \qquad (3.1.43)$$

etc.

The distribution function of the statistical ensemble is multi-valued with respect to thermodynamical and internal parameters . So far, the diagonal terms of the fluctuation matrix have been considered. There also exist the nondiagonal, or mixed fluctuations. For example, when N_A and η are fixed simultaneously (Kirkwood's ensemble), one should take into account, together with the relations (3.1.22) and (3.1.30), the following equation

$$\langle N_A\eta\rangle - \langle N_A\rangle\langle\eta\rangle = 0. \qquad (3.1.44)$$

For the ensemble of quasichemical theory (in which C_A, η and ε_{12}^{AA}) are fixed) we obtain additional relations

$$
\begin{aligned}
&\langle N_A (P_1^A - P_2^A)\rangle - \langle N_A\rangle\langle P_1^A - P_2^A\rangle = 0,\\
&\langle N_A P_{12}^{AA}\rangle - \langle N_A\rangle\langle P_{12}^{AA}\rangle = 0,\\
&\langle (P_1^A - P_2^A) P_{12}^{AA}\rangle - \langle P_{12}^{AA}\rangle\langle P_1^A - P_2^A\rangle = 0,\\
&\qquad\dots\\
&\langle N_A (P_1^A - P_2^A) P_{12}^{AA}\rangle - \langle N_A\rangle\langle (P_1^A - P_2^A) P_{12}^{AA}\rangle - \langle P_1^A - P_2^A\rangle,\\
&\langle N_A P_{12}^{AA}\rangle - \langle P_{12}^{AA}\rangle\langle N_A (P_1^A - P_2^A)\rangle + 2\langle N_A\rangle\langle P_1^A - P_2^A\rangle\langle P_{12}^{AA}\rangle = 0.
\end{aligned}
\tag{3.1.45}
$$

Thus, a large number of relations exists, several in each theory of atomic ordering, connecting different order parameters. These relations are valid irrespective to the equations of equilibrium. Strictly speaking, when calculating the order parameters in any theory, the fluctuation relations should be used together with the equations of equilibrium. However, in most cases it would be impossible because the fluctuation relations and equations of equilibrium are compatible in a limited range of changing the order parameters, and the number of equations is greater than the number of independent parameters. The partial cases that illustrate these arguments were given above. Incompatibility of equations points to the internal contradictions in the theory resulting from the restriction of the statistical ensemble. In other words, the combined application of the equations of equilibrium and fluctuation relations permits one to find not only the equilibrium order parameters but also their uncertainty due to the restriction of ensemble. Consequently, the fluctuation relationships play a double role. First, they can be used to evaluate quantitatively the uncertainty inherent in the theory. Second, one can calculate the fluctuations of the order parameters having been neglected in the theory. To do so, one should not equate the right-hand side of the fluctuation relations to zero and use on the left the equilibrium values of the order parameters. As a result, the temperature dependence of the fluctuations of the order parameters will be obtained. It is not, however, the single means of calculating fluctuations. Earlier, the theory of fluctuations in the ordering alloys has been developed by Krivoglaz [204]. The approach of this book allows one to calculate, in principle, not only the fluctuations in the ensemble but also in an individual state, in particular, those connected with the nonuniformity of the order in a specimen (i.e., the terms of the $\partial\eta/\partial x$ type). The possibilities of the Krivoglaz method are restricted by the use of thermodynamical perturbation theory and, consequently, by the assumption of the small values of fluctuations.

The method formulated in this section is free of any preliminary assumptions about the magnitude of fluctuations. Until now, however, only intraphase fluctuations have been considered. In the ordering solid solutions heterophase fluctuations may also occur, in which the arrangement of atoms corresponds to the nonequilibrium phases forming at other concentrations or completely lacking in the phase diagram. Finally, there may exist fluctuations of the domain type. The point is that all the formulas written above are related to the monodomain substance. The polydomain fluctuations may also be present in the ensemble of states. With regard to the general treatment of the problem in which the maximum account of fluctuations would be attained, the only possible way is to expand the ensemble. Certain attempts to do that have been made in the present work. The results are presented in Section 4.3.

Decouplings. The last class of relations between the order parameters includes the approximate relations called decouplings. It is appropriate to use two classes of decouplings in the theory of ordering [217]. The first is based on the assumption of the multiplicity of the probability functions; the second—on the neglect of the higher correlation parameters.

In general, the decoupling of the first type is written in the form

$$P_{i_1 \ldots i_N}^{\alpha_1 \ldots \alpha_N} \cong \frac{P_{i_1 \ldots i_N}^{\alpha_1 \ldots \alpha_{N-1}} \ldots P_{i_1 \ldots i_{N-3}}^{\alpha_1 \ldots \alpha_{N-3}} \ldots}{P_{i_1 \ldots i_{N-2}}^{\alpha_1 \ldots \alpha_{N-2}} \ldots P_{i_1 \ldots i_{N-4}}^{\alpha_1 \ldots \alpha_{N-4}} \ldots} \tag{3.1.46}$$

In particular, for N = 3,

$$P_{123}^{\alpha_1 \alpha_2 \alpha_3} \cong \frac{P_{12}^{\alpha_1 \alpha_2} P_{13}^{\alpha_1 \alpha_3} P_{23}^{\alpha_2 \alpha_3}}{P_1^{\alpha_1} P_2^{\alpha_2} P_3^{\alpha_3}} \tag{3.1.47}$$

and for N=2,

$$P_{12}^{\alpha_1\alpha_2} \cong P_1^{\alpha_1} P_2^{\alpha_2}. \tag{3.1.48}$$

The relation (3.1.47) is called the Kirkwood superposition decoupling, the relations for N>3 are the generalized decouplings. Formulas (3.1.46) - (3.1.48) may be conveniently employed to decouple the chains of the integral or integro-differential equations into the correlation functions. The relations of the kind are widely used in the theory of liquids or dense gases [216].

The decouplings of the second class have the form

$$\epsilon_{i_1\ldots i_N}^{\alpha_1\ldots\alpha_N} = 0 \tag{3.1.49}$$

or

$$P_{i_1\ldots i_N}^{\alpha_1\ldots\alpha_N} \cong P_{i_1}^{\alpha_1} P_{i_1\ldots i_N}^{\alpha_2\ldots\alpha_N} + P_{i_2}^{\alpha_2} P_{i_1\ldots i_N}^{\alpha_1\ldots\alpha_N} - P_{i_1 i_2}^{\alpha_1\alpha_2\alpha_3} P_{i_3\ldots i_N}^{\alpha_1\alpha_2\alpha_3\ldots\alpha_N} \ldots$$
$$\ldots + \ldots + (1)^{N-1}(N-1) P_{i_1}^{\alpha_1} \ldots P_{i_N}^{\alpha_N}. \tag{3.1.50}$$

The expression (3.1.50) may be written as

$$P_{i_1\ldots i_N}^{\alpha_1\ldots\alpha_N} \cong P_{i_1}^{\alpha_1} \ldots P_{i_N}^{\alpha_N}\left(1 + \frac{\epsilon_{i_1 i_2}^{\alpha_1\alpha_2}}{P_{i_1}^{\alpha_1} P_{i_2}^{\alpha_2}} + \ldots + \frac{\epsilon_{i_{N-1} i_N}^{\alpha_{N-1}\alpha_N}}{P_{i_{N-1}}^{\alpha_{N-1}} P_{i_N}^{\alpha_N}} + \right.$$
$$\left. \ldots + \frac{\epsilon_{i_1\ldots i_{N-1}}^{\alpha_1\ldots\alpha_{N-1}}}{P_{i_1}^{\alpha_1} \ldots P_{i_{N-1}}^{\alpha_{N-1}}} + \ldots + \frac{\epsilon_{i_2\ldots i_N}^{\alpha_2\ldots\alpha_N}}{P_{i_2\ldots}^{\alpha_2\ldots} P_{i_N}^{\alpha_N}}\right). \tag{3.1.51}$$

In particular, for N = 3 we obtain

$$P_{i_1 i_2 i_3}^{\alpha_1\alpha_2\alpha_3} \cong P_{i_1}^{\alpha_1} P_{i_2}^{\alpha_2} P_{i_3}^{\alpha_3} + P_{i_1}^{\alpha_1} \epsilon_{i_2 i_3}^{\alpha_2\alpha_3} + P_{i_2}^{\alpha_2} \epsilon_{i_1 i_3}^{\alpha_1\alpha_3} + P_{i_3}^{\alpha_3} \epsilon_{i_1 i_2}^{\alpha_1\alpha_2} \tag{3.1.52}$$

and for N = 2,

$$P_{i_1 i_2}^{\alpha_1\alpha_2} = P_{i_1}^{\alpha_1} P_{i_2}^{\alpha_2}. \tag{3.1.53}$$

The last equation coincides with (3.1.48), the other equations for the two classes of decouplings are not equivalent. In order to estimate the degree of approximation in the two theories, we give the relation between the expressions of the same order in both classes. In the case of N = 3, for example, the right-hand sides of (3.1.47) and (3.1.52) differ by the value

$$\frac{\epsilon_{i_1 i_2}^{\alpha_1\alpha_2}\,\epsilon_{i_1 i_3}^{\alpha_1\alpha_3}}{P_{i_1}^{\alpha_1}}+\frac{\epsilon_{i_1 i_2}^{\alpha_1\alpha_2}\,\epsilon_{i_2 i_3}^{\alpha_2\alpha_3}}{P_{i_2}^{\alpha_2}}+\frac{\epsilon_{i_1 i_3}^{\alpha_1\alpha_3}\,\epsilon_{i_2 i_3}^{\alpha_2\alpha_3}}{P_{i_3}^{\alpha_3}}+\frac{\epsilon_{i_1 i_2}^{\alpha_1\alpha_2}\,\epsilon_{i_2 i_3}^{\alpha_2\alpha_3}\,\epsilon_{i_1 i_3}^{\alpha_1\alpha_3}}{P_{i_1}^{\alpha_1}P_{i_2}^{\alpha_2}P_{i_3}^{\alpha_3}}, \tag{3.1.54}$$

which should be added to (3.1.52). Similar expressions can be written for N > 3. Hence, the decouplings of the two classes differ at least by the magnitude of the order of the last nonvanishing correlation squared. This means that the two classes of decouplings are approximately equivalent.

3.2 The ordering energy. The types of atomic order and low-temperature stability of superstructures

Stoichiometric alloys. The stability of superstructures at low temperatures when the entropy contribution is relatively small is fully determined by the ordering energy. Very important for the stability of superstructures based on a particular lattice is the dependence of the ordering energy on the distance, $V(\overline{R}) = V_{AA}(\overline{R}) + V_{BB}(\overline{R}) - 2V_{AB}(\overline{R})$, where $V_{\mu\nu}(\overline{R})$ is the energy of the interaction between atoms of μ and ν kind. The contributions into the ordering energy from the many-body interactions are also essential. In this section low-temperature conditions for stability of superstructures are formulated on the basis of the general thermodynamical relationships [218]. We show, particularly, that the stability conditions as derived earlier by Clapp and Moss [297] and Khachaturyan [21] are neither necessary nor sufficient. In its thermodynamical sense, the current analysis is similar to that of Kan et al. [219, 220] but differs from theirs by the formulation of task and by the conditions imposed on the ordering energy. Our conditions are less stringent, in particular, in that the role of many-body interactions is taken into account.

Let us consider, first, the stability conditions for the superstructures based on an f.c.c. lattice, with allowing for the pair interaction. In this case, the energy of a binary alloy has the following form

$$E = \frac{1}{2} \sum_{ij}{}' V(\mathbf{R}_i - \mathbf{R}_j) P_i^A P_j^A \tag{3.2.1}$$

or

$$E = \frac{N}{2} [V(0) C_A^2 + \sum_s \eta_s^2 A(\mathbf{K}_s) V(\mathbf{K}_s)]. \tag{3.2.2}$$

Here

$$V(\mathbf{K}) = \sum_{\mathbf{R}} V(\mathbf{R}) e^{i\mathbf{kR}} \tag{3.2.3}$$

is the Fourier transform of the ordering energy, $A(\overline{\mathrm{k}}_s)$ -- the positive coefficients defined by the symmetry of the superstructure.

The summation in (3.2.2) extends over the stars of the first Brillouin zone. On account of the distribution functions in Tables 3 - 7 describing the probabilities for atoms to occupy lattice points, expressions (3.2.2) and (3.2.3) may be written in the form quadratic in the order parameters:

$$E = \frac{N}{2} [V(0) C_A^2 + \sum_{i,s} \eta_s^2 a_{i,s} V_i], \tag{3.2.4}$$

where V_i is the ordering energy for the ith coordination shell.

If the possibility of the formation of the mechanical mixture of pure components and of different superstructures is disregarded, the stable phase will possess the minimum energy as compared to the other phases of the same composition. Under such conditions, some of the phases in a solid solution will be metastable, while the others are unstable. Further treatment will be performed for the case T = 0 K, when all the values $\eta_s = 1$. The necessary condition for the existence of a definite ordered phase L will be

$$\sum_{i,s} a_{is}^L V_i \leqslant 0, \tag{3.2.5}$$

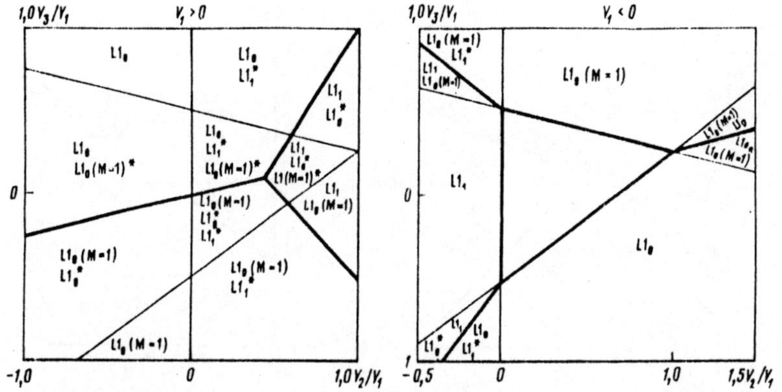

Fig. 32. The regions of superstructure stability in the interaction space for an f.c.c. lattice. AB compositions. Pair interaction. Solid lines -- stability regions, thin lines -- metastability regions.

where the sign "less" refers to the region of stability or metastability and the equality sign determines the threshold of decomposition. Sufficient conditions for stability have the form

$$\sum_{i,s} a_{is}^{L} V_i \leqslant \left\{ \begin{array}{l} \sum_{i,s} a_{is}^{L_1} V_i \\ \vdots \\ \sum_{i,s} a_{is}^{L_n} V_i \end{array} \right. \qquad (3.2.6)$$

Since the configurational energy is a linear function of the ordering energy at different coordination shells, the equation for the interphase L-L_α boundaries is written as

$$\sum_i d_i^{(L-L_\alpha)} V_i = 0, \qquad (3.2.7)$$

with

$$d_i^{(L-L_\alpha)} = \sum_s (a_i^L s - a_{is}^{L_\alpha}).$$

For the sake ot a convenient graphic representation, we confine ourselves to the consideration of three coordination shells.

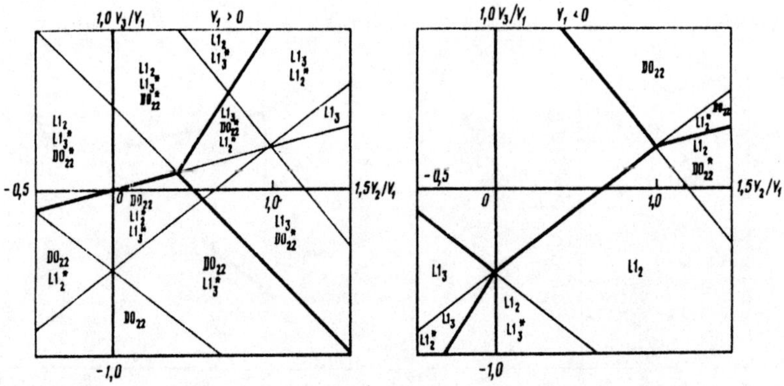

Fig. 33. The regions of superstructure stability in the interaction space for an f.c.c. lattice. AB_3 composition. See notations in Fig. 32.

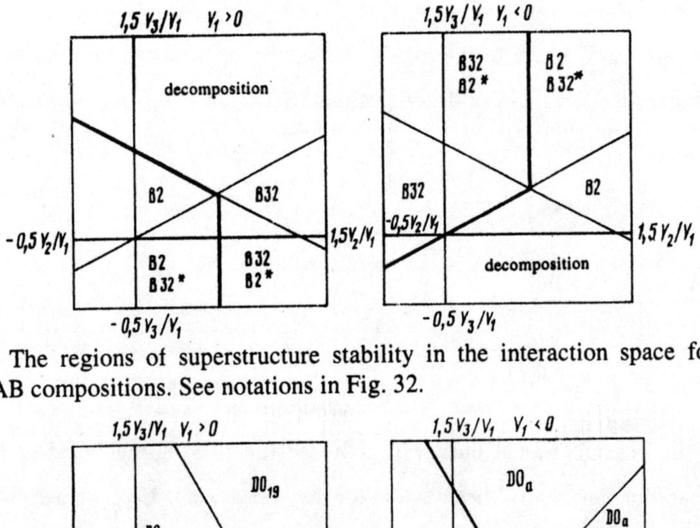

Fig. 34. The regions of superstructure stability in the interaction space for a b.c.c. lattice. AB compositions. See notations in Fig. 32.

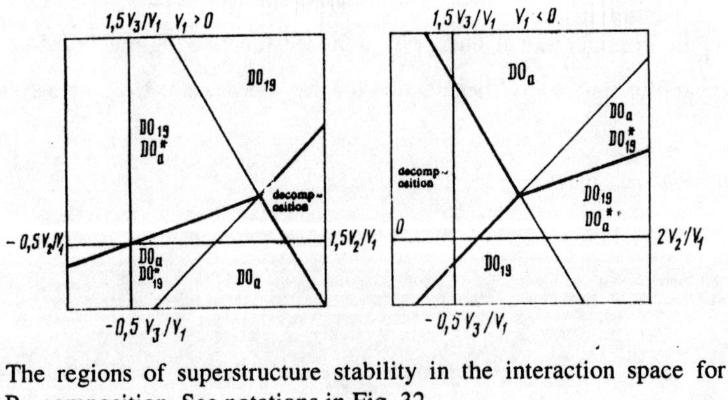

Fig. 35. The regions of superstructure stability in the interaction space for a h.c.p. lattice. AB_3 composition. See notations in Fig. 32.

The regions of stability and metastability for the superstructures of different composition based on f.c.c., b.c.c. and h.c.p. lattices are shown in Figs. 32 - 35. The antiphase structures are found to be stable on the line separating the superstructures $L1_2$ and DO_{22}.

Near the stability threshold, an excess negative energy of a superstructure decreases. The zero value of this energy at the threshold corresponds to the stability of the superstructure with periodic APBs whose limiting case is the superstructure with M = 1 lying on the opposite side of the boundary. This is explained by the fact that any superstructure can be represented by another with a definite type of the periodic APBs (see Section 1.3). Hence, the interphase boundary is simultaneously a boundary where the superstructure is stable with regard to a spontaneous formation of the APBs. Khachaturyan has formulated another criterion of stability of superstructures with respect to the formation of antiphase domains, namely, the Fourier transform of the ordering energy, $V(\bar{k})$, should have a minimum in all points of the $\bar{k}$ space corresponding to the superstructure vectors $\bar{k}_s$ for the given superstructure [19]. Unfortunately, he proved in [19] only a weaker assertion, namely, that $V(\bar{k})$ must be extremal in all points corresponding to the superstructure vectors. Let us consider this question in more detail using, as an example, the superstructures formed on the base of an f.c.c. lattice [221].

The conditions for the Fourier transform of the ordering energy $V(\bar{k})$ to be minimal ($K = \sum_{l=1}^{3} h_l \mathbf{l}_l$, $\mathbf{l}_l$ – are the reciprocal vectors of a simple cubic lattice) are written in the form

$$\frac{\partial V}{\partial h_l} = 0, \quad \frac{\partial^2 V}{\partial h_i^2} > 0, \quad \det\left\| \frac{\partial^2 V}{\partial h_i \cdot \partial h_j} \right\| > 0. \tag{3.2.8}$$

The regions of the minimum value of $V(\bar{k})$ obtained by numerical calculation are shown in Fig. 36. They can be compared with the regions where the superstructure phases exist (Figs. 32 and 33). Such comparison was first made by Kozlov et al. [35]

and later by De Fontaine [222]. It is possible for the superstructures AB that are always characterized by a single type of the superstructure vector. As is evident from Figs. 32 and 36, Khachaturyan's criterion cannot predict correctly the regions of the $L1_1$ superstructure stability with respect to its transformation to the $L1_0$ type. Obviously, Khachaturyan's condition is not sufficient. For example, in the region bounded by the lines

$$2V_3/V_1 - 3V_2/V_1 - 1 = 0; \quad 2V_2/V_1 - 1 = 0.$$

the Fourier transform of the ordering energy V(1/2 1/2 1/2) has a minimum but the energy of the $L1_0$ superstructure is lower than that of the $L1_1$ superstructure. In other words, Khachaturyan's criterion describes the regions of an absolute instability of a superstructure, i.e., the regions of the unstable state, whereas a superstructure may be unstable in a metastable sense (as in the case of the $L1_0$ or $L1_1$ superstructure). Since the coefficients $A(\overline{k}_s)$ for the superstructures of the same stoichiometry are equal, according to [218], we can compare, in place of the energies of superstructures, the Fourier transforms of the ordering energies in the corresponding points of the reciprocal space. The comparison of the superstructure stability regions with the regions where $V(\overline{k})$ reaches a minimum (Fig. 37) shows their coincidence for superstructures having a single superstructure vector. Regarding the superstructures with two superstructure vectors, the coincidence of the two above regions takes place only at the points corresponding to the superstructure vectors of the lowest symmetry.

It should be noted that for the superstructures specified by two order parameters such as DO_{22} and $L1_3$, Khachaturyan's criterion does not represent even the necessary condition for the superstructure stability because the regions with the minimum V(1 0 0) and V(1 0 1/2) and, respectively, V(1 0 0) and V(1/2 1/2 1/2) values do not overlap at all (Fig. 36). Therefore, by contrast to Khachaturyan's statement [21], the Fourier transform of the ordering energy cannot take a minimum value simultaneously at all points of the $\overline{k}$ space corresponding to the positions of the superstructure vectors.

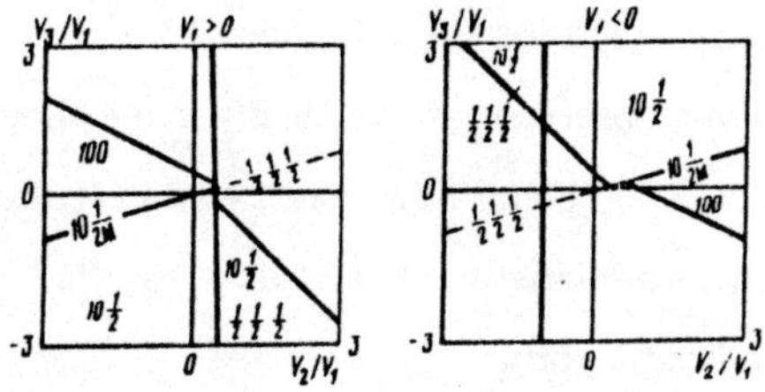

Fig. 36. The regions in the interaction space where the minima of $V(\mathbf{k})$ are realized.

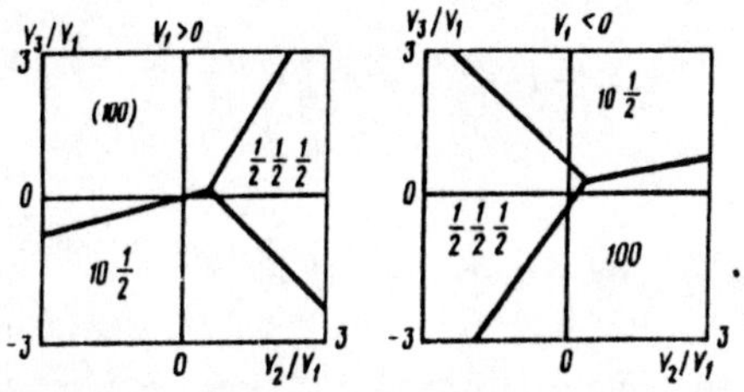

Fig. 37. The regions in the interaction space where the minimum $V(\mathbf{k})$ value is attained.

As regards the structure of short-range order in solid solutions, the Fourier transform of the ordering energy determines the structure of an alloy in the state with short-range order at temperatures far from the order-disorder transformation temperature. The regions in Fig. 36 where the Fourier transform reaches a minimum determine the maximum of the diffuse scattering connected with short-range order.

Of special interest is the analysis of stability of different structures near the A_4B compositions. In the diagram constructed in coordinates V_2/V_1, V_3/V_1, the existence of a short-range order is determined by the minimum of the Fourier transform of the ordering energy $V(\overline{k})$, while the existence of long-range order -- by the absolute minimum of the configurational energy of the alloy. The calculations showed that for the positive values of the ordering energy in the first coordination shell, as in the Ni_4Mo alloy, the region of the short-range order described by the vector (1 1/2 0) in coordinates V_2/V_1, V_3/V_1 above T_c is superimposed on the region where the superstructure $D1_a$ exists below T_c. The superstructure Ni_2Mo in this region is shown to be metastable (Fig. 38). As a result, the theoretical scheme has been found that explains, on reasonable assumptions, the existence of a variety of structures with short-range and long-range order observed experimentally in the alloy Ni_4Mo.

We can generalize the foregoing treatment by considering the case in which the three-body interaction substantially contributes to the ordering energy. The superstructure energy takes the form

$$E = N\left[\frac{V(0)}{2}C_A^2 + \frac{1}{2}\sum_s \eta_s^2 A(\mathbf{K}_s)V(\mathbf{K}_s) + \right.$$
$$\left. + \frac{1}{3}\sum_{sjs'j'} W(\mathbf{K}_s, \mathbf{K}_{s'})\eta_s\eta_{s'}\eta_{-s-s'}\alpha(\mathbf{K}_s^{j'})\alpha(\mathbf{K}_s^{j'})\alpha(-\mathbf{K}_s^j - \mathbf{K}_{s'}^{j'})\right]. \tag{3.2.9}$$

in which

$$W(\mathbf{K}, \mathbf{K}') = \sum_{i,j,k} W(\mathbf{R}_i - \mathbf{R}_j, \mathbf{R}_i - \mathbf{R}_k)e^{i\mathbf{k}(\mathbf{R}_i - \mathbf{R}_j)}e^{i\mathbf{k}'(\mathbf{R}_i - \mathbf{R}_k)},$$

and $\alpha(\overline{\mathrm{k}})$ are the coefficients determined by the superstructure type. The problem of stability is reduced to searching for the minimum of functional (3.2.9). The results of the numerical solution of this problem for the AB and A_3B superstructures based on an f.c.c. lattice are shown in Fig. 39. Allowing for the many-body interactions, the general pattern of the stability regions for different superstructures becomes more complicated but the main inferences based on the consideration of the pair interaction remain unchanged.

Consider, in conclusion, the influence of the temperature effects on the regions of stability of different superstructures. As a first approximation, we can ignore the variation of entropy for different superstructures and focus on the energy term. Near T_C, the degree of long-range order is lowered and the correlation effects become essential. The expression for the internal energy takes the form

$$E = \frac{N}{2}\sum_s N_s^l A(\mathbf{K}_s)V(\mathbf{K}_s) + \frac{1}{2}\sum_{R_i} V(\mathbf{R}_i - \mathbf{R}_j)\epsilon(\mathbf{R}_i - \mathbf{R}_j), \tag{3.2.10}$$

where ε is the correlation parameter. Account of the contribution due to the short-range order, with the sign depending on the relative magnitudes of the ordering energy for the different coordination shells, does not provide the derivation, near T_C, of an analytical

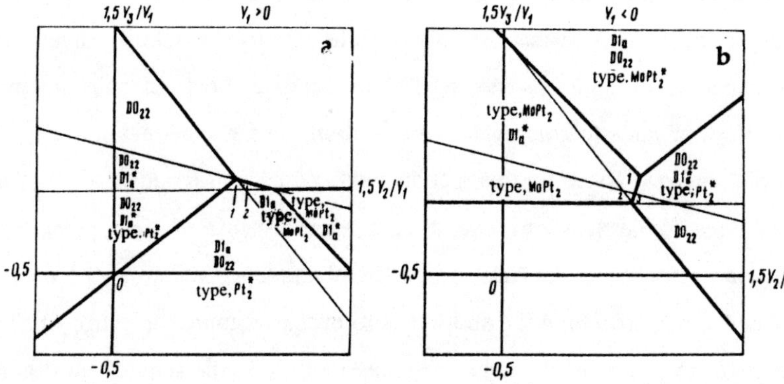

Fig. 38. Analysis of phase stability in the Ni_4Mo alloy. (a) -- $D1_a$, DO^*_{22}; $D1_a$.- (2); (b) -- $MoPt_2$ type, $D1^*_a$, DO^*_{22} (1); $MoPt_2$ type, DO^*_{22};(2); $MoPt^*_2$, type, DO^*_{22} (3). Key: type.

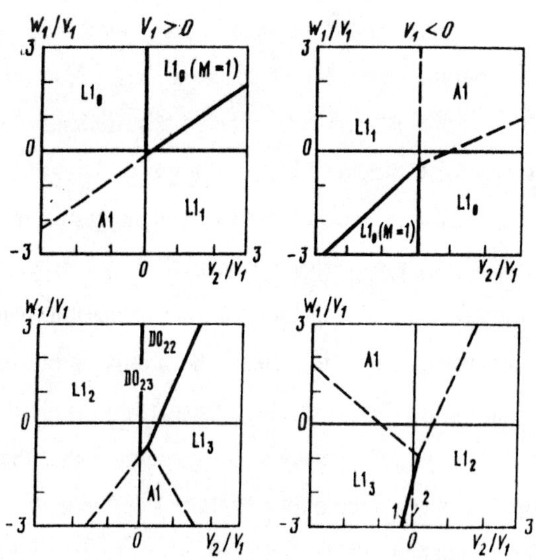

Fig.39 The regions of superstructure stability in the interaction space for an f.c.c. lattice. Compositions AB and AB_3. Pair and three-particle interactions: DO_{22} (1); DO_{23} (2).

criterion for the superstructure stability. The inclusion of the entropy effects leads, all the more, to the conclusion that the only reliable criterion of stability should be based on the comparison of the Gibbs energies for the superstructures and their mixtures.

The alloys of non-stoichiometric composition. So far, our treatment has confined to the stoichiometric alloys. In this section and Section 1.1 we outline a set of possible superstructures for such alloys and discuss some aspects of their stability. If the composition of alloys deviates from stoichiometry, situation becomes more complicated. On approaching the absolute zero of temperature, the entropy of a system tends to zero. In this case, the nonstoichiometric alloy in the state of maximum order will not satisfy the Nernst theorem because a freedom exists for excess atoms to be arranged over the sites occupied by the deficient atom species. According to the Krivoglaz concept [204], the alloys of nonstoichiometric composition should be decomposed, on cooling, into the mixture of two phases, the completely ordered alloy of the nearest stoichiometry and a pure metal in excess. Such mode of transformation may well occur in alloys but the separation of different phases requires atomic diffusion over long distances and this process may be hampered at sufficiently low temperatures.

The alternative scheme of the behavior of alloys at low temperatures can be suggested. Developing the idea of Schwartz and Cohen suggested in the discussion of the results of the work [208], the Nernst theorem can be met under the following circumstances. First, the ordering takes place, with the formation of the superstructure of the nearest stoichiometric composition; next, the atoms of the excess component undergo a complementary ordering over the sublattices occupied by the deficient component. As a result, a new structure arises, more complicated than the initial one. The emerging superstructures can have only a definite stoichiometry, i.e., the proposed scheme does not fit for all concentrations. Such behavior of nonstoichiometric alloys at low temperatures, unlike the Krivoglaz scheme, does not require diffusion over long distances. In this case, it would be sufficient for atoms to exchange their positions with the nearest neighbors. This behavior is characteristic, for example, of the Fe - Al system near the composition Fe_3Al [193]. In this case, the B2 superstructure forms initially to be followed by the complementary ordering that gives rise to the formation of the DO_3 superstructure.

In the present section such superstructures are described and the conditions are determined when the realization of such process will result in the formation of the equilibrium or metastable state, as compared to the states arising in the Krivoglaz scheme.

The treatment is given for the case in which the nonstoichiometric B2 and $L1_2$ superstructures are imperfectly ordered [35, 223]. The analytical approach and derivation of newly formed superstructures are equivalent to the scheme described in Section 1.1. The only difference is that in Section 1.1 the initial solid solution is assumed to be disordered, whereas in our case an additional ordering takes place over the sites of one of the sublattices belonging to the above listed superstructures. The results of analysis are presented in Table 10 containing, in particular, the stars of the vectors of the appropriate symmetry, the distribution functions, the stoichiometry of a new superstructure, the space group, and the number of sites occupied by atoms. The conditions can be written, under which the process described here is more probable than the scheme suggested by Krivoglaz.

In conclusion, some words should be added about the possibility of realizing the described scheme. The transformation temperatures of the initial superstructure formation and the diffusion rates of the components in the ordered phase are usually much lower than those in the disordered phase. As a consequence, these superstructures may be found experimentally only in the alloys where the initial ordered phase has a comparatively high T_c. Moreover, it is necessary that the region of the existence of the initial superstructure should cover all possible stoichiometries of the low-temperature superstructures. Given these conditions are met, it is to be hoped that the above superstructures will be found after sufficiently prolonged annealing at appropriate temperatures. There is an indication on the existence of the ordered phases Fe_7Al,

Table 10. Superstructures appearing in complementary ordering in lattices D2 and $L1_2$.

Distribution function for B2	$\frac{1}{2}+\frac{1}{2}\eta e^{i\pi z}$	$\frac{1}{2}+\frac{1}{2}\eta e^{i\pi(x+y+z)}$	$\frac{1}{2}+\frac{1}{2}\eta e^{i\pi(x+y)}$	$\frac{1}{4}+\frac{1}{4}\eta[e^{i\pi(x+y)}+e^{i\pi(x+z)}+e^{i\pi(y+z)}]$	
Composition	A_3B	A_3B	A_3B	A_7B	A_5B_5
Number of atoms per unit cell	4	16	8	16	16
Space group	P4/mmm	Fm3m	P4/mmm	Im3m	Im3m

Distribution function for $L1_2$	$\frac{1}{2}+\frac{1}{2}\eta e^{i\pi z}$	$\frac{1}{2}+\frac{1}{2}\eta e^{i\pi(x+y+z)}$	$\frac{1}{2}+\frac{1}{2}\eta e^{i\pi(x+y)}$	$\frac{1}{4}+\frac{1}{4}\eta[e^{i\pi(x+y)}+e^{i\pi(x+z)}+e^{i\pi(y+z)}$	
Composition	A_7B	A_7B	A_7B	$A_{15}B$	$A_{13}B_3$
Number of atoms per unit cell	8	32	16	32	32
Space group	P4/mmm	Fm3m	P4/mmm	Im3m	Im3m

Table 11. Description of superstructures of quasibinary section.

Distribution function		$\frac{1}{2}+\frac{1}{2}\eta e^{i\pi z}$	$\frac{1}{2}+\frac{1}{2}\eta e^{i\pi(x+y+z)}$	$\frac{1}{2}+\frac{1}{2}\eta e^{i\pi(x+y)}$	$\frac{1}{4}+\frac{1}{4}\eta(e^{i\pi(x+y)}+e^{i\pi(x+z)}+e^{i\pi(y+z)})$	
B2	Composition	A_2BC	A_2BC	A_2BC	A_4BC_3	A_4B_3C
$L1_2$		A_6BC	A_6BC	A_6BC	$A_{12}BC_3$	$A_{12}B_3C$
Space group		$P\bar{4}$/mmm	Fm3m	$P\bar{4}$/mmm	Im3m	Im3m

Distribution function		$\frac{1}{2}+\frac{1}{2}\eta e^{i\pi z}$	$\frac{1}{2}+\frac{1}{2}\eta e^{i\pi(x+y+z)}$	$\frac{1}{2}+\frac{1}{2}\eta e^{i\pi y}$	$\frac{1}{2}+\frac{1}{2}\eta e^{i\pi(x+y)}$	$\frac{1}{2}+\frac{1}{2}\eta e^{i\pi(y+z)}$
$L1_0$	Composition	A_2BC	A_2BC	A_2BC	A_2BC	A_2BC
Space group		$P\bar{4}$/mmm	Pmmm	Pmmm	$P\bar{4}$/mmm	Pmmm

Cu_5Pd_3 and Pt_7Cu [224]. If further investigations show that the phase diagrams of those alloys comply with the scheme suggested by the authors, these diagrams might serve as experimental evidence of its reality.

Similar scheme has been also applied [225] to the problem of the ordering of ternary alloys of quasibinary sections having the same superstructures at both ends of these sections. Analysis showed that in this case new intermediate ternary superstructures may arise in the quasibinary section of the phase diagram. The superstructures which may form on the base of the initial B2, $L1_0$ and $L1_2$ structures are characterized in Table 11. Most of the predicted superstructures have not been experimentally observed as yet. Only the superstructure A_2BC corresponds to the well-known Heissler alloy $L2_1$ [194]. Several works give indirect evidence of such ordering.

3.3 Long-range order and structural state of a solid solution

The long-range order parameter seems to be the only independent parameter describing the order-disorder transformation; all other parameters that characterize the structural state of a solid solution are to be functionally related to it. For this reason, it is necessary to consider in detail the experimental data showing the connection between the long-range order parameter and structural characteristics of a solid solution. The latter include the lattice constants a, b, c, the parameters of tetragonal and orthorhombic distortion, c/a and b/a, the mean square static and dynamic displacements of atoms, the Debye characteristic temperature, the thermal expansion coefficient, etc.

For cubic and "pseudocubic" superstructures, the dependence of the lattice constant upon the order parameter can be reliably measured only if the temperature range of the two-phase region is not too wide. In all cases where this dependence was reliably obtained, it was shown to be linear. In other words, where a_p and a_y are the lattice

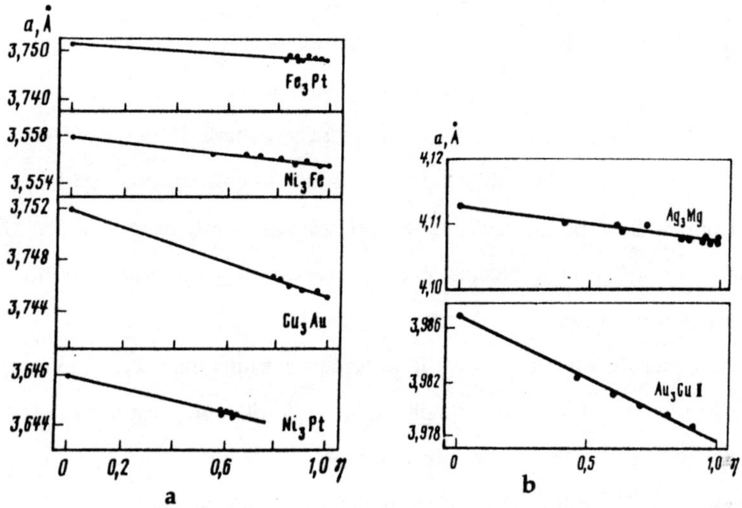

Fig. 40. Lattice constant vs order parameter for cubic (a) and quasicubic (b) superstructures.

constants in the alloys with $\eta = 0$ and $\eta = 1$, respectively. The examples are shown in Fig. 40. They provide evidence that the linear relation between a and η holds to a high accuracy, including the values corresponding to the disordered alloy. In all these drawings the lattice constants for any states of long-range order are referred to their values at room temperature (X-ray measurements were conducted specially at room temperature). In other words, in these drawings the influence of thermal expansion is excluded and the net effect of a long-range order is singled out.

It is not possible to come to any definite conclusions on the influence of the type of superstructure upon the coefficient of thermal expansion that is due to disordering. Thus, for the long-period superstructure $L1_2$(M), the value of $\partial a/\partial\eta$ (Fig. 40) can be greater or smaller than that for the $L1_2$ structure. Near the Au_3Cu composition, both the $L1_2$ superstructure and the $L1_2$(M) structure may exist. The thermal expansion coefficient for the long-period structure is two times that for the $L1_2$ superstructure. This gives evidence for a certain structural instability of the long-period superstructure in this alloy. At the same time, however, there are examples of the opposite behavior. In the Cu_3Pt alloy, the $L1_2$ superstructure and then $L1_2$(M) successively form with increasing temperature. It is the phase $L1_2$ that exhibits a marked structural instability in this alloy before the $L1_2 \rightarrow L1_2(M)$ transformation.

In general, the a(η) function for cubic superstructures is not necessarily linear. Nonlinear dependence of the lattice parameter on the long-range order parameter can be predicted on the base of the pseudopotential method (see Chapter 5). Therefore, the long-range order parameter remains the only one defining the type of transformation, even for the non-linear a(η) function.

In most noncubic superstructures, in accordance with symmetry, the crystal lattice lends itself to tetragonal or more complex (e.g., orthorhombic) distortions. The data on these distortions have been thoroughly analyzed. The lattice distortions in various superstructures differ in the sign and in the value. In the superstructures $D1_a$, $L1_2(M)$ and $C11_b$, the value of c is always less than a; in the superstructures DO_{23} and $L1_2(MM)$, c is always greater than a, and in the latter c > b as well. Only in the DO_{22} superstructures the signs of a tetragonal distortion may be different. There exists, therefore, an unquestionable correlation between the type of a superstructure and the sign of the unit cell distortion. Since we describe in this section only the alloys undergoing the order-disorder transformations, the conclusions made here are referred only to these alloys but not to the whole class of superstructures. The sign of distortion depends also on the alloy composition. In the Au-base alloys c > a and c > b; in the Cu-, Ni-, Ag- and Pd-base alloys different cases may occur, including c ≈ a.

In the long-period $L1_2$-base superstructures another interesting tendency is observed, namely, the increase in the antiphase period produces a rapid decrease in tetragonality and the transformation of the alloy into a quasicubic one. Such dependence is observed both in a series of alloys and for each alloy separately (Fig.41). This tendency can also be traced in the foregoing analysis. The increase in the antiphase period reduces the frequency of the APB occurance and approaches the alloy structure to that of the $L1_2$ type thereby decreasing tetragonality. In other words, there exists the relation between the parameters of the anisotropic Fermi surface in the disordered state, the superstructure boundaries of the Brillouin zones arising due to the contact with this surface, the superstructure unit cell parameters, and the anisotropy of the interatomic interaction and the ordering energy.

Therefore, the sign of distortion turns out to be different for various alloys and superstructures ($c > a$ or $c < a$). The reason is usually ascribed to the band structure of the material, assuming the contact of the Fermi surface with the corresponding faces of the Brillouin zone [52, 226] (see Sections 5.3 and 5.4). The structural data give evidence of an important role of the Fermi surface in changing the alloy structure due to the ordering. It follows simultaneously that the energy of the interatomic interaction and the ordering energy in such alloys become strongly anisotropic (see Sections 5.1 and 5.2). The distinct correlation for different superstructures in the dependence of the tetragonality parameter, and the a and c values upon the order parameter η has not been established. This fact confirms an important role of the peculiarities of the electronic structure in the final formation of the unit cell size and the dependence of the latter on the order parameters.

The tetragonality parameter for several alloys is given in Fig. 42. It can be seen that for many alloys the dependence of c/a on η is linear. But, as pointed out previously, it is not necessary to have a linear dependence in order to recognize the order parameter to be a single parameter specifying the order-disorder transformation. It is well sufficient to have a definite functional relation between the order and unit cell parameters. The experimental data collected in this section unequivocally indicate the existence of such dependence. The quantum theory of ordering alloys developed in Chapter 5 validates this fact. The existence of the linear relation between η and the unit cell parameters just facilitates the analysis.

Fig. 42 demonstrates, in particular, the fact that in a number of alloys the ordering is connected with a considerable anisotropy of the interactions which is most pronounced in Au-base alloys with the $D1_a$ superstructure. It should be noted that the role of the

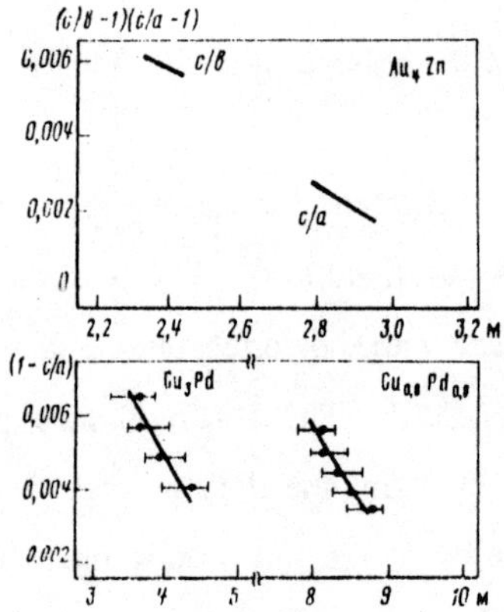

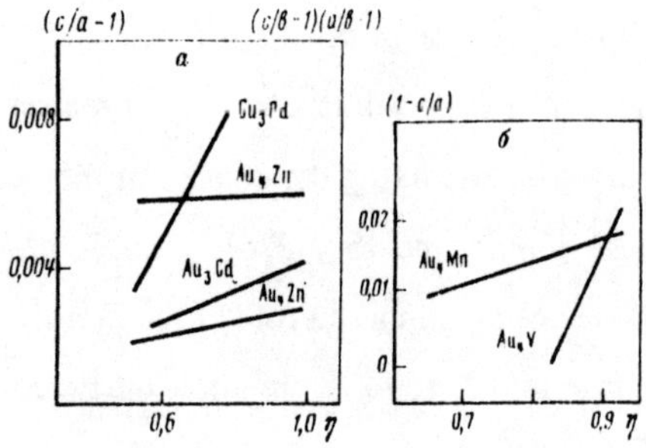

Fig. 41. Distortions of the unit cell as a function of the antiphase period.

Fig. 42. Distortions of the unit cell as a function of the order parameter in the superstructures: $L1_2$(M), $L1_2$(MM) (a); $D1_a$ (b).

anisotropic interaction in the ordering phenomenon is poorly studied by now; only experimental facts have been collected pointing to the necessity for exploring this important issue.

In Chapter 1 the description was given of the phenomenon of superstructural contraction discovered in the present work; this phenomenon is typical of the overwhelming number of superstructures. The results of this work show that this phenomenon necessarily takes place in the cubic superstructures. In the superstructures of lower symmetry, superstructure contraction is observed in most cases. The observed deviations (Ni_4W, Au_4Zn, Au_4V) provide evidence of a strong influence of the anisotropic interactions on the volume effects in the atomic ordering.

Along with the change in the unit cell parameters, finer variations in the arrangement of atoms in an alloy occur upon ordering. Among them are, particularly, oriented (anisotropic) static and dynamic displacements of atoms from the lattice sites. Consider them one by one. Oriented displacements may be of such type that their occurrence does not change the symmetry of the superstructure. Such displacements take place when, in ordering, atoms occupy the positions on symmetry axes or symmetry planes or, in general, in arbitrary points of the lattice [227]. These displacements are "legal" for the superstructure in which they arise. For example, in the $C11_b$ superstructure the atomic coordinates contain the free parameter Z (see Section 1.1). The dependence of this parameter on the order parameter for the $AlCr_2$ alloy is close to linear. The oriented displacements increase with an increase in the order parameter. The free parameters occur in the superstructures $L1_2$(M), $D1_a$ and others [228] depending on the order parameter in a similar way. This fact confirms the conclusion about the existence of a single order parameter in the described class of transformations.

Another type of oriented displacements have been observed in the alloys CuPd, Fe_3Pt, NiTi and some others [229–232]. These displacements are oriented in such a way that they are not inscribed in the symmetry of the superstructure and distort it. In this case, we deal, therefore, with the lattice instability preceding the structural transformation occurring in these alloys together with the order-disorder transformations. The structural instability in the ordering alloys represents a very interesting phenomenon [233].

In high-symmetry superstructures, such as $L1_2$, the coordinates of atoms do not contain free parameters. However, the displacements of atoms from lattice sites take place in these structures. These displacements proceed in a random manner. Only in the stoichiometric alloy, in the state of perfect order, all atoms find themselves exactly in the lattice points. With the decreasing order parameter (due to the temperature or concentration disordering), atoms of different components with unequal size occupy the foreign sites, pushing away one another from the lattice sites. As a result, static random displacements arise in the lattice [205]. These are complemented by dynamic random displacements caused by the thermal atomic vibrations. The phenomenological theory of static displacements in ordering alloys has been developed by Krivoglaz [23]. He predicted the following dependence of the mean square static displacements

on the order parameter η

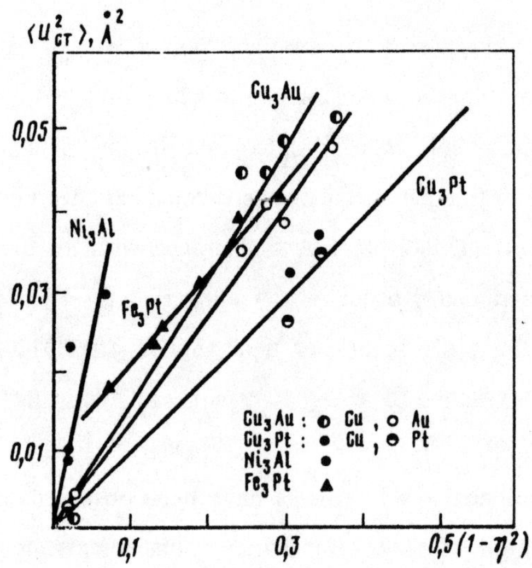

Fig. 43. Dependence of static displacements on the order parameter. Superstructure $L1_2$.

$$\overline{(\delta R)^2} \sim 1 - \eta^2 + \epsilon(\rho) \tag{3.3.1}$$

where ε is the pair correlation parameter. If the correlation is disregarded, we have

$$\overline{(\delta R)^2} \sim (1 - \eta^2). \tag{3.3.2}$$

The experimental study of the problem of static displacements in the ordering alloys has been conducted in [234]. In Fig. 43 the dependence of rms static displacements on η is given in coordinates corresponding to the Krivoglaz theory. The measurements were carried out both for the temperature disordering (Cu_3Au, Cu_3Pt) and for the concentration disordering (Ni_3Al). For the alloys Cu_3Au and Cu_3Pt, the displacements were sorted out according to different sublattices. The results are in accord with the inferences of the Krivoglaz theory: the displacements are equal to zero at $\eta = 1$ and proportional to $1 - \eta^2$ as η decreases.

Therefore, the atomic ordering creates premises for a variety of structural changes. When a low-symmetry superstructure is formed, the resulting atomic arrangement allows only particular distortions and atomic displacements from the lattice sites. In high-symmetry superstructures, the volume of alloy changes, superstructural contraction is observed and chaotic displacements of atoms emerge. The experimental data described in this section, provide evidence that these structural variations always occur in a way dictated by the ordered arrangement of atoms in the crystal lattice.

The magnitude of the corresponding distortions or displacements turns out to be related to the long-range order parameter by a simple functional relation (for the most part, linear). All this maintains the thesis stated above that the change in the atomic arrangement on ordering is the dominant process in such transition. The remaining changes in the alloy structure arise as a direct result of variations in the atomic arrangement. The only quantitative transition parameter is the long-range order parameter η. Changing in other parameters proceeds in accordance with η and depending on η. This property is one of the underlying factors in the separation of an individual class of transitions, i.e., the order-disorder transformations.

4. ORDER-DISORDER PHASE TRANSITIONS IN ALLOYS

4.1 Thermodynamics of order-disorder phase transitions

Any classification of the phase transitions from the thermodynamical standpoint is based on the well-known Erenfest approach [24] which distinguishes between the 1st and 2nd-order transitions (Fig. 44). Although the Erenfest classification provides the basis for any other classification, it is not sufficiently elaborated. The intention to make the thermodynamical description of the phase transitions more detailed has evoked a large number of attempts. Among them, Mayer's classification [235] should be mentioned in the first place; it is based on the account of the order of the correlation function undergoing a jump at the phase transition point. An alternative approach has been employed by Mayer and Stritter (see the review in [236] and Fig. 45). It is based on a detailed analysis of the mathematical features of the behavior of the internal energy and heat capacity C_p near T_c.

The next step in this line represents the Pippard classification (see the review in [237]) which provides virtually the treatment of the isocritical phenomena [298]. It has been customary to classify the order-disorder phase transitions in accord with the Erenfest scheme. However, the experimental data accumulated by this time [221] allow a more detailed classification of the order-disorder phase transitions. It has an original character and represents the generalization of the approaches developed by Mayer and Stritter and Pippard for the case of order-disorder transitions. This scheme is represented in Fig. 46.

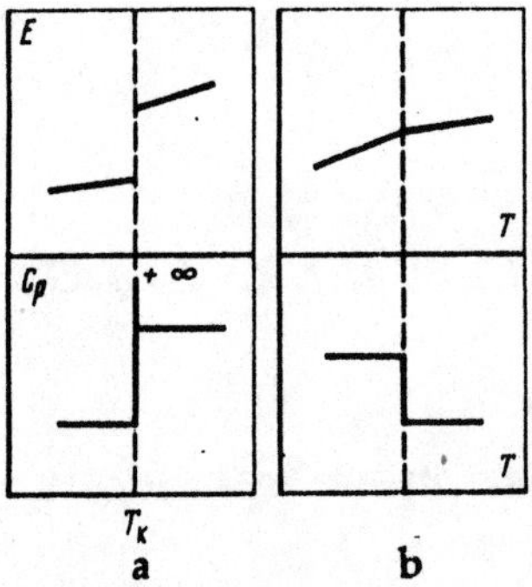

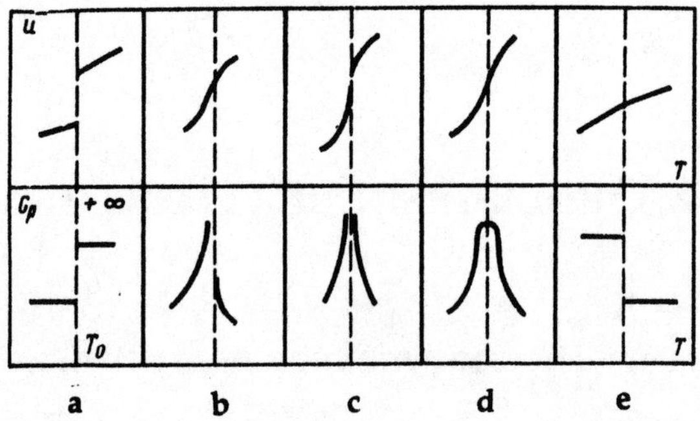

Fig. 44. The Erenfest classification of phase transitions: 1st-order (a); 2nd-order (b).

Fig. 45. Classification of phase transitions after Mayer and Stritter: 1st-order (a); λ-transition (b); anomalous 1st-order (c), diffuse (d); 2nd-order (e); dashed line indicates T_0.

Let us draw attention to the fact that in our classification, along with the assessment of the behavior of the internal energy E and heat capacity C_p, the order parameters η and α are also used. It should be reminded that, to the first approximation, the internal energy is proportional to η^2 and Cowley's parameter α. The use of the latter is necessary because, in contrast to the long-range order parameter η, the short-range order parameter α describes the behavior of both the ordered and disordered phases. The mathematical properties of the behavior of E, C_p, η, and α in the vicinity of T_c are given in Table 12. It is these properties that constitute the basis of the proposed classification.

We shall illustrate first the necessity of using the Cowley-Warren parameter. Various kinds of α(I) curves are shown in Fig. 46. Owing to the presence of low- and high-temperature phases, we obtain two branches of $\alpha(T)$ on which T_c and T_{sl} (the temperature corresponding to the loss of stability of the low- and high-temperature phases) are denoted. The jump value $\Delta\alpha|_{T_c}$ is determined by the proximity of the 1st-order transition to the 2nd-order transition. In the limiting case of the 2nd-order

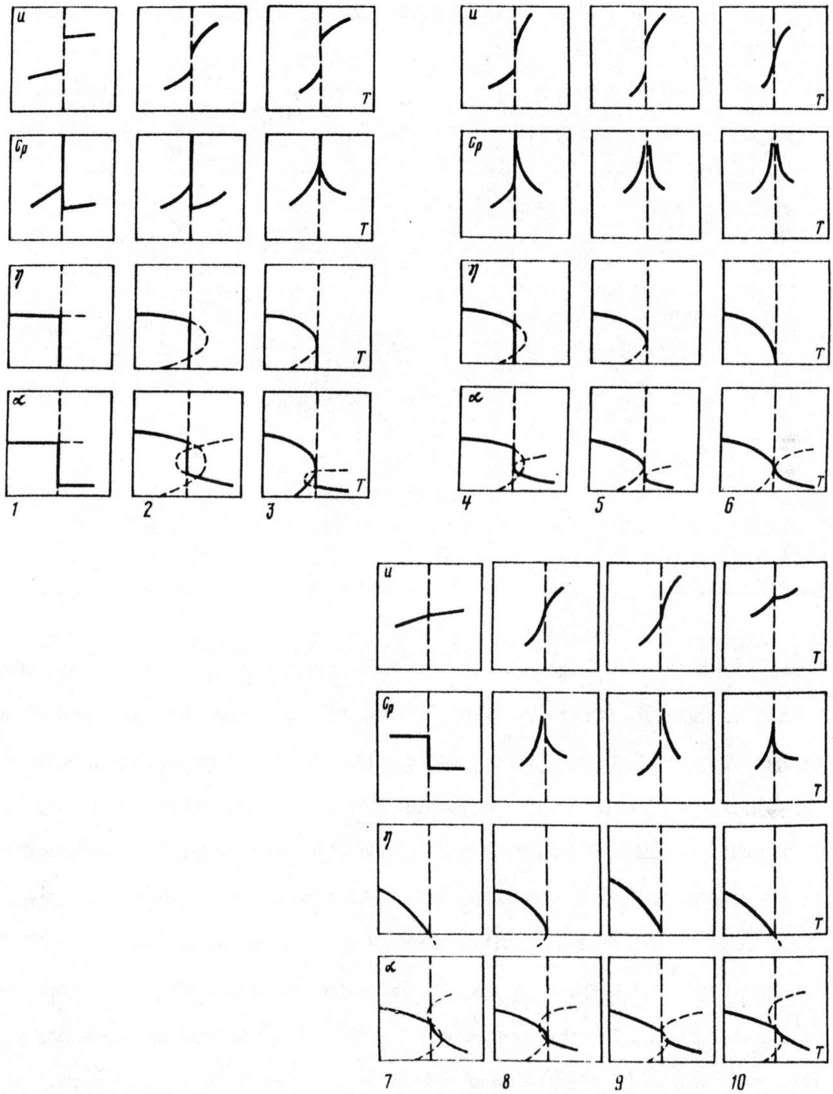

Fig. 46. Classification of phase transitions: 1st-order (1); order-disorder transition of the 1st-order (2); critical disordering of the 1st order (3); critical ordering of the 1st order (4); critical of the 1st order (5); anomalous of the 1st order (6), 2nd-order (7); λ-transition (8); inverse λ-transition (9); anomalous of the 2nd order (10).

Table 12. Characteristic features (gaps and derivatives) of thermodynamical parameters in phase transition region.

Notations in Fig. 46	ΔU	$\left.\frac{dU}{dT}\right\|_{T_K}$	Δc_p	$\frac{dc_p}{dT}$	$\Delta\eta$	$\left.\frac{d\eta}{dT}\right\|_{T_K}$
а	$\neq 0$	const	∞	const	$\neq 0$	const
б	$\neq 0$	const	∞	const	$\neq 0$	const
в	$\neq 0$	$T_K - 0$ ∞ $T_K + 0$ const	∞	$T_K - 0$ ∞ $T_K + 0$ const	$\neq 0$	const
г	$\neq 0$	$T_K - 0$ const $T_K + 0$ ∞	∞	$T_K - 0$ const $T_K + 0$ ∞	$\neq 0$	const
д	$\neq 0$	$T_K - 0$ $T_K + 0$ ∞	$c_p(T_K) = \infty$	$T_K - 0$ $T_K + 0$	$\neq 0$	const
е	0	$T_K - 0$ ∞ $T_K + 0$	$c_p(T_K) = \infty$	$T_K - 0$ $T_K + 0$ ∞	0	∞
ж	0	$T_K - 0 \neq T_K + 0$	const	const	0	const
з	0	$T_K - 0$ ∞	$T_K - 0$ ∞	$T_K - 0$ ∞	0	∞
и	0	$T_K + 0$ ∞	$T_K + 0$ ∞	$T_K + 0$ ∞	0	const

transition, both branches merge and three specific temperature points reduce to a single point. Turning the α(T) curves through 180° around the temperature axis and through 90° around the axis orthogonal to the drawing plane represents the α(T) functions in the Van-der-Vaals coordinates. Such projection illustrates well common features and, simultaneously, the distinction between the phase transitions strongly differing in their nature. The regions of stable, metastable and unstable states are especially clearly seen. Let us proceed now to a detailed consideration of the new classification.

The first type of transition is the net 1st-order transition: the order parameters remain equal to unity up to the temperature T_C. It is this type that corrresponds to the 1st-order transition in Erenfest's sense. The order-disorder transitions of such type do not occur; they may occur only in allothropic and martensitic transformations. The second type of transition is a typical order-disorder transition of the first order. If one of the phases (ordered or disordered) loses its stability at T_C, there exist the transitions of the third type (critical 1st-order disordering) or the sixth type (critical 1st-order ordering). The fifth and sixth types of transitions would be found in the case when the stability loss temperatures are the same for both phases. The critical 1st-order transition implies the abrupt change in η and α at $T = T_C$, whereas an anomalous transition corresponds to the absence of these changes. By "anomalous" it is meant that, in principle, this transition may be assigned both to the 1st-order and the 2nd-order types. The seventh type describes the 2nd-order transition following Erenfest. Such order-disorder transitions do not occur, but the transformations of the λ type (eighth and ninth types) exist in which one of the phases loses stability at T_C. Then, anomalous 2nd-order transitions begin, passing to the hypercritical phenomena and the 3rd-order phase transitions. An example of the anomalous 2nd-order transition is presented as the tenth type. We are not in position to consider in detail this part of our classification because it is rather cumbersome, the more so that the order-disorder phase transitions are mostly referred to the 1st-order phase transitions.

The classification described illustrates a variety of order-disorder phase transitions. It is not a comprehensive one and, at the same time, underlines the significance of using the configurational effects in the classification of order-disorder phase transitions, at least of the second order (parameter α).

4.2 Experimental data on the equilibrium behavior of the long-range order parameter

The nature of long-range atomic order can hardly be understood without detailed information on the nature of the order-disorder transformation and on the behavior of the long-range order parameter at temperatures close to the transformation temperature. The main reason to study the order-disorder transformation for a variety of alloys is that only this way leads to the understanding of the general laws governing the phase transitions of the order-disorder type and permits one to refer them to an individual class. It would be impossible to gain such understanding by selective studying of a

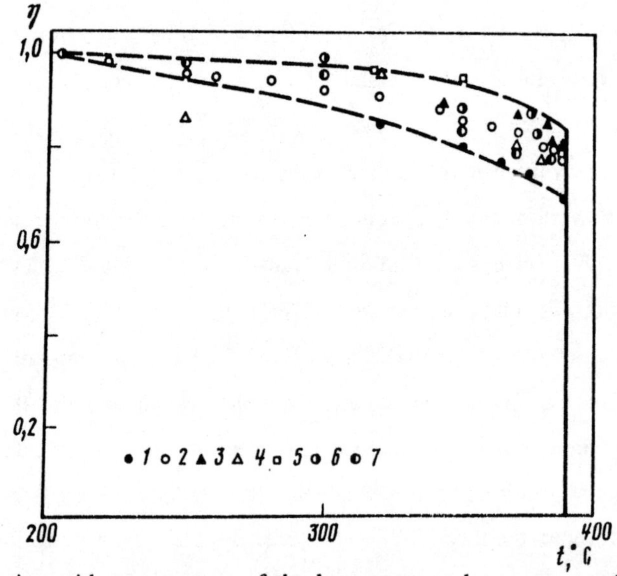

Fig. 47. Variation with temperature of the long-range order parameter for the Cu_3Au alloy according to the literary data: Cowley (1); Feder (2); Buinov (3); Wilchinsky (4); Anguetti (5); Warren and Keating (6, 7).

limited number of alloys without generalizing the facts obtained. The authors of this book, together with their colleagues, have studied the order-disorder transformation by using a group of alloys which included the most part of alloys accessible at present for the experimental investigation. The exception constituted the alloys with very high transformation temperatures or those containing easily sublimating component. The material of the present section is given accordingly to the type of superstructure.

L12 superstructure. Previously, the temperature dependence of the long-range order parameter and the nature of transformation were investigated on the alloys Cu_3Au [239 - 241], Ni_3Fe [242 - 244], Ni_3Mn [245 - 247], and Pt_3Co [248]. The data for the Cu_3Au alloy are shown in Fig. 47. It is evident that the scatter of the results obtained by different authors exceeds the experimental uncertainty. The most reliable results for the Cu_3Au alloy have been obtained by Keating and Warren [241], whereas no choice is available for Ni_3Fe.

Table 13. Characteristic features of $L1_2 \rightarrow A1$ transformation.

Alloy	T_c, K	Boundaries of two-phase regions, K;	$\Delta\eta$ (or $\Delta\eta_{max}$ and $\Delta\eta_{min}$)	$\frac{\partial(T/T_K)}{\partial\eta}$ in critical point	$\frac{\partial(T/T_K)}{\partial\eta}$ in two-phase region	ΔT in two-phase region, K	Occurrence of short-range order at $T>T_c$	Correlation radius (half-size of domains), A		References
								$T > T_K$	$T > T_K$	
Cu_3Au	659 ∓ 2	Нет	0,77	0,90	–	–	yes	350	5–15	[221, 249]
Cu_3Pt	958 ∓ 3	893–958	0,68; 0,62	–	2; 0,7	65	yes η = 0,3–0,4	400	10–15	[221, 250]
Au_3Cu	481 ∓ 3	469–481	0,69; 0,62	–	0,5; 0,5	12	yes η = 0,3	35	20	[221, 251]
Ni_3Fe	808 ∓ 2	Нет	0,44	0,05	–	–	nc	70	7	[221, 249, 252]
Ni_3Pt	825 ∓ 3	805–821	0,55; 0,50	–	1; 0,25	16	yes η = 0,15–0,2	80	7,5–10	[221, 249]
Ni_3Mn	793 ∓ 3	733–793	0,86; 0,66	–	0,6; 0,14	60	no	95	–	[221, 252, 253]
Pd_3Fe	909 ∓ 3	803–909	0,98; 0,63	–	2; 0,3	106	no	420	10	[252]
Pt_3Fe	1565 ∓ 3	1501–1565	0,85; 0,60	–	0,2; 0,03	64	no	1000	10	
Fe_3Pt	1068 ∓ 3	Нет	0,36	1	–	–	no		10	[254, 234]

The measurements of the long-range order parameter against temperature for the Ni_3Mn alloy have been made for compositions far from stoichiometry [245]. For the stoichiometric alloy, Vintaikin et al. [247] have measured the temperature dependence of the intensities of diffraction peaks. The measurements on the Pt_3Co alloy [248] have been performed carefully. It should be noted that the transformation in the $L1_2$ superstructure represents a favorable object for specialists who devise statistical theories of this phenomenon. These theories are usually compared with the experimental data of Keating and Warren for the Cu_3Au alloy [241]. Meanwhile, as may be seen from the works prior to the present investigation, the characteristic features of the transformation occurring in the $L1_2$ superstructure cannot be understood if we restrict ourselves to the description of the transformation in a single alloy.

In the present work the transformation in the $L1_2$ superstructure was studied in the alloys Cu_3Au, Cu_3Pt, Au_3Cu, Ni_3Mn, Ni_3Fe, Ni_3Pt, Pd_3Fe, and Fe_3Pt. The basic characteristics of the alloys and the details of their behavior near T_c are presented in Table 13. This table incorporates the data on the alloy compositions, the transformation temperatures, the temperature and long-range order parameter ranges of the two-phase regions, the jump of the order parameter η at T_c (including the case where the transformation proceeds through the two-phase region), the correlation radii at temperatures above and below T_c, and the proximity of the alloys to the point where the ordered phase loses its stability (which is determined by the deviation of $\partial(T/T_c)/\partial\eta$ from zero).

The variations with temperature of the long-range order parameter η (the volume-averaged value and that in the ordered phase) are shown in Fig. 48. In all alloys, except Fe_3Pt, the net transformation $L1_2 \rightarrow A1$ occurs (i.e., the order-disorder transition in a classical sense). In the Fe_3Pt alloy, two structural transitions -- to a long-period phase and to the $L1_2$ phase with the extraordinary large volume -- superimpose on the $L1_2 \rightarrow A1$ transformation. This fact may explain the difference between the $\eta(T)$ curves for Fe_3Pt and other alloys. In most alloys the transformation proceeds through the two-

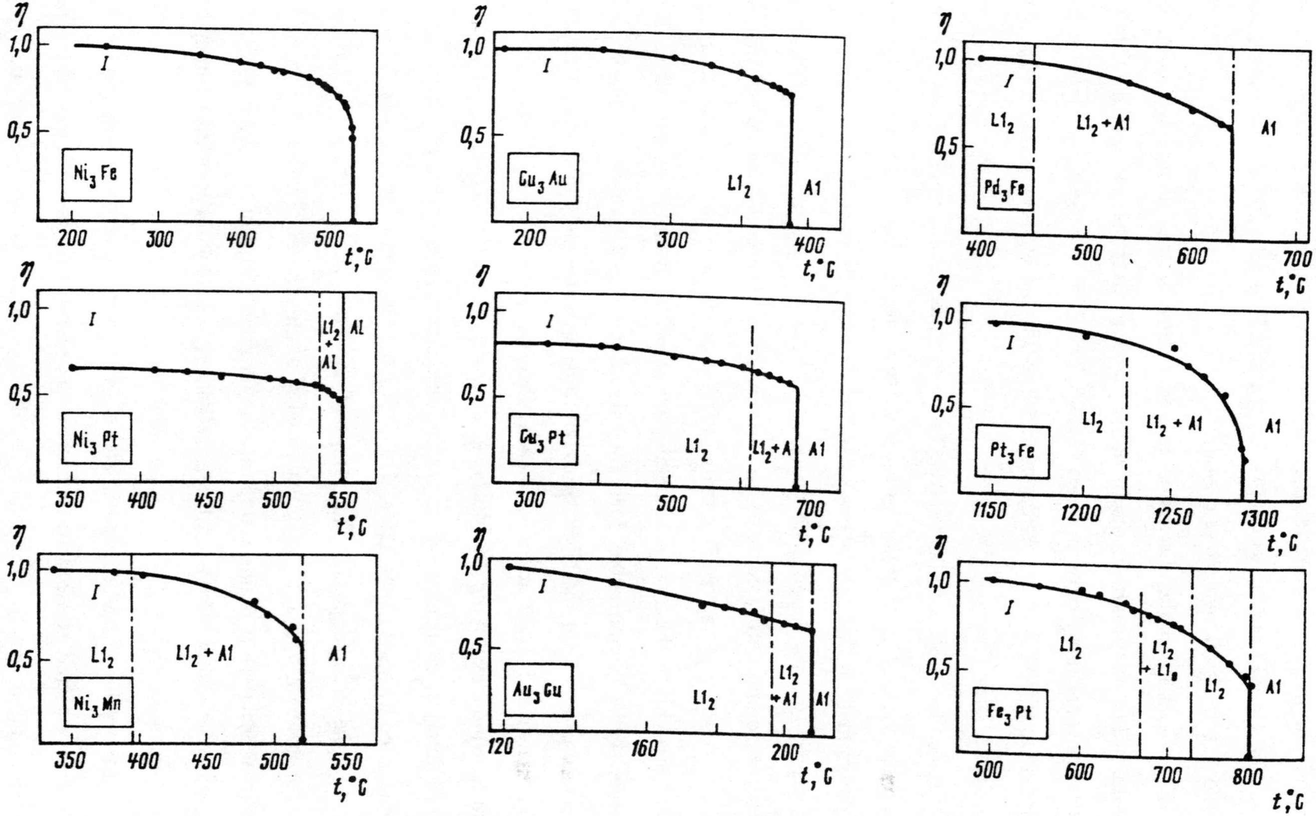

Fig. 48. Variation with temperature of the long-range order parameter in the $L1_2$ superstructure (experiment) for different phases.

phase region with the temperature range diminishing in the line: Pd_3Fe, Cu_3Pt, Ni_3Mn, Ni_3Pt, Au_3Cu. Only in the alloys Cu_3Au, Ni_3Fe and Fe_3Pt the phase transition takes place without forming the two-phase region.

Before proceeding to the discussion of the results of the present work, let us compare the $\eta(T)$ curves with the results of the previous investigations. For the Cu_3Au alloy, our results are fairly close to those of [240, 241, 255]; for the alloy Ni_3Fe in the low-temperature region -- to the results of [256], and in the high-temperature region -- to the results of [243].

The first fact to consider in the analysis of the obtained data is nonfulfillment of the law of corresponding states. For the 1st-order phase transition that is typical of the $L1_2 \rightarrow A1$ transformation, this fact is not surprising; it only shows that the origin of the ordering energy and its dependence on size and other factors are different in these alloys. Had a common $\eta(T)$ curve existed in the alloys undergoing the transformation of the $L1_2 \rightarrow A1$ type, all $\eta(T)$ curves in T/T_{sl} coordinates would have been the portions of this universal curve. A detailed analysis of the available experimental data showed that such common curve does not exist, but the behavior of the alloy pairs Cu_3Pt and Cu_3Au, Au_3Cu and Ni_3Pt turns out to be approximately the same. The $\eta(T)$ curves for Ni_3Mn and Ni_3Fe are close to each other only in the vicinity of the transition temperature while the alloy Fe_3Pt displays a suprisingly unique behavior.

It is absolutely clear that the $L1_2$ superstructure is not characterized by a common $\eta(T/T_{sl})$ curve, both volume-averaged and related to the behavior of the long-range order parameter in the ordered phase. The jumps of the order parameters at T_c are also different for different alloys, ranging within wide limits (from 0.9 ± 0.02 for Pt_3Fe to 0.45 ± 0.02 for Ni_3Fe). It is interesting that the terminal alloys in this series display a point 1st-order transition, whereas all intermediate alloys show the transition through a two-phase region.

The smallest value of the order parameter is an important quantity. It determines the stability of the ordered phase with respect to the phase with short-range order. The smaller the jump $\Delta\eta$ at T_c, the higher is this stability. The alloys with the $L1_2$

superstructure may be positioned in a series starting from the most unstable and ending with the most stable: Pt_3Fe - Cu_3Au - Ni_3Mn - Pd_3Fe - Au_3Cu - Cu_3Pt - Ni_3Pt - Fe_3Pt - Ni_3Fe.

Along with the minimum degree of order persisiting in the alloy near T_c, the maximum value of η at which disordering begins can also be used as an important parameter. This quantity corresponds to the onset of the two-phase region and characterizes the stability of the ordered state. The highest degree of order at which disordering begins (the maximum jump) is observed in Pt_3Fe where $\Delta\eta = 0.95$. The series of alloys arranged in accordance with the maximum η values correlates with that in which the alloys get in line according to the minimum η values.

As the alloy approaches the point of the stability loss (retaining, however, the 1st-order transition), the possibility of the occurrence of the temperature hysteresis and its value diminish; in other words, the region of the existence of the metastable states decreases. At the point of the stability loss $\partial(T/T_c)/\partial\eta = 0$. The nearest to this point is the Ni_3Fe alloy, for which $\partial(T/T_c)/\partial\eta = 0.05$. The comparison of the jumps of the order parameters with the slope of the $\eta(T)$ curve at this point (see also Table 13 where, for convenience, the reciprocal quantity $\partial(T/T_c)/\partial\eta$ is given) reveals the absence of the correlation between these values. This fact confirms the conclusion made before that the dependences of η on T are substantially different for various alloys (even if they are referred to the point where the ordered state loses its stability).

The loss of stability for different alloys occurs at varying η values. From the standpoint of the statistical theory of order-disorder transitions (even if the theory operates with the constant ordering energy independent of the lattice parameter, the state of order, temperature, and so on), this implies that the variation of the ordering energy with the number of the coordination shell is different for various alloys. If the statistical theory takes into consideration at least the pair correlation, $\eta(T/T_c)$ curves will be different for different contributions V_1, V_2, V_3 into the ordering energy. Thus the availability of a series of curves $\eta(T/T_c)$ gives evidence that in the investigated alloys (1) the pair ordering energy depends in different ways upon the distance (if the correlation is substantial), or (2) the dependence of V on η and T is different for various alloys, or (3) the contributions of the three-particle, four-particle and higher interactions into the ordering energy are different.

The choice between these options cannot be made by examining only the $\eta(T/T_C)$ functions; it is necessary to involve several other characteristics. We must underline, however, the difference in the behavior of the alloys based on transition metals (Ni) and a noble metal (Cu). In the alloys based on transition metals (TM) the temperature range of the two-phase region exceeds, as a rule, that in the alloys based on noble metals (NM). In addtition, the two-phase region in TM-base alloys appears at higher η values. Regarding the jump of the long-range order parameter at T_C, or, equivalently, the minimum η value at which the ordered phase is still stable, it may be said that, in general, in NM-base alloys, especially in Cu-base ones, this value is higher than in the alloys based on transition metals Ni or Pd.

It is of interest to note that such behavior correlates with a stronger tendency for the antiphase boundaries to lie on cube planes in the Cu-base alloys compared with to the Ni-base alloys. Besides, in Cu- and Au-base alloys in the neighboring temperature or concentration ranges and, occasionally, at the same composition but at the other ordering mode, the formation is observed of superstructures with periodic antiphase boundaries in cube planes, namely $L1_2(M)$ and $L1_2(MM)$. At the same time, such structures never occur in Ni-base alloys while in Pd-base ones they were found only in a few systems. Such behavior, as was noted above, is connected with the dependence of the ordering energy on the distance, $V(\overline{R})$, and with many-particle interactions. If one compares the present results with those from [42], it will be clear that in Cu-base alloys the contribution of the third shell into the ordering energy is small and the ordering energies for the first and second shells have opposite signs, whereas in nickel alloys the contribution of the third shell must be significant. If the contribution of the three-particle interaction into the ordering energy is significant, the contribution of the second shell must be small in Cu-base alloys while in Ni-base alloys it is significant and has the opposite sign.

In terms of the electronic structure of the alloys of interest (see Chapter 5), such behavior is closely related to the wave vectors of the singularities which determine the formation of the ordered state. It is evident that in Cu-base alloys these singularities will coincide with the general points in the Brillouin zone (or lie on symmetry lines or planes) while in Ni-base alloys they coincide with the high-symmetry point (1 0 0). This effect is determined to a great extent by the shape of the Fermi surface and by the degree of filling of the d band. From the comparison of the results of this section with those in Subsection 6.4 it follows that, with successive filling of the d band in the line Ni - Pd - Cu, the singularity wave vector responsible for the formation of the superstructure, moves in the direction (1 0 0) $\rightarrow$ (1 0 1/2M).

At temperatures above that of the transition, one observes usually in an alloy the state of short-range order. Sometimes, the degree of short-range order is so small that this state does not practically manifest itself in the experiments on polycrystalline specimens. However, in some alloys a significant short-range order was detected above T_C; which can be recalculated into the long-range order using the microdomain theory. As can be seen, in some alloys the magnitudes of η may reach 0.2 - 0.3. This poses the question: what does it happen in the point of the order-disorder phase transition? In such cases, it would be more accurate to define this transition as the "long-range to short-range order" transformation. The order parameters decrease abruptly, at least by a factor of two. But this change does not reflect the essence of the transformation. The main effect is the abrupt decrease in the correlation radius r_C. The latter characterizes the maximum distance at which one can still observe a regular intermittence of atoms of different kind on the successive coordination shells. The correlation radius sharply decreases at $T = T_C$. Below T_C it must be principally $r_C = \infty$ but, owing to the presence of the antiphase domains, the value of r_C amounts approximately one half of the domain size. Above T_C, the correlation radius is limited by the value of the order of several interatomic distances (typically, ~ 2–6 nearest interatomic spacings). This limit does

Table 14. Characteristic features of $L1_2(M)$, $L1_2(MM) \rightarrow A1$ transformation.

Alloys	Superstructure type at T_c	T_c, K	Occurrence of c/a	Order of phase transition	Jump in η_{mean} at T_c	$\frac{\partial(T/T_K)}{\partial\eta}$	M range	$\frac{\partial M}{\partial\eta}$	References
Au_3CuII	$L1_2$ (MM)	204 ± 4	quasicubic	close to II	0,1	0,05	M_1 = 19 ÷ 17 M_2 = 6,7 ∓ 0,5	0,6	[252, 262]
Cu_3Pt	$L1_2$ (M)	630 ± 5	quasicubic	close to II	0,1	0,1	8,5–4,5	2,5	[263, 250]
Au_4Zn	$L1_2$ (MM)	308 ± 5	orthorhombic	I	0,5	0,3	M_1 = 2,78 ÷ 3,10 M_2 = 2,37 ÷ 2,45	−0,4 −0,18	[262]
Au_3Mn	$L1_2$ (MM)	658 ± 3	orthorhombic	I	0,5	–	M_1 = 1,20 ± 0,02 M_2 = 2,07 ± 0,07	0 0	[262]
Cu_3Pd	$L1_2$ (MM)	468 ± 5	tetragonal	I	0,54	0,5	M_1 = 3,5 ÷ 4,5 M_2 = 5,5	−4	[252]
Cu – 22ат.% Pd	$L1_2$ (MM)	488 ± 3	tetragonal	I	0,50	0,25	5,5–6,5	–	–
Cu – 20ат.% Pd	$L1_2$ (M)	478 ± 3	tetragonal	I	0,54	0,4	8–9	–	–

not depend virtually on the degree of order. As for the decrease in the η value at $T = T_c$, it seems to relate mainly to the increase in the density of the APBs, in accordance with the microdomain model.

The essence of the order-disorder phase transition (the long-range to short-range order transition) lies in a jumplike decline of the correlation radius [221]. In principle, it drops from infinity to the magnitude of the order of several interatomic distances.

We can summarize now some results of studying the order-disorder $L1_2 \rightarrow A1$ transformation. This is the 1st-order phase transition with an abrupt disappearance of long-range order in a point or in the two-phase region. Alloys exhibit a wide spectrum of properties near T_c, markedly differing both in the stability with respect to the disordered phase and in the proximity to the point of the stability loss of the ordered phase. The degree of the long-range order changes near T_c with the different rate for various alloys. All this difference in the behavior of alloys is in turn determined by the difference in the contributions of the pair and many-particle interactions into the ordering energy and in their dependence on the distance.

$L1_2$(M) and $L1_2$(MM) superstructures. The order-disorder phase transformations leading to the formation of long-period superstructures as well as the variations of long-range order parameters and antiphase periods with temperature are comparatively little studied. Before the present work, information on the nature of transformations related to the formation of long-period superstructures as well as some details of the temperature variations of the long-range order parameters and antiphase periods have been obtained by studying the alloys CuAu [258, 259] and Cu + 30% Au [260, 261] in which the $L1_0$ - $L1_0$(M) - A1 and $L1_2$ - $L1_2$(M) - A1 transformations occur. Both transformations are of the 1st order, but the temperature range of the existence of the long-period superstructure is rather narrow and, consequently, many details of the behavior cannot be studied on these alloys. Besides, in the case of the $L1_2$(M) superstructure only one order parameter was measured in these references. For this reason, there has been an urgent necessity of investigating the order-disorder transformation on the alloys with an extensive region of existence of the long-period superstructures.

In the current work experiments were conducted on the alloys Au_3Cu, Au_4Zn, Au_3Mn, Cu_3Pd, and Cu_3Pt. The main experimental characteristics and information about the details of the transformations occurring in these alloys are presented in Table 14. It comprises, in particular, the compositions of the studied alloys, the types of the superstructures formed, the critical temperatures T_c, the order of the phase transition, the variation range of the antiphase period and the sign of its temperature dependence, the jumps of the mean order parameters associated with the phase transitions, the type of the unit cell distortion, and the values of $\partial(T/T_c)/\partial\eta$. To describe the structure of the ordered state, in addition to the antiphase period and the volume-averaged order parameters, we also employed the long-range order parameters for the transformation stars and the order parameters describing atomic order directly at the APBs and far apart from them. The relationships between these order parameters depend on the antiphase period and cannot be introduced in a general way. They have been introduced separately for each superstructure.

The question about the number of parameters in the long-period superstructures $L1_2(M)$ and $L1_2(MM)$ is not trivial for two reasons. First, a large number of lattice sites of different kind in these structures (see Section 1.1) necessitates a great number of long-range order parameters which are difficult to measure experimentally. The difficulties stem from the insufficient number of superstructure lines and their satellites as well as from the overlap of these satellites. It is necessary, therefore, to determine the number of independent long-range order parameters for each value of the antiphase period M. Second, the question about the number of independent order parameters is closely related to a finer problem: whether there exists a sound basis to recognize these superstructures as individual phases. It should be noted that many authors give them the status of the conventional ordered phases with their own region of existence and a definite type of the phase transition into the ordinary superstructures and introduce for them in a usual way the order parameters. At the same time, some authors do not agree with this approach supposing that the behavior of such superstructures is determined not only by the long-range order parameter alone [19].

Crystallography of long-period superstructures has been investigated in the works [9, 10]. The superstructures DO_{22}, DO_{23} and $L1_2(M)$ with the integer M values belong to the space group D_{4h}^{12}. When M is fractional, the $L1_2(M)$ superstructure can be assigned to one of the four space groups. The determination of the actual space group for the $L1_2(M)$ superstructures is important because it defines the type of X-ray diffraction pattern (the number and the type of satellites). The total number of long-range order parameters is defined by formula (3.1.9). The number of different types of lattice sites Q is determined by the magnitude of the antiphase period. Analysis taking into account crystallography of the superstructure and the availability of free parameters in the atomic arrangement shows that the number of independent sublattices is equal to (3M+2)/2 if M is even and M -- 3(M+1)/2 for odd M. Thus an increase in the antiphase period gives rise to an increase in the number of long-range order parameters. For illustration, we show in Fig. 49 the unit cell of the $L1_2(M = 4)$ superstructure of the Cu_3Pd alloy with seven lattice types denoted. The displacements of atoms from the sites are not indicated. The structure factor is expressed in this case by the formula:

$$F_f = (C_A f_A + C_B f_B)\,[1 + \cos\pi\,(h + k + l)]\left\{\sum_{n=0}^{(M-2)/2} \cos\frac{(2\pi + 1)\pi l}{2m} + \right.$$
$$+ 2\sum_{n=\frac{M}{2}}^{M} \cos\frac{(2n+1)\pi l}{2M} + \cos\frac{\pi}{2}(2h + l) + \cos\frac{\pi}{2}(2k + l) +$$
$$\left. + (\cos\pi h + \cos\pi k)\left(1 + \sum_{n=1}^{(M-2)/2} \cos\frac{\pi n l}{M}\right)\right\} \tag{4.2.1}$$

for the fundamental reflections and

$$F_s = (f_A - f_B)[1 + \cos\pi\,(h + k + l)]\left\{\frac{1}{4}\eta_1 (\cos\pi h + \cos\pi k) - \right.$$
$$- \cos\frac{\pi}{2}(2h + l) - \cos\frac{\pi}{2}(2k + l)] +$$
$$+ \frac{3}{4}\sum_{i=2}^{(M+2)/2} \eta_i \left[2\cos\frac{(2i - 3)\pi l}{2M} - \cos(\pi h + \cos\pi k)\right] -$$
$$- \frac{1}{4}\sum_{m=(M+2)/2}^{M} \eta_m \left[2\cos\frac{(2m - 3)\pi l}{2M} - (\cos\pi h + \cos\pi k)\right] -$$
$$\left. - \frac{1}{4}\sum_{m=1}^{(M-2)/2} \eta_m \left(2\cos\frac{\pi m l}{M} - 1\right)(\cos\pi h + \cos\pi k)\right\} \tag{4.2.2}$$

for the superstructure lines. The displacements of atoms are not introduced into the structure factor.

In the real superstructures, the number of order parameters is limited for several reasons. First, at great values of M the regions remote from APBs are similar to the $L1_2$ superstructure with a single order parameter. This is due to the limited value of the radius of the interatomic interaction and cannot be proved in general. Second, the reflections of the (0 0 1/M) type arising from the structure factor are always absent in the long-period structure. This fact points to the degeneracy of some part of the order parameters. Third, the satellites of different diffraction orders are often absent in X-ray patterns from the long-period superstructures which increases the degeneracy of the long-range order parameters.

It should be noted that the satellites of different superstructure reflections frequently overlap, especially in the case of the two-dimensional superstructure. A small size of thermal antiphase domains due to a large number of domains of the translational type in these superstructures enhances the effect of the satellite overlap. For this reason, for some alloys we had to confine ourselves to the determination of the mean order parameter. Nevertheless, in these alloys the behavior of the quantities characterizing the intensity maximum give evidence of different changes in the degree of order for different stars.

The types of the APBs forming a long period in an alloy were determined in accordance with the character of diffraction pattern by using the well-known scheme suggested by Sato and Toth [55, 264]. In this scheme, a set of indexed satellite lines uniquely determines a conservative or nonconservative type of the periodic antiphase boundaries. In addition to X-ray diffraction, the superstructures were identified by using electron diffraction. The variations of the order parameters and antiphase period with temperature are shown in Figs. 50 and 51. As an example, Fig. 50 shows, for the Cu_3Pd alloy, the long-range order parameters related to different transformation stars.

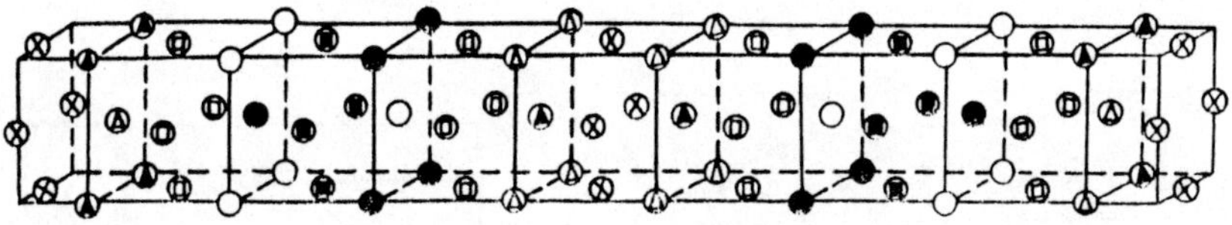

Fig. 49. The unit cell for the $L1_2(M)$ (M = 4) superstructure, the Cu_3Pd alloy.

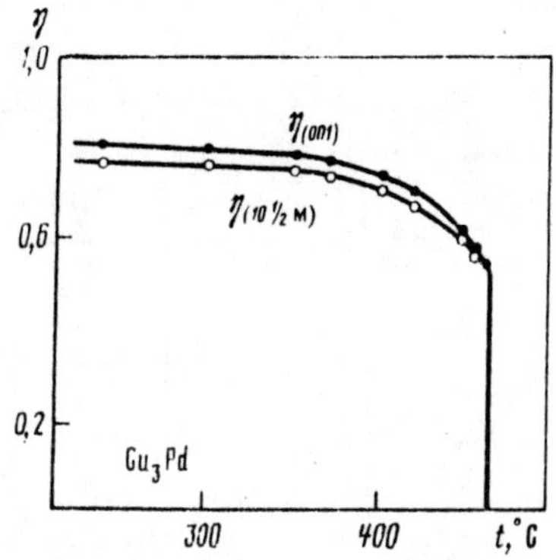

Fig. 50. Long-range order parameter as a function of temperature. The $L1_2(M)$ superstructure. The stars are indicated for order parameters.

The unit cells of the two-dimensional long-period superstructures with the antiphase periods corresponding to the most ordered state are depicted in Figs. 52 and 53. The appearance of these unit cells is rather complex and not accustomed to an eye. Nevertheless, we single out the most extraordinary unit cell, namely, that for the Au_3Mn alloy (Fig. 52). The antiphase period in this cell coincides with the period of the DO_{22} superstructure in one direction and with the period of DO_{23} in the other. However, this unit cell cannot be reduced to these superstructures, being typically a two-dimensional periodic structure. It is of interest that just in this alloy the transformation of the long-period superstructure into the A1 structure proceeds through a two-phase region.

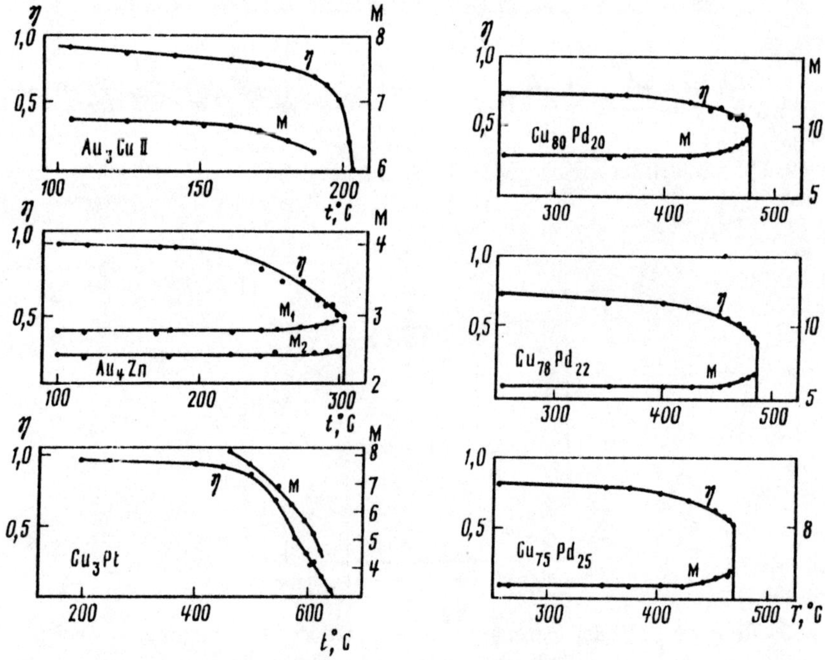

Fig. 51. Long-range order parameter vs temperature plots for different phases. The $L1_2$(M) superstructure. Mean order parameters are given.

It is worthwhile noting among the general regularities that in Au-base alloys only the two-dimensional periodic antiphase structure is observed, while in Cu-base alloys the one-dimensional structure is also found. The type of the periodic APBs in the one-dimensional structure is always conservative, whereas in the two-dimensional one the boundaries of the deficient and interstitial type are usually observed along the second period of the structure. The $L1_2$(MM) → A1 and $L1_2$(M) → A1 transformations may be of the 1st or the 2nd order. In this case, we have the first difference of the $L1_2$(M) and $L1_2$(MM) superstructures from the $L1_2$ superstructure. The presence in the structure of periodic APBs makes possible the transition that is close, at least, to the 2nd order.

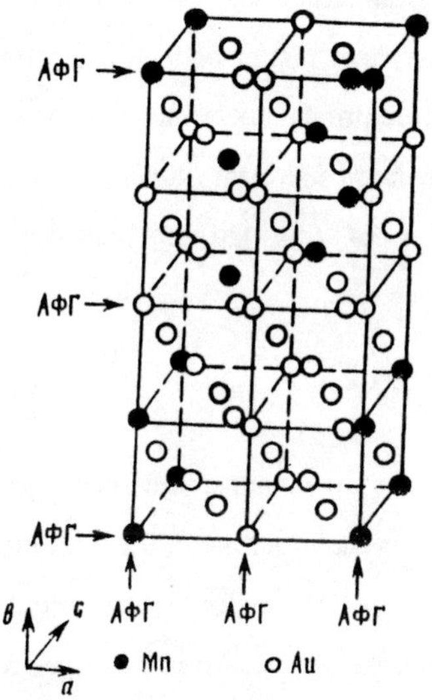

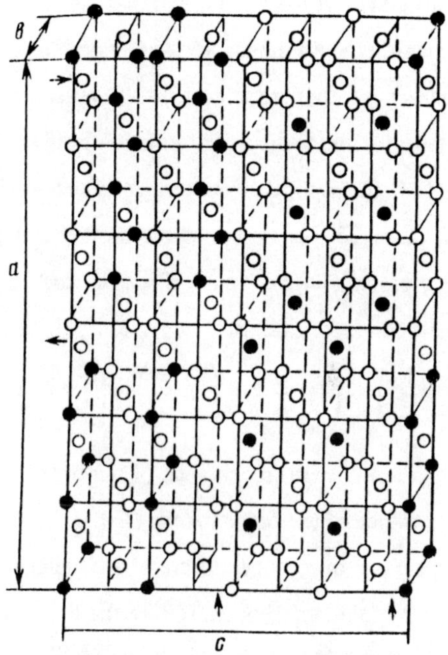

Fig. 52. The unit cell of the Au_3Mn alloy. The $L1_2(MM)$ superstructure; $M_1 = 1$, $M_2 = 2$.

Fig. 53. The unit cell of the Au_4Zn. The $L1_2(MM)$ superstructure; $M_1 = 2$, $M_2 = 3$.

The more the differencies between elements of alloys in the Periodic system, the less the antiphase period. With a drop in M factor, its dependence on temperature decreases or disappears altogether. Finally, M is nearly always a fractional quantity. It gives evidence that the mixture of at least two unit cells is present in the alloy, differing in length by one unit cell of the disordered alloy.

The order-disorder transformation occurs in all the alloys simultaneously for all transformation sites, i.e., at the same temperature. The long-range order parameters $\eta_{0\,0\,1}$, $\eta_{1\,0\,1/2M}$, $\eta_{1\pm1/2M\,0\,0}$ vanish at the same temperature. This phenomenon of the degeneracy of the transition temperature for different stars has been discovered in the present work. It is due to the mechanism of the formation of the long-period superstructures. It has been found that the superstructure reflections determining various order parameters in the $L1_2(M)$ superstructures must disappear simultaneously, i.e., the degeneracy of the order parameters must be accompanied by the degeneracy with respect to the order-disorder transition temperature. As a matter of fact, the reflections of the (1 0 1/2M) type characteristic of the superstructure cannot occur without the (0 0 1) reflection. This is explained by the fact that the sum of several translation vectors in the reciprocal lattice must also be the vector of this lattice. For the DO_{23} superstructure, for example, we have

$$\left(10\,\frac{1}{4}\right)+\left(10\,\frac{1}{4}\right)+\left(\bar{1}0\,\frac{1}{4}\right)+\left(\bar{1}\,0\,\frac{1}{4}\right)=(001).$$

Therefore, the disappearance of the order for the (0 0 1) star entails necessarily the disappearance of the order for the (1 0 1/2M) star. On the other hand, the conditions for the existence of a superstructure are determined by the region where the Fourier transform of the ordering energy for the low-symmetry star has a minimum (see Section 3.2). If V(1 0 1/2M)<V(0 0 1), the transition temperature or, at least, the temperature of the stability loss for the $\eta_{1\ 0\ 1/2M}$ parameter is higher than for $\eta_{0\ 0\ 1}$ (the rigorous proof has been obtained, however, in the simplest version of the statistical theory [19]). In this case, the order for the $\eta_{1\ 0\ 1/2M}$ parameter will always coexist with the order for the parameter $\eta_{0\ 0\ 1}$. Hence, in the long-period superstructures the degeneracy of the transition temperature for all order parameters is to be expected. This is what is really observed experimentally.

The order parameters for various stars at low temperatures are somewhat different so that $\eta_{0\ 0\ 1} > \eta_{1\ 0\ 1/2M}$, but upon reaching T_c they turn out to be close to each other. The degree of order on planes adjacent to the APBs is, as a rule, low. As a consequence, in a perfectly ordered alloy the volume-averaged long-range order parameter for alloys of different concentration is linearly dependent on the APB density (for example, in the Cu_3Pd alloy). The highest degree of order is attained inside the antiphase domains while at the boundaries it is much lower. That is why in the long-period structures the complete ordering cannot be reached. Therefore, the order parameter oscillates in the space with the period M. Diffraction pattern shows that with the same period oscillate the atomic concentration (segregations on the APBs) and directed statistical atomic displacements (i.e., the value of the free parameter in the coordinate of the lattice site). The reflections of the (0 0 1/M) type in diffraction pattern were not recorded (at least, in the state of perfect order). This results from the degeneracy of probabilities and the decrease in the number of the types of lattice sites really existing in the structure.

The distinction of the long-period superstructure from $L1_2$ is connected not only with the appearance of the phase transitions close to the 2nd-order type. Typical of long-period phases are also the transformations without the two-phase region (exception is the Au_3Mn alloy) and almost constant value of the jump of the mean order parameter ($\Delta\eta = 0.5$) pointing to a high stability of the superstructure with respect to disordering. The latter is also connected with the presence of the APBs in the structure. Long-period superstructures come, on disordering, closer to the point of the absolute loss of stability (compare the $\partial(T/T_C))\partial\eta$ values in Tables 13 and 14).

An interesting observation concerns the Au_3Cu alloy. The variations of the order parameters with temperature in the $L1_2$ superstructure and a long-period superstructure are similar, except for the region immediately adjacent to T_C. The critical transition temperatures are also close to each other. Since such similarity in the behavior of the structures is determined by the same nature of the ordering energy in both alloys,the electronic structures of these alloys seem to be close to each other. This conclusion can be reached by comparing the present results with thoses described in Chapter 5 (see, for example, Section 5.4), where the electronic structure of the alloys possessing the periodic and quasi-periodic APBs is considered and the calculations are presented of the transformation energies for the Au_3Cu alloy with and without a long period. These calculations showed that the energies are quite close to each other in accord with the experimental evidence and the phase diagram. Regarding the similarity of the electronic structures of the alloys, it should be noted that, according to the experimental data, in the Au_3Cu alloy with $L1_2(MM)$ superstructure the conservative APBs occur after each 7 unit cells and the nonconservative -- after about 14 unit cells. There are no periodic APBs in the Au_3Cu alloy with the $L1_2$ superstructure -- only closely positioned thermal APBs exist. At the degree of long-range order attained in our work, the conservative APBs were found to occur after each 33 unit cells and the nonconservative -- after 13. These facts, together with the material presented in Chapter 5, confirm the resemblance of the band structures of these alloys. A similar behavior of the ordering energies and, respectively, $\eta(T)$ curves follows from this reasoning.

We should draw attention to the interesting fact, which points to a deep connection existing between the ordering energy (understandable in the interatomic sense) and the volume contribution into the transformation energy. In the Au_3Cu and Cu_3Pt alloys undergoing the phase transition close to the 2nd-order type, the unit cell in the ordered phase is pseudocubic (with the tetragonality parameter $c/a = 1$). Conversely, in the other alloys showing 1st-order transition, one can observe considerable tetragonal or orthorhombic distortions. It is also of interest that the antiphase period in these alloys is greater, on the average, than that in the alloys showing the 1st order transition. It may be supposed that the APBs with a greater period cause a smaller distortion of the unit cell.

Another important fact is the existence of the linear relation between the long-range order parameter and the antiphase period (Fig. 51). In addition to our results, the data of other authors are plotted in this drawing.

First, as mentioned above, this gives evidence for the substantial role the APBs and the degree of order in planes adjacent to APBs take part in the disordering of the alloy. The degree of order is lowered at the antiphase boundaries; in some (but not in all!) alloys an increase in the density of APBs gives rise to disordering. In most alloys with long-period superstructures the value of M either does not change on ordering or increases. It means that a simple increase in the density of APBs cannot be regarded as

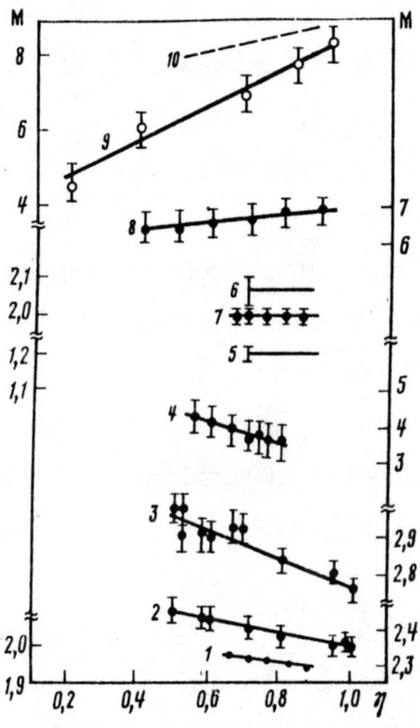

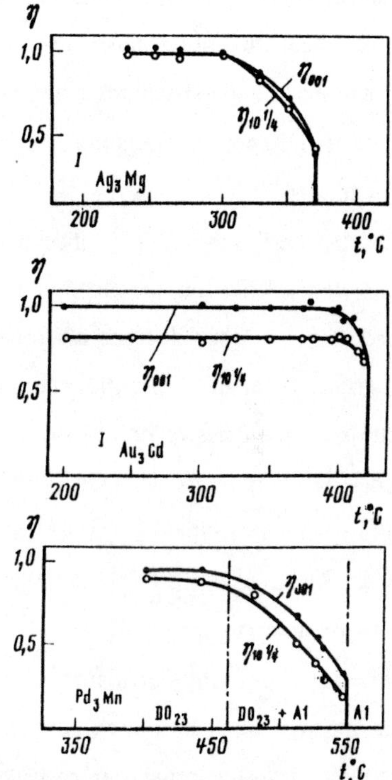

Fig. 54. Antiphase period as a function of the degree of long-range order. Ag_3Mg (1); Au_4Zn(2,3): M_1--(2),M_2--(3); Cu_4Pd (4); Au_3Mn (5,6): M_1--(5),M_2--(6); Au_3Cd, Au_3Zn (7); Au_3Cu (8); Cu_3Pt (9); Cu_3Au (10);

Fig. 55. Variation of the long-range order parameter with temperature for different phases. The DO_{23} superstructure.

the principal mechanism of disordering of the long-period superstructures. In the aforementioned Cu-Pd alloys, but considered separately, in a definite temperature range the density of APBs decreases on ordering.

Second, the linear relation between M and η shows that no special transition parameter, besides those introduced for the stars of the superstructure vectors, is needed to describe the $L1_2(M) \rightarrow A1$ transition. The classical long-range order parameters completely specfy this kind of transition.

Third, the data in Fig. 54 refute, in essence, the known Khachaturyan's assertion [19] that the temperature dependence of the antiphase period allows one to single out the long-period superstructures from a general class of the ordered phases. Obviously, this is not the case. Incidentally, this is evidenced also by the appearance of the η (T/T_C) curves.

Fourth, the following correlation has been established between the antiphase period and its dependence on the degree of order: the longer the period, the stronger its dependence on the degree of order. Thus, the behavior of the long-period superstructures at the order-disorder transition fully justifies the assignment of these structures to the class of order-disorder phase transitions. In addition, the presence of periodic APBs gives rise to a series of interesting features which differ the $L1_2 \rightarrow A1$ transformation from the $L1_2(M) \rightarrow A1$ and $L1_2(MM) \rightarrow A1$ transitions. Many of the observed features can be explained with the use of the geometric and band theories of stability of superstructures.

It is worth pointing out that there exists a series of the long-period superstructures based on the $L1_2$ structures (the one-dimensional $L1_2(M)$ and two-dimensional $L1_2(MM)$). If we confine our consideration primarily to the treatment of the one-dimensional superstructures, the periodic APBs in them will be conservative and oriented in cube plane. The neighboring APBs are more frequently spaced through the one lattice period in the DO_{22} superstructure in which the antiphase period M = 1. Next in this line is the DO_{23} superstructure with M = 2. The superstructures with M > 2 have no special name. The antiphase period may often be fractional due to the alternate occurrence of different spacings between the APBs. It turns out to be fractional even in some alloys with the DO_{23} superstructure. The $L1_2$ superstructure terminates cyclically a series of the long-period superstructures: it can be represented as the superstructure with $M = \infty$ or M = 0.

Another interesting feature of the long-period superstructures is the region of their existence. Its dimension, on account of only the pair interaction, is equal to that of the region where the $L1_2$ superstructure exists minus one (see Section 3.2). Such behavior will hold if one takes into account the three-particle interaction. It might be seen that the long-period superstructures are more complicated. First, this may be connected with the symmetry aspects and, if so, the corresponding contributions into the ordering energy are not reduced to the interparticle interactions. This assertion was substantiated and the relevant example of the $L1_2$(M) superstructure was given in Section 1.1.

Second, when a long-period superstructure is stabilized by the pair interaction, such stabilization takes place due to the presence on the Fermi surface of a disordered solid solution of the section with a large radius of curvature whose center is located at the halfdistance from the nearest site of the $L1_2$ superstructure. The Fermi surface of several alloys, in particular, based on noble metals, is characterized by he presence of such sections (the Fermi surface topology and its role in stabilizing superstructures will be discussed in detail in Section 5.4).

The DO_{23} superstructure. Of the transformations occurring in alloys with the DO_{23} superstructure, only the transformation in Ag_3Mg has been investigated so far [265]. The DO_{23} superstructure represents an interesting object in view of the following considerations. This superstructure takes up an intermediate position between the long-period structures $L1_2$(M) in which the antiphase period may strongly change and the DO_{22} superstructure with a stable value of the antiphase period. The DO_{23} superstructure is obviously closer to the latter. It is shown, according to our scheme, in Fig. 1. The structure factor takes the form

$$F(hkl) = [1 + \cos\pi(h+k+l)]\left\{(P_A^1 f_A + P_B^1 f_B)\, 2\cos\left(\frac{\pi l}{4} + 2\Delta Z_A \pi l\right) + \right.$$
$$+ (P_A^2 f_A + P_B^2 f_B)\, 2\cos\left(\frac{3}{4}\pi l + 2\Delta Z_B \pi l\right) + (P_A^3 f_A + P_B^3 f_B) \times$$
$$\left. \times \left[\cos\pi\left(h + \frac{l}{2}\right) + \cos\pi\left(k + \frac{l}{2}\right)\right] + (P_A^4 f_A + P_B^4 f_B)[\cos\pi h + \cos\pi k]\right\}. \qquad (4.2.3)$$

Owing to the existence of the atomic displacements ΔZ depending on the order parameter, the intensities of the fundamental and superstructure lines also depend on this parameter. Only the hk0 reflections are divided in the usual way into the fundamental and superstructure ones. If we neglect displacements at all ($\Delta Z=0$), the structure factor takes the form typical of the fundamental and superstructure lines:

$$|F_{\text{осн}}| = 4(f_A + 3f_B), \qquad \left|F_{10\frac{1}{4}}\right| = 2\sqrt{2}\eta_{10\frac{1}{4}}(f_B - f_A),$$

$$|F_{001}| = 4\eta_{001}(f_B - f_A), \qquad \left|F_{00\frac{1}{2}}\right| = \eta_{00\frac{1}{2}}(f_B - f_A). \tag{4.2.4}$$

Since the reflections of the (0 0 1/2) type are absent in diffraction pattern, two long-range order parameters, $\eta_{0\,0\,1}$ and $\eta_{1\,0\,1/4}$, suffice to specify the long-range order. In the present work, the $DO_{23} \rightarrow A1$ transformation and the variation of the long-range order parameters with temperature have been studied on the alloys Au_3Cd, Au_3Zn, Pd_3Mn, and Ag_3Mn. The characteristics of the order-disorder $DO_{23} \rightarrow A1$ transformations are presented in Table 15. The variations of the long-range order parameters calculated for different transformation stars are given in Fig. 55.

Before proceeding to the general regularities observed for this superstructure, one substantial feature should be noted. In Au_3Zn below 200°C there exists the superstructure DO_{23}' differing from DO_{23} by introducing periodically along the axis the rotatory boundary. Above this temperature both DO_{23} and DO_{23}' coexist. The last circumstance did not allow all characteristics of the transformation in this alloy to be measured.

In all the alloys the long-range order for both stars disappears at the same temperature. As in the $L1_2(M)$ superstructure, the transformation temperature is degenerate. The phase transition in all alloys is of the 1st order for both stars. This fact distinguishes the disordering in the DO_{23} superstructure from transformations in the long-period superstructures. The order for the fractional reflection is always lower than for the integer. However, the enhanced disordering at APBs is not observed. The jumps of the order parameter at T_C are rather different in magnitude, but for each superstructure are close to each other. The alloys come close to the point of the absolute loss of stability. Obviously, all these signs markedly distinguish the $DO_{23} \rightarrow A1$ transformation from the $L1_2(M) \rightarrow A1$ transition.

And finally, the difference between these two superstructures lies in the behavior of the antiphase period. In Ag_3Mn this value slightly changes near T_c, in other alloys it is equal to 2. Therefore, the superstructure DO_{23} represents an individual class in a series of superstructures and cannot be regarded as the limiting case of the $L1_2(M)$ superstructure as supposed before. At the same time, this superstructure also differs from DO_{22}. The distinction between the superstructures $L1_2(M)$, DO_{22} and DO_{23} is evidenced by the results of the theoretical derivation of possible superstructures in Section 1.1.

The DO_{22} superstructure. This superstructure is the limiting case of the structures with the periodic antiphase boundaries with M = 1. It is characterized by three sorts of lattice sites and two order parameters $\eta_{0\ 0\ 1}$ and $\eta_{1\ 0\ 1/2}$.

Before the current investigation, no attempts have been made to measure the temperature dependence of the order parameter in this superstructure. In this work the $DO_{22} \rightarrow A1$ transformation was investigated on the alloys Ni_3V and Pd_3V [268]. The case of Ni_3V faced with nontrivial difficulties due to the close values of atomic scattering factors. The application of neutron diffraction turned out to be impossible for the bulk specimens in which we could not prevent the ordering on quenching. In Fig. 56 and Table 16 the results obtained are given. The table comprises, in particular, the transformation temperatures and the order of transformation for different stars, the jump of the long-range order parameter at T_c, and the degree of proximity to the point of stability loss. The long-range order vanishes simultaneously for both stars. This phenomenon of degeneracy of the transition temperature is common for all structures with periodic antiphase boundaries. The order for the fractional index M is always lower than for the integer one. This fact and a rigorous constancy of the antiphase period (for the DO_{22} superstructure $M \equiv 1$) are characteristic of both DO_{22} and DO_{23} superstructures. However, these are the only common features of the superstructure. Although the phase transitions for both order parameters refer to the first order type, the transition corresponding to the fractional reflection is obviously closer to the second

Table 15. Characteristic features of $DO_{23} \to A1$ transformation.

Alloy	T_C, K	Order of phase transition		$\Delta\eta$ at T_C		$\partial(T/T_K)/\partial\eta$		M range	$\frac{\partial M}{\partial \eta}$	References
		η_{001}	$\eta_{01\frac{1}{4}}$	η_{001}	$\eta_{01\frac{1}{4}}$	η_{001}	$\eta_{01\frac{1}{4}}$			
Ag_3Mg	645 ∓ 3	I	I	0,50	0,50	0,12	0,25	2–2,01	0,04	[221, 266]
Au_3Cd	695 ∓ 3	I	I	0,70	0,65	0,05	0,16	≡ 2	0	[221, 267]
Pd_3Mn	820 ∓ 3	I	I	0,30	0,18	0,16	0,12	≡ 2,03	0	[221, 252]
Au_3Zn	698 ∓ 3	I	I	–	–	–	–	≡ 2	0	–

Table 16. Characteristic features of $DO_{22} \to A1$ transformation.

Alloy	T_C, K	Occurrence of c/a:	Order of phase transition		$\Delta\eta$ at T_C		$\partial(T/T_K)/\partial\eta$		References
			η_{001}	$\eta_{10\frac{1}{2}}$	η_{001}	$\eta_{10\frac{1}{2}}$	η_{001}	$\eta_{10\frac{1}{2}}$	
Pd_3V	881 ∓ 3	0,9966	I	I	0,69	0,50	0,3	0	[268]
Ni_3V	1318 ∓ 3	1,017	I	close to II	0,10	0,02	0,12	0,25	[221, 268]

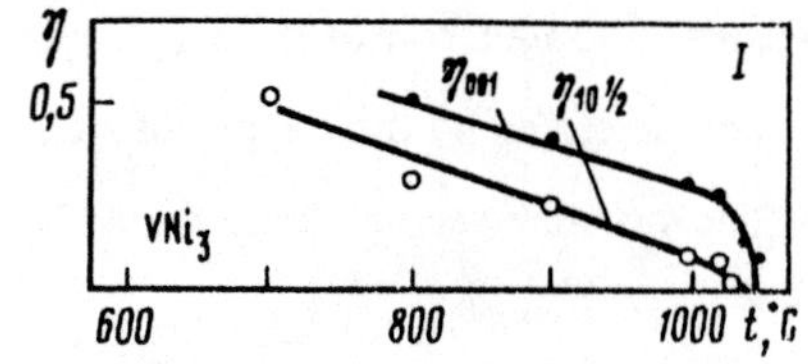

Fig. 56. Variation of the long-range order parameter with temperature for different phases. The DO_{22} superstructure.

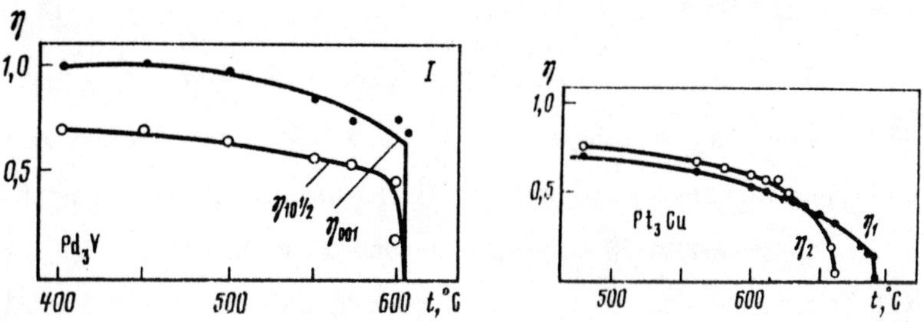

Fig. 57. Variation of the long-range order parameter with temperature. The $L1_3$ superstructure.

order. The jumps of the order parameters at T_C in each alloy differ significantly for various stars. As regards the degree of proximity of alloys to the point of stability loss, the experimental difficulties did not allow us to reliably determine this value for the DO22 superstructure. The values in Table 16 must be regarded, therefore, as only tentative.

To summarize, it should be noted that the transformations in the superstructures $L1_2$(M), DO_{22}, DO_{23} have a number of similar features. Nevertheless, these transformations are different, if we take into consideration all the signs revealed. These differencies are connected with variations in the interatomic interactions leading to the formation of each of these superstructures. The experimental data obtained in the present work confirm the conclusion about the existence of different classes in a series of long-period superstructures.

The L1$_3$ superstructure. The unit cell of the L1$_3$ superstructure is shown in Fig. 57. The superstructure is characterized by three types of sites and two order parameters $\eta_{0\,0\,1}$ and $\eta_{1/2\,1/2\,1/2}$. The structure factor is determined by the formulas:

$$F_{\text{осн}} = (C_A f_A + C_B f_B)[1 + \cos \pi h + \cos \pi \left(\frac{h}{2} + \frac{k}{2} + l \right) + $$

$$+ \cos \pi \left(\frac{h}{2} + \frac{3k}{2} + l \right)] [1 + \cos \pi (h + k)];$$

$$F_{\text{сверхстр}}(\eta_1) = \frac{1}{4} \eta_1 (f_A - f_B) \left[1 + \cos \pi h - \cos \pi \left(\frac{h}{2} + \frac{k}{2} + l \right) - \right.$$

$$\left. - \cos \pi \left(\frac{h}{2} + \frac{3k}{2} + l \right) \right] \times [1 + \cos \pi (h + k)];$$

$$F_{\text{сверхстр}}(\eta_2) = \frac{1}{2} \eta_2 (f_A - f_B) [1 + \cos \pi (h + k)] [1 - \cos \pi h]. \tag{4.2.5}$$

The superstructure L1$_3$, like those of the L1$_2$ and DO$_{22}$ types, belongs to the set of principal superstructures formed from an f.c.c. lattice at the AB$_3$ composition (see Section 1.1). In contrast to the L1$_2$ superstructure formed in a great majority of alloys and the DO$_{22}$ superstructure detected in several alloys, the L1$_3$ superstructure is known to occur just in one alloy, Pt$_3$Cu. This exclusiveness is fully determined by the ordering energy: the contribution of the second coordination shell must be great enough and have the same sign as the contribution of the first shell. It is only possible if the wave vector of the singularity responsible for ordering is close to the vector (1/2, 1/2, 1/2), see Chapter 5. Such cases occur very rarely because they call for the specific shape of the Fermi surface. Before this work, the order-disorder transformation in the L1$_3$ superstructure and the η versus T dependence have not been investigated. Our results are presented in Table 17 and in Fig. 57. The table indicates, in particular, the temperature and the type of the transformation for both stars as well as the jump of the long-range order parameter at T_c and the parameter defining the proximity of the alloy to the point of an absolute loss of stability.

The transformation proceeds in two stages: L1$_3$ → L1$_0$ and then L1$_0$ (in the alloys with the AB$_3$ composition) → A1. The critical temperatures differ by 30°. The types of transformation are also different: the L1$_3$ → L1$_0$ transformation is quite close to the 2nd-order transition, while the L1$_0$ → A -- to the 1st-order one. The latter passes

Table 17. Characteristic features of $L1_3 \rightarrow A1$ transformation.

Transformation star	T_c, K	$\Delta\eta$ at T_c	Order of phase transition	$\frac{\partial(T/T_K)}{\partial\eta}$	References
(001)	963 ∓ 3	0,17	I	0,17	[221]
$(\frac{1}{2}\frac{1}{2}\frac{1}{2})$	933 ∓ 3	0	II	0	[221]

through a two-phase region and is accompanied with a small jump of the long-range order parameter. Above T_C short-range order for both stars persists but for the (1/2 1/2 1/2) star it disappears rapidly. Like in the long-period superstructures, the degree of order for the low-symmetry star is lower than for the other stars. The unit cell of the Pt_2Cu alloy in the ordered state is quasicubic. Orthorhombic distortions are very small. The important and almost unique feature of the $L1_3$ superstructure is the lack of degeneracy of the transition temperatures for different stars. The order for the (1/2 1/2 1/2) star loses its stability at a lower temperature, before the disordering for the (0 0 1) star is achieved. To reveal and confirm such behavior, the measurements of order parameter were carried out on alloys upon quenching and directly at the temperatures of interest. In the last case, upon disappearing of the $\eta_{1/2\ 1/2\ 1/2}$ order, the long-range order parameter $\eta_{0\ 0\ 1}$ behaves in an anomalous way, but the critical temperatures differ by 30°. This is, in fact, the second case of finding different transition temperatures for various stars in the same alloy. The first case is related to the DO_{23} superstructure (the Fe_3Al alloy).

The $D1_a$ superstructure. The order-disorder phase transition in alloys with this superstructure has been investigated experimentally, with the measurement of the degree of order only in our works. Five alloys have been studied: Ni_4Mo, Ni_4W, Au_4Cr, Au_4Mn, and Au_4V [269 - 272]. The plots of the long-range order parameter against temperature are given in Fig. 58. In all the alloys the $D1_a \rightarrow A1$ transformation

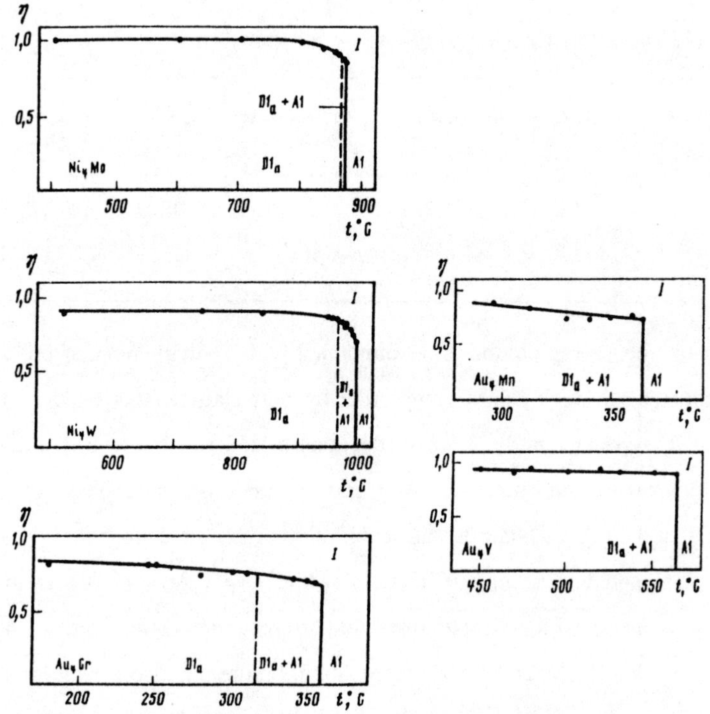

Fig. 58. Variation of the long-range order parameter with temperature for different phases. The $D1_a$ superstructure.

represents the 1st-order transition that proceeds through a two-phase region. This is a characteristic distinction of the $D1_a$ superstructure from the others. The main data on the transformation in the investigated alloys are presented in Table 18.

Consider specific features of the $D1_a \rightarrow A1$ transformation. The phase transition in this structure starts at high values of the order parameter. The mean jump value $\Delta\eta$ is considerably greater than that in the $L1_2$ superstructure. In other words, the $D1_a$ superstructure exhibits a weak stability with respect to the transition to the disordered state. The transformations in all the alloys, except Ni_4W, take place far from the temperature of the absolute loss of stability of the $D1_a$ phase. Analysis of the $\eta(T/T_c)$ curves for the $D1_a$ superstructure shows that the same curve presumably exists for all the alloys, except for Au_4Mn.

Table 18 Characteristics features of $D1_a \rightarrow A1$ transformation

Alloy	T_c,K	Boundaries of two-phase region	Jump in the degree of order		ΔT in two-phase region, K	$\delta(T/T_k)/\delta\eta$		C/a	References
			at the beginning of two-phase region	at the end of two-phase region		at the beginning of two-phase region	at the end of two-phase region		
Au_4Cr	633±3	588-633	0.74	0.66	45	1.00	0.70	0.999	[269]
Au_4Mn	638±5	568-638	0.89	0.70	70	0.70	0.50	0.981±0.001	-
Au_4V	838±5	713-838	0.95	0.90	125	5.00	1.00	0.982±0.001	-
Ni_4Mo	1148±3	1133-1148	0.93	0.85	15	0.50	0.33	0.986±0.001	[270,271]
Ni_4W	1268±3	1228-1268	0.86	0.66	40	0.70	0.09	0.982±0.001	[272]

All the alloys with the $D1_a$ superstructure undergo a tetragonal distortion. The behavior of the tetragonality parameter in the course of the order-disorder transition is typical of the 1st-order phase transitions. The distortion parameters are generally small. Therefore, the $D1_a \rightarrow A1$ transformation is characterized by considerable size and dimension effects. An interesting anomaly is observed in this superstructure: the volume per one atom in the Ni_4W alloy is smaller in the disordered state than in the ordered one. This is the only case in which a pure order-disorder transformation is accompanied by the volume change with the sign opposite to the effect of contraction.

Another problem is also connected with the $D1_a \rightarrow A1$ transformation: the type of short-range order in the A1 structure does not conform with the superstructure type. A characteristic superstructure reflection from **The Pt_2Mo and $C11_B$ superstructures** $D1_a$ superstructure is (2/5 4/5 0), while for the A1 phase in the state of short-range order the maximum of diffuse scattering corresponds to the (1 1/2 0) reflection. From the energy point of view, this problem was discussed in Section 3.2. Here we shall consider some experimental data. On approaching T_c from above as well as in the quenched state, one can find the reflection (1 1/2 0) in diffraction pattern. On passing through T_c from below, i.e., on disordering, the reflection (2/5 4/5 0) is observed just above T_c. No doubt, one of the corresponding short-range order configurations is metastable, and they are divided by the energy barrier. At temperatures above T_c stable is the order of the (1 1/2 0) type. These experimental findings correlate with the energy scheme described in Section 3.2.

The occurrence of the energy barrier between these states is also confirmed by the kinetic investigation of the earlier stages of ordering in Au-base alloys. The transition from the (1 1/2 0) order to the (2/5 4/5 0) order is connected with the incubation period testifying to the transition over the energy barrier. The difference between these two states consists not only in the type of order but also in the value of the correlation radius. The order of the (1 1/2 0) type is stable at low values of the correlation radius. The stability of the order of the (2/5 4/5 0) type needs in extending the correlation over much longer distance.

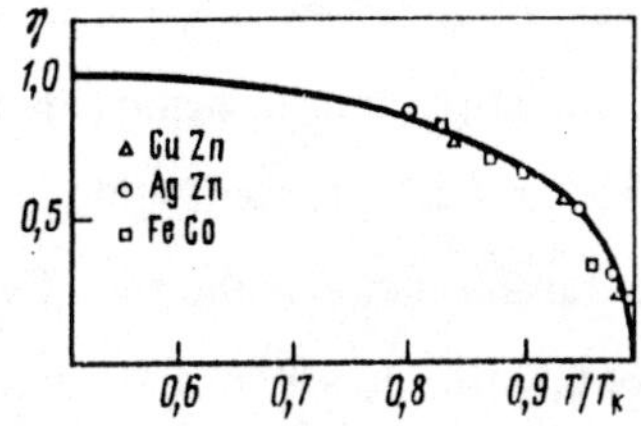

Fig. 59. Temperature dependence of the long-range order parameter during the A2→B2 transition.

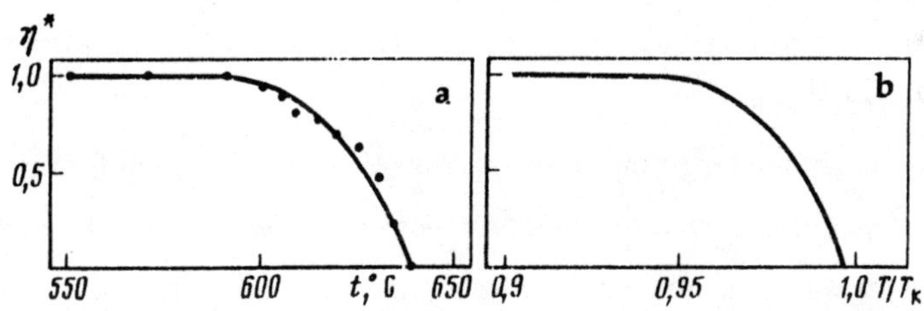

Fig. 60. Variation of the long-range order parameter with temperature for the CuPd alloy. The B2 superstructure.

These observations can be illustrated by some figures for the Au_4Cr alloy. The order of the (1 1/2 0) type corresponds to the correlation length of 8 - 10Å, the order of the (2/5 4/5 0) type -- to the length of 35 - 50 Å, and the long-range order $D1_a$ -- to the length of 200 - 250 Å. As a matter of fact, the (1 1/2 0) order in this alloy represents short-range order, whereas the (2/5 4/5 0) corresponds to the microdomains with a long-range order of the size close to a critical, minimally stable size. Thus, the $D1_a \rightarrow A1$ transformation, like $D1_a \rightarrow A1$, is characterized by an abrupt change in the correlation radius. In the last case, however, it is complicated by changing of the correlation type.

The B2 superstructure. The phase transformation accompanied by the formation of the B2 superstructure is well studied. The dependence η on T is reliably measured for CuZn [273], AgZn [274], and FeCo [275] alloys. The results expressed in the relative temperature are shown in Fig. 59. As is evident, the law of the corresponding states holds well for these alloys. The phase transformation in all the alloys is close to the 2nd-order type. On disordering, an A2 b.c.c. solid solution forms. We have investigated [229, 230] the order-disorder transformation in the alloy Cu+40 at.% Pd. The distinctive feature of this alloy is that the order-disorder transition does not proceed through the B2 → A2 reaction but through B2 → A1, i.e., the disordering in the alloy is accompanied with the allotropic b.c.c. → f.c.c. transformation. The investigation showed that the B2 → A1 transformation in the Cu-Pd alloy includes four phase transitions. Three of them are the intraphase transformations in the B2 and A1 phases connected with the 2nd-order process of "softening" the superstructure. The b.c.c. → f.c.c. transformation is the 1st-order phase transition.

The variation of the long-range order parameter with temperature for the alloy CuPd is shown in Fig. 60. Although the disappearance of the long-range order similar to that in B2 structures was found to be the 2nd-order transition, as in other alloys, the slope of the $\eta(T)$ curve in Cu - Pd is steeper than that in the B2 → A2 transformation. Such behavior distinguishes the Fe_3Pt alloy in which the disordering of the $L1_2$ phase is accompanied by the structure transformation. The arising mechanical instability of the lattice is likely to affect the stability of the state with long-range order.

The Pt_2Mo- and $C11_B$-type superstructures. These two superstructures form at the AB_2 composition; $C11_B$ is based on an f.c.c. lattice, that of the Pt_2Mo type -- on a b.c.c. lattice.

The superstructure of the Pt_2Mo type. The unit cell is depicted in Fig.1. The transformation star is (2/3 2/3 0). The order-disorder transition in this superstructure has been studied earlier on the Ni_2Cr alloy [276] and Ni_2V (Fig. 61). The characteristics of the transformations in these alloys are presented in Table 19. In the Ni_2Cr alloy the two-phase region is very narrow and, as a consequence, has not been found. The two-phase region in Ni_2V extends over 25°C. In both alloys the value of the jump of the order parameter at T_c is significant. It was also shown that the order-disorder

Table 19. Characteristic features of transformations of the $Pt_2Mo \rightarrow A1$ type.

Alloy;	T_C, K;	$\Delta\eta$ at T_C	$\partial(T/T_K)/\partial\eta$
Ni_2V	1173	0,745	0,23
Ni_2Cr	863	0,725	0,25

transformation in both alloys ends at the temperature where the ordered phase is far from the state of the stability loss. According to [277], a small short-range order corresponding to the type of the emerging superstructure is present above T_C.

The C11$_B$ superstructure. A number of alloys possesses this superstructure at the AB_2 composition. However, only in one of them, Cr_2Al, the order-disorder transformation proceeds below the melting temperature. The investigations of the C11$_B$ $\rightarrow$ A2 transformation have not been conducted earlier. Our results are given in Fig. 62 [277]. The unit cell of the C11$_B$ superstructure is shown in Fig. 1 The structure is characterized by one order parameter and by the low-symmetry star corresponding to the transition (0 0 2/3). The measurements of the temperature dependence of the order parameter in such superstructures are difficult due to the presence of free parameters involved in the atomic coordinates. Displacements of atoms from the lattice sites depend on the degree of order (see Section 3.3).

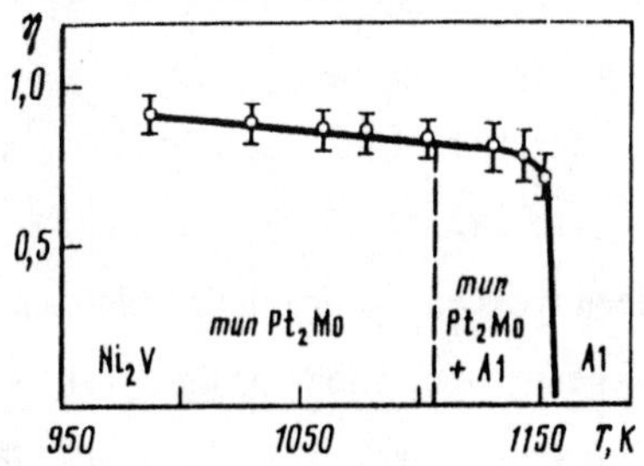

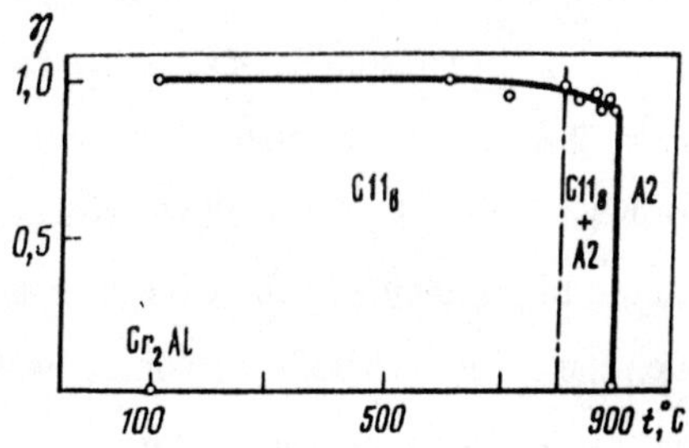

Fig. 61. Variation of the long-range order parameter with temperature for the Ni_2V alloy. The Pt_2Mo superstructure.

Fig. 62. Variation of the long-range order parameter with temperature. The C11$_b$ superstructure.

The order-disorder transformation in the $C11_B$ superstructure is similar to the $D1_0 \rightarrow A1$ transformation. The 1st-order phase transition proceeds through a two-phase region showing a great jump in the order parameter. The transformation begins far below the temperature of the absolute loss of stability. The value of $\partial(T/T_c)/\partial\eta$ is equal to 0.3. These features of the $C11_B$ transformation are probably typical of the order-disorder transformations in which the superstructures with a single low-symmetry star take part.

4.3 Statistical theories of atomic ordering

Fundamentals. The problems to be encountered in the statistical theory of order-disorder transitions are entirely determined by the partition function Z of the alloy. In the most general case, the partition function for the equilibrium generalized statistical ensemble is written as [278, 279]

$$Z = \sum_{\{k\}} \exp\left[-\beta(E - \mu N - pV)\right]. \tag{4.3.1}$$

where the summation extends over all possible configurations, $\beta = 1/kT$, p is the pressure, V is the volume of the system, E, μ and N are the internal energy, the Gibbs free energy and the number of particles of the system in the given configuration, respectively. In the generalized ensemble, the fluctuations of the energy, of the number of particles and of the volume of the system are possible.

The problem of the evaluation of the partition function (4.3.1) for a solid solution has not been solved as yet and all theories employ approximations of different kind. In most cases, the problem of atomic ordering has been reduced to the Ising problem [280] in which the volume alterations, atomic displacements, fluctuations of these values as well as changes in the phonon spectrum on ordering are not treated. Let us consider first the development of statistical theories within the framework of the Ising model. It should be noted that the term "Ising model" is often used to denote quite different models. In a classical sense, this is the model of a rigid lattice with the pair interaction, usually on the first coordination shell. In the present work we rely upon the conception of the generalized Ising model, i.e., that of the rigid lattice with the pair and many-particle interaction of an arbitrary radius. In this case, the ensemble corresponds to the grand canonical one, and expression (4.3.1) takes the form

$$Z = \sum_{\{k\}} \exp\left[-\beta(E - \mu N)\right]. \tag{4.3.2}$$

The solution to the statistical problem is reduced to obtaining the equations that describe adequately the equilibrium state of the system and include the long-range order parameters and the correlations or the correlation functions (see Section 3.1). The ways of obtaining these equations can be divided into three groups.

The first relates to the derivation of the equilibrium equations without a direct calculation of the partition function. In these methods, the chain of integro-differential equations is being constructed with the use of the Liouville theorem [281] to decouple this chain into the correlation functions or Green functions [282]. The solutions of the set of these equations represent the required averages over the statistical ensemble. The central problem in these methods is the choice of the procedure for decoupling the infinite chains of equations and its verifications.

The second group employs a direct representation of the correlation function as the average over the ensemble. The solution is attained by the expansion of the partition function into a power series. This expansion may be a certain version of the high-temperature series, including the group expansions [283] and those in the random field. In applying a high-temperature expansion, the convergence problem arises (with respect to the radius and the velocity). In the random field technique, the main problem is the approximate representation of the distribution function (in an analytical form or according to the number of the moments retained).

The third group of methods is based on finding the solutions providing the maximum of Z. This is attained by restricting the partition function, passing to the free energy F and subsequent minimization of F. The restriction of Z consists in selecting the states with the given values of the order and correlation parameters. The problem inherent in this method is the calculation of the free energy which may be carried out with the aid of the temperature (cumulant [282]) expansion of the logarithm of the restricted partition function or through the transition to an approximate expression for F = E - TS. The first approach is adopted in Kirkwood's method [193] and in the random field techniques. The entropy methods [191] relate to the second case.

Traditional methods of the theory of atomic ordering can be considered within the framework of the classification presented above. For example, Gorsky-Bragg-Williams (GBW) approximation [284], the quasichemical method [285], the cluster variation technique [288], the methods of Li and Yang [286, 287], Cowley [289], Christy-Hall [290] as well as their modofications in the works of Borelius, Dalinger, Shockley, Takagi and others are related to the entropy methods. They represent a consistent development of a combinatory technique for calculating entropy, ranging from the use of the one-particle cluster to the four-particle (tetrahedrical) one. The third group of methods also includes those worked out by Kirkwood [291], Krivoglaz [189, 204], Gurov [292], Khachaturyan and Badalyan [19] who used the temperature expansion technique.

The methods of the secon group have not been so widely used in the theory of atomic ordering. There should be mentioned, among others, the works of Zernike [293], Ivon and Fourne [294] based on the group expansion technique, and the theoretical investigations of Bethe [295], Peierls [296], Clapp and Moss [297] which may be assigned to different modifications of the random field approximation. In fact, the development of the statistical theory of atomic ordering was connected with the refinement of the approximate expression for the free energy, or the expression for entropy (combinatory methods), or with allowing for the fluctuation contributions into the configurational energy (temperature expansions). Since the entropy of a system is a measure of fluctuations in this system, these approximations are, in essence, quite close to each other though the authors may not realize this.

The methods of the first group have not been used, in practice, in the theory of atomic ordering. The exception is the works of Tachir-Khelli [298], who used the first approximation of the Green function method.

The important role of the works carried out by the GBW method (the general behavior of the $\eta(T)$ function), the quasichemical and Kirlwood's methods (the elucidation of the role of correlation and fluctuations), and Li and Yang (the essential part of higher corellations) methods for understanding of the statistical regularities in phase transitions should also be emphasized. Most of the interesting calculations have been carried out by the GBW method (note, especially, the contribution of Smirnov's school), the cluster variation technique (Golosov's school [299] and Gan's calculations [300, 301], and the group expansion method (Fourne).

The progress in the theory of atomic ordering has been frequently reviewed. Among others, it should be noted the works of Nix and Shockley [192], Muto and Takagi [193], Guttman [194], Krivoglaz and Smirnov [189, 204], Domb [302], and Green and Hurst [303]. We shall not give here a detailed survey and the assessment of these works. At this point we want to emphasize only that the above described classification of various approaches to the solution of the statistical problem and the separation, accordingly, of the works devoted to the theory of atomic ordering, are original and differ substantially from the classification suggested in mentioned reviews.

It is seen from the above classification that most of the works devoted to the theory of atomic ordering was carried out by combinatory methods. Although an obvious success has been achieved in the application of these methods, considerable difficulties and limitations in their use should be noted. First, it is difficult to derive a reliable expression for the combinatory factor. Second, the criterion is lacking in these methods for an unambiguous choice of the optimum cluster. Third, different approximations are used to describe various superstructures by the same method. And, finally, there is no reliable criterion for evaluating the accuracy of the approximation involved.

Now we will formulate the requirements the statistical theory of ordering should meet, in our opinion. Based on definite initial approximations, such theory should be able to describe the transformations occurring in any superstructure and to provide, in a single scheme, the treatment of the correlation and the interactions of any order without limitations imposed on the radius. In addition, the physical nature of approximations used in constructing the theory should be apparent. The latter is hardly possible within the framework of one particular method. Only the use of different techniques allows one to evaluate comprehensively the sense of approximations adopted in the theory. In the modern statistical physics, a variety of efficient methods has been worked out to meet the above criteria. At the beginning of this section, the classification of these methods was described. Historically, they have not been used practically in the theory of atomic ordering. As a consequence, a pressing need existed to apply various methods of statistical physics to the problem of atomic ordering. And what is more, a radical progress could not be achieved without this approach.

There exists another nontrivial cause to apply the methods related, according to our classification, to various groups to the problem of atomic ordering. In the case of the first or second group, the solution is the average values, over the ensemble, of the physical entities, while in the third group -- the most probable. These characteristics are known not to coincide in the case of asymmetric distributions. The distinction may be essential if the approximate solution to the statistical problem is obtained. Knowledge of the mean and the most probable order parameters provides information on the distribution of states over the statistical ensemble that was actually used for the solution to the problem. For example, the long-range order parameter in the B2 superstructure at the constant energy may suffer the quadratitic fluctuations (the variance of the distribution) which are proportional to the square root of the pair correlation parameter ($\sqrt{\varepsilon_{12}}$). The asymmetry of the distribution is related in a similar way to the three-body correlation parameter ($\sqrt[3]{\varepsilon_{123}}$). The estimates show that near the critical temperature of the phase transition the variance and asymmetry of the distribution of the η values may be of the order of the mean value.

All unsolved general problems of the statistical theory of phase transitions persist in the theory of order-disorder transformations. This concerns, in particular, the problem of singularities in the behavior of thermodynamical functions near the point of the stability loss and the problem of accounting for fluctuations (especially heterophase) in constructing the Gibbs free energy. The former is a global problem that can hardly be solved within the framework of the theory of order-disorder transitions. The significance of the second problem depends on a particular task and the model chosen to solve it. Kirkwood was the first to show the importance of intraphase fluctuations in the theory of order-disorder transitions. Nowadays, it becomes clear, however, that the account of the domain and heterophase fluctuations is also important. In the following, we shall suggest several ways of handling these effects.

The variation with temperature of the equilibrium long-range order parameter has a general form shown in Fig. 7. At temperatures below T_C, the phase with long-range order is stable. In the range between T_C and T_{Sl} it is metastable. Above T_{Sl} the phase with long-range order loses its stability. It can be shown that if the free energy depends not only on the long-range order parameter but also on the correlation parameter, the point of absolute loss of stability will be defined by the criterion $\partial\eta/\partial T = \infty$ which is analogous to that obtained without correlations (see Section 4.1).

We showed in the previous chapters that the energy behavior and stability of alloys in the phase transition are determined not only by the cooperative redistribution of atoms over the lattice sites but also by the volume changes. There have been several attempts to take these effects into account in the statistical theory of alloys without using of the Ising model. They have been based, as a rule, on the model of elastic spheres [304] with the expansion of the energy of pair interaction into a series in atomic displacements [306]. It can be shown that the approximations in these works correspond to the harmonic model of a crystalline solid [305]. This model enables one to allow for the volume changes in an alloy upon ordering. It has been successively employed by Dines, Ter-Haar, Vesnik and, particularly, by Smirnov and co-workers [188, 307]. Of course, this is the simplest approximation in taking the volume effects into account. The next step consists in the use of the anharmonic interatomic potential or in the account of the dependence of the ordering energy on the degree of long-range order as predicted by the quantum theory of solid solutions (see Chaper 5). Both these approaches will be described below.

The concept of the statistical ensemble. We have already mentioned that approximations used in constructing the statistical theory of atomic ordering impose the constraints on the statistical ensemble. The choice of the ensemble comprises eventually all the assumptions involved, thus determining all achievements and potentialities of the given version of the theory. It is important, therefore, to strictly define the selected ensemble in constructing the theory. This has not been done usually in the works on the statistical theory of atomic ordering.

In general, the choice of the ensemble is a nontrivial task. It depends, particularly, upon the object under investigation and its model. For example, gaseous systems and crystalline solids without vacancies and interstitial atoms are described by different generalized ensembles, if one takes into consideration the fluctuations in the number of particles. In the first case, the variation of the number of particles includes the pressure effect, while in the second it is not so. In the subsequent discussion, the total number of particles or sites N is assumed to be constant (unless the presence of vacancies, interstices or other defects is specially stated). The number of component atoms, say N_A, in a binary solid solution can be variable. Consider in more detail the problem of the ensembles in metallic solid solutions.

First, the reader should be reminded about the classical definition of the ensemble in the statistical theory. The generalized ensemble is being constructed so as to take into account the equilibrium fluctuations of all conceivable thermodynamical and internal parameters. To the former we refer the energy E, the volume V, the number of particles N_A; to the latter -- the long-range order parameter η, the correlation parameters ε, the correlation functions D, the displacements of atoms from the lattice sites and so on. A grand canonical ensemble does not involve the volume fluctuations. In the foregoing chapters the important role of the volume effects in atomic ordering has been demonstrated. Meanwhile, in the great many works devoted to the statistical nature of atomic ordering, the volume effects have not been taken into account. We shall proceed now to the expression for the partition function of the generalized ensemble represented by the integral

$$Z = \int_0^\infty dE \int_0^{N_A} dN_A \int_0^V dV \omega(E, N_A, V) e^{\beta[-E+\mu N_A+PV]}. \tag{4.3.3}$$

Here $\omega(E,N_A,V)$ is the density of configurations. We introduce in a usual way the entropy of the microcanonic distribution

$$S(E, N_A, V) = k \ln \omega(E, N_A, V). \tag{4.3.4}$$

Then we obtain for (4.3.3)

$$Z = \int_0^\infty dE \int_0^{N_A} dN_A \int_0^V dV e^{\beta[TS(E, N_A, V) - E + \mu N_A + PV]}. \tag{4.3.5}$$

In principle, any ensemble can be expressed through the microcanonical one by expanding it in fluctuations. Examples of this kind are given in [280]. Such treatment can be generalized to the ensemble considered above. To do this, we expand the exponential into the Taylor series around the most probable value (E, $\tilde{N}_A$, V):

$$TS(E, N_A, V) - E + \mu N_A + PV = TS(\tilde{E}, \tilde{N}_A, \tilde{V}) - \tilde{E} + \mu\tilde{N}_A + $$
$$+ P\tilde{V} - \frac{1}{2}\frac{1}{TC_V}(E - \tilde{E})^2 - \frac{1}{2}\frac{P}{T\alpha}(V - \tilde{V})^2 - \frac{1}{2}\left(\frac{\partial\mu}{\partial N_A}\right)_{\tilde{E}, \tilde{N}_A, \tilde{V}} \times$$
$$\times (N_A - \tilde{N}_A)^2 - \frac{1}{\alpha T}(E - \tilde{E})(V - \tilde{V}) + \ldots, \tag{4.3.6}$$

where α is the volume expansion coefficient, C_V is the heat capacity. The expression on the left side gives the Gibbs free energy of a particular state in the ensemble. It would be pertinent to remind the reader that two ways of averaging are available for metallic solid solutions [308]: over the state and over the ensemble of states. The values E, V and N_A in expressions (4.3.3) to (4.3.5) are already averaged over the given state. Series (4.3.6) includes only the fluctuations of the parameters of different states in the ensemble. Combining (4.3.5) and (4.3.6) gives:

$$
\begin{aligned}
Z = {} & e^{\beta[TS(\widetilde{E}, \widetilde{N}_A, \widetilde{V}) - \widetilde{E} + \mu \widetilde{N}_A + P\widetilde{V}]} \int_0^{\infty} dE \int_0^{N_A} dN_A \int_0^{V} dV \times \\
& \times \exp\left\{ \beta \left[-\frac{1}{2} \frac{1}{TC_V} (E - \widetilde{E})^2 - \frac{1}{2} \frac{P}{T\alpha} (V - \widetilde{V})^2 - \right.\right. \\
& \left.\left. - \frac{1}{2} \left(\frac{\partial \mu}{\partial N} \right)^0_{\widetilde{E}, \widetilde{N}, \widetilde{V}} (N_A - \widetilde{N}_A)^2 - \frac{1}{\alpha T} (E - \widetilde{E})(V - \widetilde{V}) + \ldots + \right] \right\}.
\end{aligned}
\tag{4.3.7}
$$

From this expression it can be seen that in the considered ensemble the distribution declines from the Gaussian form. If we neglect the higher terms in the expansion, the distribution over the ensemble will become a three-dimensional Gaussian. Its variance is determined by the heat capacity, the thermal expansion coefficient and by the dependence of the chemical potential on the number of particles. As pointed out in numerous handbooks on statistical physics, this distribution is very narrow and, consequently, the most probable values E, Ñ, and V calculated in different ensembles for the real systems coincide. This also implies the smallness of fluctuations of these quantities for macroscopic systems. However, we shall confine our consideration to what distinguishes these ensembles. On approaching the temperatures of phase transitions and especially near the critical points for C_V and α, the thermal expansion coefficients will diverge and, therefore, the transition to the microcanonical ensemble becomes impossible. In other words, near these points the ensembles are not equivalent. From the physical point of view, this is due to the fact that near the critical points fluctuations increase infinitely and can be adequately treated only with account of the properly chosen ensembles [309]. In the foregoing consideration we have dealt with the fluctuations of thermodynamical parameters. If we proceed to the internal parameters which do not depend on N, their fluctuations are not small even far apart from the critical points (see Section 3.1). Therefore, the ensembles will not be equivalent in the whole temperature range even if the consistent calculation of the internal parameters such as η, ε and others is carried out.

Expressions (4.3.6) and (4.3.7) are quite general. The partial cases of the expansion of the canonical and grand canonical ensemble near the microcanonical one can be obtained from these expressions by omitting in them the appropriate variables. The above reasoning remain valid in this case.

Because the problem of the ensemble used in the theory, as we can see, is quite serious and important, we shall consider its application to the models used in the statistical theory of atomic ordering. At this point, we must single out two branches in the forthcoming classification: the theories accounting for the volume alterations on ordering and those neglecting such effects. The latter correspond to the generalized Ising model. The relation between these two groups of theories consists in that those in the second one may always be generalized by introducing the volume terms into the Hamiltonian.

Most common in the theory of atomic ordering is the assumption about the occurrence of the degeneracy in the system. This assumption enables one to go over to the restricted partition function. In this case, the partition function Z is expressed by

$$Z = \sum_{\{\alpha\}} Z_{\{\alpha\}}, \tag{4.3.8}$$

where $\{\alpha\}$ is a set of some parameters specifying the state. The summation in (4.3.8) extends over all possible values of the parameter involved:

$$Z_{\{\alpha\}} = \sum_{\{K\}}^{\{\alpha\}} e^{-\beta(E - \mu N - PV)}, \tag{4.3.9}$$

In equation (4.3.9), the summation is accomplished over all possible configurations at the fixed values of the parameters $\{\alpha\}$. The use of (4.3.8) means the break of the ensemble into the sum of subensembles with a fixed set of the parameters $\{\alpha\}$ in each. If, for example, $\{\alpha\} \equiv E$, then (4.3.8) implies that the subensembles are microcanonical. In general, $\{\alpha\}$ may include the number of particles, the energy, the long- and short-range order parameters, etc. If the volume or the lattice constant of a cubic alloy is fixed ($\{\alpha\} \equiv V$ or a), we obtain

$$Z_V = \sum_{\{K\}}^{V} e^{-\beta(E - \mu N_A - PV)} = e^{+\beta pV} \sum_{\{K\}}^{V} e^{-\beta(E - \mu N_A)}. \tag{4.3.10}$$

It is evident that the restricted ensemble corresponds to the grand canonical one. The partition function of the grand canonical ensemble differs from (4.3.10) by the factor $\exp\{\beta PV\}$ that is constant in a given ensemble. Addition to (4.3.10) of the series in the volume fluctuations of the form expressed by (4.3.6) and (4.37) implies the transition to the expansion of the generalized ensemble near the grand canonical one.

Before leaving the generalized ensemble, we notice two points. First, if the lattice of an alloy belongs to the tetragonal or orthorhombic singony but not to the cubic one, the relation between the ensembles becomes more complicated. The fixation of the volume in this case retains the opportunities for the fluctuations of the parameters of tetragonality or orthorhombicity and the corresponding restricted ensemble will be wider than the grand canonical one. Second, in a solid solution consisting of atoms of different size, static displacements are possible, while in some structural types the displacements from the lattice sites are admissible by symmetry and by the equal sizes of atoms. These displacements may also be present at the constant volume. Account of them also makes it necessary to work with the ensemble that is wider than the grand canonical one.

If, in addition, we fix the number of particles in the system, we shall come to the consideration of the canonical ensemble. The restricted canonical sum Z_{V,N_A} will then be

$$Z_{V,N_A} = e^{\beta(PV + \mu N_A)} \sum_{\{k\}}^{N, V} e^{-\beta E}. \tag{4.3.11}$$

This expression differs from the partition function of the canonical ensemble by the factor $\exp(\beta\mu N_A)$ that is constant for all configurations in the canonical ensemble. The overall energy of the grand canonical ensemble can be expanded into a series in the vicinity of the canonical ensemble in the fluctuations of the number of particles. It should also be noted that the canonical ensemble can be expanded so as to allow for the volume effects, by choosing $\{\alpha\} \equiv N_A$.

We can classify the methods employed in the theory of atomic ordering in the above-described scheme of the restricted partition function. To do this, the classification needs in refining. It is particularly important because the authors of the works devoted to the statistical theory of atomic ordering have not discussed this issue. In the Kirkwood--Khachaturyan method we have $\{\alpha\} \equiv V, N_A, \eta$. Hence, the statistical ensemble used in this method (Kirkwood's ensemble) is narrower than the canonical one but, at the same time, is wider than the microcanonical ensemble. Indeed, the Kirkwood expansion is that in the energy fluctuations at the fixed value of the order parameter. It is narrower than that of Kerson Huang in which the fluctuation of energy is calculated without the fixation of η.

The entropy methods are formally related to the narrower ensemble than that of Kirkwood. The introduction of a combinatory factor into the free energy requires the fixed value of the configurational energy.This means the transition to the microcanonical ensemble. Besides, in the combinatory methods all internal parameters describing the state of order in a system (η, ε_{12}, ε_{123}, etc.) are also fixed. As a consequence, the ensembles used in these methods must be assigned to the narrower ones than the microcanonical ensemble. The distinction between combinatory methods consists in the way the order parameters were fixed. They can be equated to zero (a rigid condition) or simply made constant (not so rigid condition). From this point of view, the narrowest is the ensemble adopted in the GBW approximation, the less narrow is that used in the cluster variation method.

Two important subtle points should be elucidated in this context. First, the expression for the total partition function, both in Kirkwood's approximation and in entropy methods, is formally written in the canonical ensemble. For example, in the entropy methods

$$Z_{N_A,V} = \sum_{\eta,\epsilon_{12}\ldots}^{N_A,V} Z_{\eta,\epsilon_{12}\ldots} \tag{4.3.12}$$

But one should always remember that the equilibrium conditions for $F=kT\ln Z_{\eta,\varepsilon_{12},\ldots}$, i.e.,

$$\frac{\partial F}{\partial \eta} = \frac{\partial E}{\partial \epsilon_{12}} = \ldots = 0 \tag{4.3.13}$$

choose from the canonical ensemble the restricted subensemble that is smaller than the microcanonical one; a complete description of the system is virtually achieved just in this subensemble.

The restriction of the ensemble means the neglect of the fluctuations of definite entities. The implication of this fact, consisting in the appearance of new relations between the order and correlation parameters (fluctuation relations), has been considered in Section 3.1. Here we must draw attention to the fact that the large bulk of the theory of atomic ordering has been performed on the base of the highly restricted ensembles.

So far, we have classified the statistical ensembles used in the works of the third group. The works of the second group have "started" from the canonical ensemble. Provided the group expansion technique was used in these works, the canonical ensemble was, in fact, expanded into a series in the energy fluctuations and a certain part of them was taken, in some way, into account. In this case, the ensemble is narrower than the canonical one but wider than the microcanonical ensemble. The expansion into a series in the moments of a random field also leads to the reduction of the canonical ensemble. In this case, the fluctuations of energy are allowed for indirectly, through the fluctuations of a random field. The choice of the distribution function determines the choice of the ensemble. The distribution described by δ function leads to the microcanonical ensemble. In the other cases, the ensemble is intermediary between the canonical and microcanonical.

It would be more difficult to give the classification of the works of the first group because the partition function is not explicitly employed in these works. The technique developed by Tahir-Khelly neglects the fluctuations of the occupation numbers relative to their averages. In the model chosen by Tahir-Khelly, the average occupation number is connected with the order parameters and energy. So, this approximation corresponds to the microcanonical ensemble.

The classification hierarchy of the ensembles used in the statistical theory of ordering can be represented, along with the main ensembles, in a well defined scheme. Besides the ensembles discussed above, one should single out in this scheme the Smirnov ensemble in which the GBW method is generalized to take account of the volume effects. One more detail is worth to be noted. Kirkwood's and the quasichemical methods in the proposed hierarchy proved to belong to different ensembles. This question was considered in Section 3.1.

The above classification gives evidence that the available theories of atomic ordering are constructed on the base of fairly restricted ensembles. As a consequence. an account of fluctuations of certain quantities turns out to be impossible (or incorrect). It may be said that the theory of ordering should be advanced along the following lines. It is necessary (1) to find out all the consequences of neglecting the fluctuations and, (2) to work out statistical theories of atomic ordering using the approximations relied upon wider ensembles.

New approaches in the theory of atomic ordering. All that has been stated above provided the grounds for developing a new technique in the theory of atomic ordering based upon the up-to-date methods of statistical physics. This approach employs the method of integral equations by Bogolyubov-Born-Green-Kirkwood-Ivon (BBGKI), Green function method, the random field approximation, various schemes of temperature expansions and, finally, the entropy methods. In the present monograph we shall not expound the details of the complex and cumbersome transformations in the statistical theory developed in the works [215, 217 ,278 ,310 - 319] since they have been discussed in the recently published monograph [320]. Our task is to give only some final equations of different theoretical approaches that may be used in practical computations of the $\eta(T/T_C)$ curves. The results of the computations of this kind are presented in the next section.

The equilibrium equations in the BBGKI method. In applying the superpositional decoupling, the equations for the B2 superstructure take the following form:

$$\ln\frac{\left(C_A+\frac{\eta}{2}\right)\left(1-C_A+\frac{\eta}{2}\right)}{\left(C_A-\frac{\eta}{2}\right)\left(C_A-\frac{\eta}{2}\right)}=\frac{ZV_1\eta}{2kT}\left\{1+\frac{\epsilon_{12}^{AA}}{4}\,\frac{2[C_A^2+(1-C_A)^2]-\eta^2}{(C_A^2-\eta^2/4)[(1-C_A)^2-\frac{\eta^2}{4}}\right\} \tag{4.3.14}$$

$$\ln\left\{(\epsilon_{12}^{AA})^2+\epsilon_{12}^{AA}\left[C_A^2+(1-C_A)^2-\frac{\eta^2}{2}\right]+\left(C_A^2-\frac{\eta^2}{4}\right)\left[(1-C_A^2)-\frac{\eta^2}{4}\right]\right\}+$$

$$+\,(Z+1)\,\frac{V_1}{kT}=(Z-1)\,\frac{V_1}{kT}\,\frac{(-\epsilon_{12}^{AA})}{\left(C_A+\frac{1}{2}\right)\left(1-C_A-\frac{\eta}{2}\right)}\,.$$

For the $L1_2$ superstructure, this system is written as

$$\ln\frac{P_1}{P_2}+4\,\frac{V_1}{kT}\left[\frac{3P_{12}}{P_1}-\frac{2P_{22}}{P_2}-\frac{P_{12}}{P_2}\right]=0$$

$$\ln\frac{P_{12}}{P_{22}}+\frac{V_1}{kT}\left[\frac{P_{12}P_{22}}{P_1P_2P_2}-2\,\frac{P_{12}P_{22}}{P_2P_2P_2}+7\,\frac{P_{12}}{P_1}-2\,\frac{P_{12}}{P_2}-5\,\frac{P_{22}}{P_2}\right]=0$$

$$\ln\frac{P_{12}P_1}{P_{22}P_1}+\frac{V_1}{kT}\left[6\,\frac{P_{12}P_{22}}{P_1P_2P_2}-3\,\frac{P_{12}P_{12}}{P_1P_2P_2}-3\,\frac{P_{22}P_{22}}{P_2P_2P_2}+17\,\frac{P_{12}}{P_1}-\right.$$

$$\left.-\,12\,\frac{P_{22}}{P_2}-5\,\frac{P_{22}}{P_2}\right]=0. \tag{4.3.15}$$

where 1 and 2 denote the types of the lattice sites. All probabilities are written for the A atoms. In the square brackets one can easily identify the contributions due to the decoupling of three-particle probabilities, i.e., the three factors in the denominator. The occurrence of such terms confers a clear advantage to the theories of atomic ordering based on the use of the infinite chains of equations over the combinatory methods. The latter contain only the terms of the P_{ji}/P_i type. Owing to the presence of the indirect contributions of higher correlations, equations (4.3.14) differ from those of the quasichemical theory.

The Green function method. Various decouplings of the chains of the equations into the Green functions for the B2 superstructure in the nearest neighbor approximation lead to the following system of the equilibrium equations:

$$\epsilon_{12}=-\frac{1}{32}\,(1-\eta^2);\quad \ln\frac{(1+\eta)^2-4\epsilon_{12}}{(1-\eta)^2-4\epsilon_{12}}=7\beta V_1\eta,$$

$$\ln\frac{1-\eta}{1+\eta}=-\frac{4\beta V_1\eta(1-\eta^2+4\epsilon_{12})}{1-\eta^2}\;; \tag{4.3.16}$$

$$\ln \frac{(1+\eta)^2 - 4\epsilon_{12}}{(1-\eta)^2 - 4\epsilon_{12}} = \frac{7}{4} \ln \frac{1+\eta}{1-\eta}, \tag{4.3.17}$$

$$\ln \frac{(1+\eta)^2 - 4\epsilon_{12}}{(1-\eta)^2 - 4\epsilon_{12}} = 7\beta V_1 \eta \frac{1-\eta^2+4\epsilon_{12}}{1-\eta^2}; \quad \epsilon_{12} = -\frac{(1-\eta)^2}{100+28\eta^2}. \tag{4.3.18}$$

The two first decouplings yield the same system, (4.3.16), due to the nearest neighbor approximation in the theory. The first equation of the system links the degree of order and correlation (irrespective to temperature) and coincides with one of the fluctuation relations in Section 3.1. This means that the neglect of the fluctuation in the occupation numbers around the mean value corresponds, at least, to the transition to the canonical ensemble. The remaining decouplings do not use such strong restrictions and give rise to different equations. The second equation in system (4.3.17) coincides with one of the equations of the quasichemical approximation.

The random field method. Various approximations in this method are determined by the cluster size and the form of the distribution function describing a random field or the number of moments retained in the expansion of this function into a series. For example, in the case of the one-particle cluster and δ-shaped distribution function, one can obtain both the equations of the GBW approximation and those of Cowley's theory. Choosing the one-particle cluster and the normal distribution leads, in the case of the $L1_2$ superstructure, to the following equilibrium equation:

$$\ln \frac{P_\alpha^A P_\beta^B}{P_\alpha^B P_\alpha^A} = 3\left(\frac{V}{kT}\right)^2 [P_\beta^A P_\beta^B + 2(P_{\beta\beta}^{AA} - P_\beta^A \cdot P_\beta^A)] -$$
$$- \left(\frac{V}{kT}\right)^2 \Big[P_\alpha^A \cdot P_\alpha^A + 2P_\beta^B \cdot P_\beta^B + 4(P_{\alpha\beta}^{AA} - P_\alpha^A P_\beta^A) +$$
$$+ 4(P_{\beta\beta}^{AA} - P_\beta^A \cdot P_\beta^A)\Big] + \frac{4V}{kT}(P_\alpha^A + 2P_\beta^A) - 12\frac{V}{kT}P_\beta^A\Big]. \tag{4.3.19}$$

In this equation the left side and two last terms on the right correspond to the GBW approximation; the remaining terms determine the extension of this approximation due to the use of the normal distribution. These terms involve the pair probability functions.

The simplest way to solve (4.3.19) is to neglect the correlation which gives rise to the conversion of the equation into that with one unknown quantity, the long-range order parameter.

High-temperature expansions. For the B2 superstructure, the equation of equilibrium based on the use of the extended statistical ensemble and on retaining the terms up to that of the eighth degree of order, has the form:

$$\ln\frac{1+\eta}{1-\eta} = 112\eta u^2 - (976\eta + 816\eta^3)\,u^4 - \left(4400\eta - 67989\,\frac{1}{3}\,\eta^3 + 27712\eta^5\right)u^6 - \left(68183\,\frac{1}{3}\,\eta + 268010\eta^3 + 1495990\eta^5 - 1091360\eta^7\right)u^8. \qquad (4.3.20)$$

This equation enables one to calculate the η versus T dependence. It approximately accounts for the contribution of rather high correlations, up to the 16-particle one. Moreover, equation (4.3.20), as a single equation with one unknown quantity, the long-range order parameter, is very convenient for computations. In equation (4.3.20), U = --th(βV_1).

The entropy methods enable one to formulate the general approach to the determination of the free energy and entropy with account of the correlations taken in advance. This approach is rigorous due to the use of the canonical ensemble and the consistent derivation of the formulas from the exact expression for entropy. In doing so, the results of the other combinatory theories are obtained simultaneously, as partial cases. The particular versions of the theory are being constructed with the aid of decoupling of the correlation functions discussed in Section 3.1. The previously known entropy methods can be obtained as the particular cases of the general approach described here. For this reason, we shall not give here the concrete theoretical formulas and proceed further to the consideration of the newly developed methods of statistical theory. In working out these methods, we assumed that new approaches in the theory of ordering have to be more general than traditional theories: they must provide (1) the possibility of obtaining the same approximation for different superstructures, (2) the account, without principal complications, of the long-range and many-particle interactions and, finally, (3) the opportunity to generalize theory with taking into account of higher order correlations. Furthermore, our objective was to extend, as far as possible, the ensemble in which the theory operates. As is seen, this problem has been solved, on the whole, and the devised methods have brought an unexpected result.

In particular, in all newly developed methods, except for the random field technique, the involvement of the long-range and many-particle interactions does not face any difficulty. As for the higher correlations, their evaluation does not require any changes in the theoretical formalism in all known methods. Moreover, the expansion into a series in correlations (the entropy methods) enables one to select automatically the best cluster within the given approximation. The contribution of higher correlations into the lower ones is also accounted for automatically in decoupling the chain of the integral equations involved in the BBGKI and Green function methods. At last, if the proper approximations are chosen in the devised methods, all superstructures are described in the same way. The exception is the method of high-temperature expansions in which the structure of the series depends on lattice geometry.

All our methods enable one to work in rather wide ensembles: the Green function method in the grand canonical ensemble, the temperature expansion method -- in the ensemble wider that the canonical, the entropy and BBGKI methods -- in the canonical ensemble, and the random field method - in the ensemble restricted by a particular form of the random field distribution function. The limitations connected with different approximations lead to the narrowing of the statistical ensemble. One should keep in mind that the decouplings used in constructing the concrete versions of the theory are equivalent to the neglect of all higher correlations. This gives rise to the fixation of all disregarded correlations (they are made equal to zero) and the corresponding narrowing of the ensemble.

It should be reminded that this narrowing leads to the appearance of the fluctuation relations (see Section 3.1). These relations are, in essence, the complementary (surplus) equations in each version of the theory, which may coincide with the basic equations of the theory, yield similar numerical results, or do not coincide with them at all.

Examination showed that the complete analytical involvement of the fluctuation relations into the theory takes place in the methods based on the use of the temperature expansions or Green functions. The remaining versions prove to be more and more accurate as the accuracy of approximation is improved.

As noted above, the correlation functions describing the order in alloys, can be found as the averages over the ensemble or as the most probable values. These two approaches are equivalent only in lower approximations. The methods based on the temperature expansions or the random field approximation provide the determination of both the averaged and the most probable correlation functions. In other methods, the solution to the problem represents the averages over the ensemble. Such variety of approaches provides appreciable advantages over traditional techniques in which one can find, as a rule, the most probable values of quantities.

Another feature common to each method developed in the present work is the physical clearness of the assumptions involved. In the BBGKI and entropy methods the character of assumption is determined by the mode of decoupling into the correlation functions, while in the random field approximation -- by the the number of moments retained in the expansion or by a particular form of the distribution function. More complicated is the situation with the Green function method and the temperature expansions. In the latter method, the physical sense of the neglected terms is evident, but the problem of convergence of the expansion remains unsolved. In the Green function method, decouplings are performed at the level of Green functions themselves. In this case, the occupation number operator is replaced by a certain conventional average, so that the analytical assessment of the assumptions is not achieved. Nevertheless, in all the methods developed in this work, the physical sense of all approximations made at each step of the theoretical treatment is very clear, whereas the combinatory theories and Cowley's method need in additional arguments to evaluate the degree of approximation.

In this respect, the results presented above are of methodical value. The point is that almost every method outlined here, yields, in lower approximations, the results which coincide with or are close to the achievements of traditional theories. In other words, the formulas derived here in the simplest approximations are equivalent to those obtained in the conventional theories. How and where that occurs is written in [217].

4.4 Effect of the superstructure type and the size factor on the characteristic features of the order-disorder transformation

In this section we shall consider the behavior of order parameters in the order-disorder phase transition for different classes of alloys to reveal common trends of such behavior. After the fulfillment of this work, the conditions for such generalization are ready because we have studied the order-disorder phase transformations in as many as twenty alloys. Combining these results with the previously obtained data on the order-disorder transitions provides a sufficient experimental base for the generalization of the results. Thus, by this time the order-disorder phase transformations have been investigated at least on forty alloys including those of the AB, A_2B, A_3B, and A_4B composition. The alloys comprise the superstructures of various symmetry formed from an f.c.c. solid solution -- $L1_0$, $L1_0$(M), $L1_1$, the structure of the Pt_2Mo type, $L1_2$, $L1_2$(M), $L1_2$(MM), $L1_3$, DO_{23}, DO_{22}, and D1a, as well as those formed from a b.c.c. solid solution -- B_2, $C11_b$, and DO_3.

Analysis of the $\eta(T/T_c)$ curves, represented in Section 4.2 of this chapter, together with the literary data showed the occurrence of the three types of such curves (Fig. 63). The first two types relate to the alloys with a linear behavior: the $\eta(T/T_c)$ dependence is described, in fact, by a single curve falling smoothly with temperature. The curves I and II in Fig. 63 differ in that the first describes the nearly linear dependence of η on T/T_c, usually with a comparatively slow decrease of the degree of order with temperature, while the second one is characterized by a continuous drop of the long-range order parameter with increasing temperature. The term "linear behavior" referred to the first two types of the $\eta(T/T_c)$ curves is not connected directly with the real form of the η (T/T_c) function; it was taken from the thermodynamical theory of phase transformations [14] and characterizes the independence of the moving force of the transformation, the ordering energy in our case, upon the long-range order parameter and temperature. It is this type of the $\eta(T/T_C)$ curves that results from the theoretical calculations performed in the statistical theory of atomic ordering based on the models with a constant pair ordering energy V(r). In such calculations, the first type is obtained if the role of correlations in the alloy structure is overestimated, the second type corresonds to the neglect of the correlation effects.

The $\eta(T/T_C)$ curves of the third type are characterized by the nonlinear behavior. These curves appear in the theoretical calculations provided the dependence of the pair ordering energy on the degree of order $V(\eta)$ becomes substantial above a certain temperature, or when the contribution of many-particle interactions is commensurate with that from the pair interaction. The curve $\eta(T/T_C)$ of the third type consists of the two parts: a linear one and that where the drop of the degree of order sharply accelerates with increasing temperature. The curve with two nonlinear portions separated by a marked interface is also possible.

Such classification of the phase transition $\eta(T/T_C)$ curves is accomplished for the first time. It relates primarily to the 1st-order phase transitions and based mostly on experimental data but is supported in part by the results of theoretical calculations. It should be noted that the experimental curves obtained for an alloy cannot always be referred to that or other class; the intermediate cases also occur. Not all experimental η values do necessarily define the equilibrium state, especially in the cases when the ordering in an alloy proceeds slowly and the temperatures at which the equilibrium is attained are low. Nevertheless, we believe that the available information on the order-disorder phase transitions confirms the classification proposed.

One more remark should be made at this point. The second type of the $\eta(T/T_C)$ curve is in a certain sense intermediary between those showing the linear and nonlinear behavior. This can be seen from the comparison of the curves of type I and type II. The nonconfigurational terms connected with the size and phonon effects, the presence of

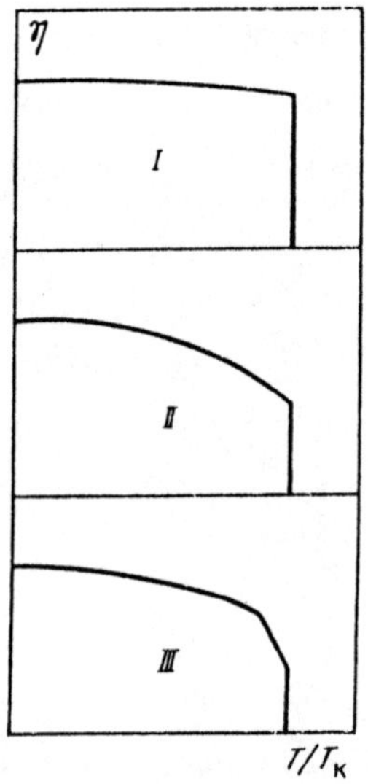

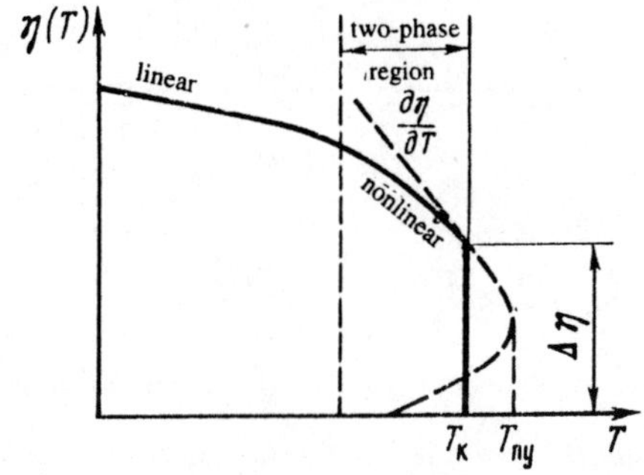

Fig. 63. Three types of experimental long-range order parameter vs temperature dependence (schematically).

Fig. 64. The main characteristics of the order-disorder phase transition.

APBs and the correlation effects, especially complicated by their interaction, may diminish, to some extent, the distinction between the two curves. Let us consider this question in more detail.

The behavior of the $\eta(T/T_c)$ curves can be analyzed from the general thermodynamical standpoint by considering the characteristics affecting the value of the free energy of an alloy such as the correlation parameter ε, the dependence of the lattice constant a on the degree of order, the characteristics of the phonon spectrum, and the periodical APBs. It may be inferred from the general thermodynamical considerations [263] that the equation of the tangent to the $\eta(T/T_c)$ curve is determined in part by the behavior of the derivatives $\partial\varepsilon/\partial\eta$ and $\partial a/\partial\eta$ and the difference in the atomic masses of the alloy.

As a consequence, for an alloy showing a stronger dependence of the lattice constants and the correlation parameters upon the degree of order and at a greater difference in the masses of the component atoms, the curves $\eta(T/T_C)$ are less inclined to the horizontal axis at higher η values and are steeper near T_C than the curves for the alloy where these factors are less pronounced. The curves $\eta(T/T_C)$ represent virtually the equation of equilibrium

$$\frac{\partial(E - TS)}{\partial\eta} = 0, \tag{4.4.1}$$

where E is the configurational energy and S is the configurational entropy. The tangent to the curve (4.4.1) is expressed by

$$\frac{\partial\eta}{\partial T} = \frac{\partial S}{\partial\eta} \bigg/ \left(\frac{\partial^2 E}{\partial\eta^2} - T\frac{\partial^2 S}{\partial\eta^2}\right). \tag{4.4.2}$$

The numerator is always negative here while the denominator is positive.

If we single out explicitly the dependence of the unit cell volume Ω on the degree of order (see Section 3.3), then (taking into account the fact that this volume is always in equilibrium) the equation (4.4.2) will be written as

$$\frac{\partial\eta}{\partial T} = \frac{\partial S}{\partial\eta} \bigg/ \left\{\frac{\partial^2 E}{\partial\Omega^2}\left(\frac{\partial\Omega}{\partial\eta}\right)^2 + \left[\frac{\partial^2 E}{\partial\eta^2} - T\frac{\partial^2 S}{\partial\eta^2}\right]\right\}. \tag{4.4.3}$$

An additional term in expression (4.4.3), compared to (4.4.2), is always positive for an equilibrium solid solution. Therefore, a stronger dependence a on η will diminish the slope of the $\eta(T/T_C)$ curve.

To take the effect of short-range order into account is more difficult. In this case, on account of the equilibrium condition and a linear dependence of η^2 on ε (see Sections 3.1 and 4.3), expression (4.4.3) takes the form

$$\frac{\partial\eta}{\partial T} = \left(\frac{\partial S}{\partial\eta} + \frac{\partial S}{\partial\epsilon}\frac{\partial\epsilon}{\partial\eta}\right) \bigg/ \left(\frac{\partial^2 E}{\partial\eta^2} - T\frac{\partial^2 S}{\partial\eta^2} + A(\epsilon,\eta)\right). \tag{4.4.4}$$

Here $A(\varepsilon,\eta)$ is a function that tends to $(\partial E/\partial\varepsilon)(\partial^2\varepsilon/\partial\eta^2)$ at low temperatures. It can be shown that both factors are positive. Hence, when the correlation in an alloy is more pronounced, the $\eta(T/T_C)$ curve shows a lesser slope to the T/T_C axis. At higher temperatures, the dependence of entropy on the correlation parameter becomes the decisive factor. This dependence is expressed by the numerator in (4.4.4). Since both factors in the second term of the numerator have different signs, analysis of the behavior of the $\eta(T)$ curve is ambiguous.

Next, we shall proceed to the consideration of the experimental trends common to all investigated alloys and superstructures. In all these alloys, the transition to the phase with a short-range order is preceded by a partial disordering in the phase with a long-range order. This phenomenon is universal, without exceptions. The cause lies in a cooperative nature of the ordering and disordering processes, which is reflected in all statistical theories of atomic ordering predicting an immediate decrease of the equilibrium degree of long-range order as soon as the temperature rises from absolute zero. But in various alloys this process proceeds in different ways which is reflected in the classification of the $\eta(T/T_c)$ curves as proposed above. The reason for such distinction lies in a variety of interatomic interactions in alloys and their different variations with temperature and the state of order. Thus, the order-disorder phase transition is always preceded by a partial disordering of the alloy. The transformation never occurs directly in a completely ordered phase. This fact distinguishes drastically the order-disorder transformation from some other transformations in metallic alloys, for example, from most of the structural phase transformations, diffusion or martensitic. The latter give an example of purely 1st-order transition. In ordering alloys there may be pure 1st-order transitions, those close to the 2nd-order and, finally, purely 2nd-order transitions.

A great majority of the experimentally investigated order-disorder transformations relates to the 1st-order. This fact distinguishes the order-disorder transformation, as a cooperative phenomenon, from the magnetic phase transformations where the 2nd-order transitions frequently occur.

In the case of the 1st-order phase transition, usually just before disordering, the decline in the degree of long-range order speeds up. This gives evidence that in a majority of alloys the nonlinear interactions are quite important. Nevertheless, this fact was ignored in most of the works devoted to the statistical theory of order-disorder transitions. In the present investigation, we made an attempt to fill this gap by carrying out a certain part of calculations using the models which take into account the dependence of the ordering energy on the order parameter. The choice of such models is based on the quantum theory of solid solutions developed in Chapter 5.

In spite of the features common to all transitions, in most alloys one can find specific trends characteristic of different superstructures. In view of a small number of investigations performed on the ordering alloys with initial b.c.c. and f.c.c. lattices, the comparison of different superstructures is restricted to the alloys with initial f.c.c. lattice. Before making such comparison, we must pay attention to the existence of definite relations between various superstructures. The $L1_2$ and $L1_0$ superstructures can be defined as the basic ones for the alloys of the A_3B and AB compositions, not only due to their high symmetry. The point is that the superstructures $L1_2$(M), DO_{23}, DO_{22}, and $L1_2$(MM) are deduced from the $L1_2$ structure by introducing the simplest periodic APBs. In this case, in addition to the main star (0 0 1), we have the stars (1 0 1/2M) specifying the long period of the structure. The phase transitions related to the remaining stars in such superstructures are degenerate with respect to the phase transition corresponding to the star (0 0 1). The $L1_2$ and $L1_3$ superstructures have one common star (0 0 1) and are characterized each by more complex type of periodic APBs. Only the $D1_a$ superstructure and that of the Pt_2Mo type are not connected directly with $L1_2$, but they have different stoichiometric composition. Similarly, the $L1_0$ superstructure is the basic one for the $L1_0$(M) and $L1_0$(MM) superstructures, being not connected directly with $L1_1$. Let us compare now the main features of the phase transitions involved (Fig. 64).

The transformation temperature. Most superstructures are characterized by the transitions which occur in various alloys at different temperatures, i.e., in a wide temperature range (except for the superstructures with a single representative, for which no specific trends can be established). Two superstructures, $L1_2$(M) and $L1_2$(MM), are to be assigned to the low-temperature ones, since the order of this kind does not occur at temperatures higher than 600 K. In comparison to them, the DO_{23} superstructure and especially DO_{22} are undoubtedly the high-temperature ones. The region where the $L1_2 \rightarrow A1$ transformations occur is also shifted toward higher temperatures (the more so, if we bear in mind the alloys in which the transformation has not been studied as yet in detail due to the high transformation temperature). The comparison of these observations to the data on changing the number of bonds on different coordination shells caused by disordering shows that one of the reasons for such shift is a relative energetical "benefit" of different superstructures. If we focus on the pair interaction model and ignore lattice distortions, it can be seen from Table 2 that the difference in the temperature ranges of the existence of the superstructures is due to the different contributions of the distant coordination shells into the ordering energy. Furthermore, taking the distortions of a crystal lattice into account enables one to see another cause for the decrease in stability of the long-period superstructures, namely, the competition of the size factor and the factor of electron structure.

It should be pointed out that the order of the $L1_2$ type, as well as those of the $L1_0$ and B2 types, occur at the highest possible temperatures. The high-temperature ordered phases seem to be more high-temperature ones not only due to the definite sequence of the temperatures of the phase transitions between different superstructures but also in view of the maximum absolute temperature at which the order corresponding to the given superstructure type can still be observed. And finally, the $D1_a$ superstructure should also be assigned to the comparatively high-temperature one.

The jump in the order parameter, or the maximum disordering attained just before the transformation, characterizes the maximum stability of the ordered phase with respect to disordering, i.e., the maximum decrease in the degree of long-range order at which the phase with long-range order remains stable with respect to the phase with short-range order. The disordering takes place if the free energies or the Gibbs energies of the ordered and disordered phases are equal (the external loss of stability) or when the temperature is attained at which $\partial\eta/\partial T=\infty$ (the internal loss of stability). This division has a certain sense only for the 1st-order transitions; in the case of the 2nd-order transitions, the temperatures of the external and internal losses of stability are coincident.

The high-symmetry superstructure $L1_2$ can exist in various alloys with different degree of disordering. However, the minimum value of the degree of order usually exceeds 0.6. The $L1_2$-base long-period superstructures are definitely capable of retaining the ordered state at lower values of the order parameter. This is especially so for the $L1_2$(M) and $L1_2$(MM) superstructures in which the phase transition does not begin before the degree of order is lowered to 0.5. Also capable of significantly disordering is the $L1_3$ superstructure. Such distinction between the $L1_2$ and the long-period superstructures is connected with the complex combination of the configurational and nonconfigurational contributions into the transformation entropy.

The superstructures of the A_4B, A_2B and AB composition are definitely less stable to disordering than the $L1_2$ superstructures. In many of them, a slight decrease in the degree of order suffices to initiate the phase transition. When comparing these superstructures with $L1_2$, the following criterion is valid: lower symmetry needs in higher order for its stabilization. The realization of this criterion implies the complex interaction of the configurational and nonconfigurational effects, and the band-structure effects. Finally, the relation between the $L1_0$ and $L1_0$(M) superstructures is the same as between the $L1_2$ and $L1_2$(M) superstructures.

The proximity to the point of the stability loss is determined by the slope of the η (T/T_c) curve at the phase transition point (or by the reciprocal value). This closeness determines, in turn, how close is the order-disorder phase transition to the "spinodal" disordering ($\partial\eta/\partial T=\infty$, $\partial(T/T_c)/\partial\eta=0$) or, in other words, determines the distance between the temperatures of the external and internal stability loss. For the superstructures with several order parameters, we can compare the above characteristic (in relation to the order parameter) with their common star (0 0 1). The widest variations of $\partial(T/T_c)/\partial\eta$ exhibit the alloys with the DO_{23} superstructure. Along with the alloys belonging to the $L1_2$(M), $L1_2$(MM), and DO_{23} superstructures, they approach most closely to the point of the absolute (internal) loss of stability of the ordered phase. It is worth noting that, on the average, for all long-period superstructures, in contrast to that of the $L1_2$ type, disordering is closer to "spinodal". Conversely, all superstructures of other stoichiometry undergo disordering at higher values of $\partial(T/T_c)/\partial\eta$ than that for the $L1_2$ superstructure, i.e., the transformation in these alloys bears certain features of the external loss of stability.

Attention should be drawn to the fact that the trends pointed out in two last sections of this chapter are virtually the same (with the exception of the alloy with the $L1_3$ superstructure). The smaller the jump in the order parameter, the closer, as a rule, the alloy to the temperature of the internal loss of stability (Fig. 65). The superstructure

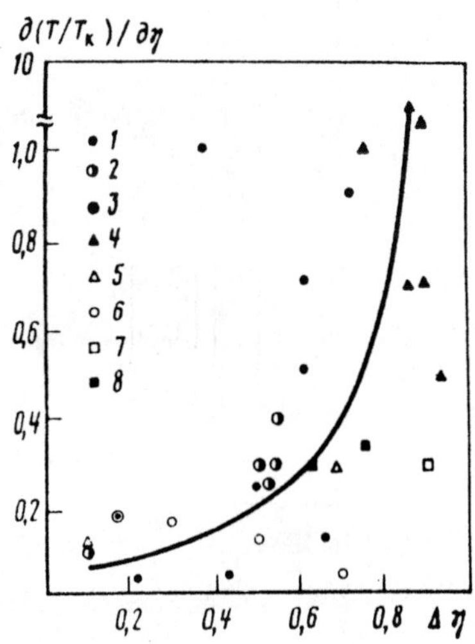

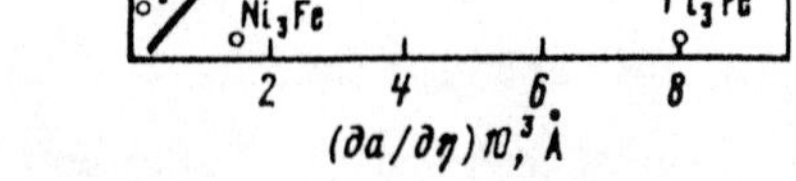

Fig. 65. Dependence of the jump in the long-range order parameter on the transformation temperature and the slope of the $\eta(T/T_c)$ curve in the jump point. Superstructures: $L1_2$ (1); $L1_2$(M), $L1_2$(MM) (2); $L1_3$ (3); $D1_a$ (4); DO_{22} (5); DO_{23} (6); $C11_b$ (7); Pt_2Mo-type (8).

Fig. 66. Correlation between the configurational expansion coefficient $\partial a/\partial\eta$ and the proximity of the alloy to the point of stability loss for different alloys. The Pt_3Fe alloy undergoing the structure transformation together with disordering falls out of the common curve.

Table 20. Three superstructure groups differing in structures and phase transition

Group	Superstructure	Alloy	Type of $\eta(T/T_c)$ curve	Closeness to 1st or 2nd order
first	$D1_a$	Au_4Cr, Au_4V, Ni_4Mo, Ni_4W	I	I
		Au_4Mn	III	I
	$L1_0$	CuAu, NiPt	I	I
	$L1_0$ (M)	CuAuII	I	I
	$L1_1$	CuPt	II	I
intermediate	$L1_2$	Cu_3Au, Au_3Cu, Ni_3Mn, Pd_3Fe	III	I
		Pt_3Fe, Fe_3Pt	III	to anomalous 1st order
		Ni_3Fe	II	to anomalous 1st order
		Cu_3Pt, Ni_3Pt	I	I
		Pt_3Co	II	I
third	$D0_{22}$	Pd_3V	III	I
		Ni_3V	III	to anomalous 1st order
	$D0_{23}$	Au_3Cd, Ag_3Mg	III	I
		Pd_3Mn	III	to anomalous 1st order
	$L1_2$ (M)	Cu_3Pd	II	I
		Cu_3Pt	III	to
	$L1_2$ (MM)	Au_4Zn	III	to anomalous 1st order
		Au_3CuII	II	to

$L1_3$ divides all superstructures into two classes. The first includes the superstructures of the AB, A_2B, and A_4B stoichiometry that are characterized by a considerable jump in the order parameter and a typical external loss of stability. The second class comprises all long-period superstructures with a small jump in the order parameter and manifest closeness to the complete loss of stability of the ordered state.

The occurrence of the two-phase region and its characteristics. It is also interesting that in alloys of the first group the transformation always passes through a two-phase region whereas in alloys of the second group this phenomenon occurs rarely. Naturally, the degree of long-range order at which the two-phase region begins is always higher in alloys of the first group and, particularly, in alloys of the second group than in alloys with the $L1_2$ superstructure.

If the transition proceeds through a two-phase region, it exhibits, according to Mayer's and Stritter's classification (improved recently by Rolov [236]), the features of a diffuse phase transition. It should be stressed at this point that we analyze here the behavior of the long-range order parameter η in an ordered phase but not the volume average of η, as in Rolov's works. Specially for this purpose, the appropriate technique has been devised in the present investigation. After the removal of the effect of the long-range order parameter for the two-phase region, the phase transition in alloys of the first group proved to be purely of the 2nd-order.

To summarize, we may note that the alloys of the first group undergo the 1st-order phase transition in a pure form, while the phase transition in the alloys of the second group, if it is of the 1st order, is closer to the 2nd-order transition.

To be more accurate and by using Mayer's and Stritter's classification [236] instead of Erenfest's one, it can be said that in the second group of alloys the 1st-order phase transition occurs which is close to an anomalous 1st-order transition. In some alloys of the second group, namely in Au_3Cu II and Cu_3Pt, the phase transition of the 1st order is observed which cannot be distinguished from the 2nd-order transition or an anomalous 1st-order transition (the terminology depends on the classification used).

The types of the $\eta(T/T_c)$ curves are indicated in Table 20. In most alloys, the third type of the long-range order versus temperature dependence usually exists. In other words, the nonlinear behavior is typical of most alloys. In the case of the $\eta(T/T_c)$ curves of the second type, the linear correlationless behavior occurs rarely (this is so, no doubt, for Ni_3Fe and Pt_3Cu, but NiPt and Au_3Cu II represent disputable cases). Most of the $\eta(T/T_c)$ curves relate either to the linear type with a strong correlation (the first type) or to the ordinary nonlinear type (third type).

In alloys of the first group, with a pure 1st-order phase transition, the first type of the $\eta(T/T_c)$ curves (linear behavior with a high correlation) occurs most frequently. The slope of the $\partial\eta/\partial(T/T_c)$ curve in a wide temperature range is close to zero or constant and small, i.e., the degree of long-range order does not decrease appreciably with temperature or declines slowly. The intermediate group of alloys, with the L12 superstructure, displays rather diverse behavior: all the four types of the curves may occur. The slope of $\partial\eta/\partial(T/T_c)$ in the "linear" region of the curve may be small (0 - 0.3), moderate, or large (0 - 0.5). In the third group of alloys, the third type of the $\eta(T/T_c)$ curves (i.e., an ordinary nonlinear behavior) is mostly observed. The linear part of the curves is nearly horizontal. In alloys of the third group there also may be the $\eta(T/T_c)$ curves intermediary between the first and the third types. For such alloys, the linear part of the curve shows an appreciable slope to the T/T_c axis ($\partial\eta/\partial(T/T_c)$ = 0.2 - 0.4). Nonlinear parts of the $\eta(T/T_c)$ curve in alloys of the third group have a high slope, i.e., they show a steep fall of the long-range order before the beginning of the transformation ($\partial\eta/\partial(T/T_c) \sim$ 1.7 - 1.6). The intermediate group of alloys with the $L1_2$ superstructure is characterized by a lesser nonlinearity ($\partial\eta/\partial(T/T_c) \sim$ 1 - 1.6). Thus, various groups of alloys show different tendency to the nonlinear behavior of the $\eta(T/T_c)$ curves.

It should be reminded that the ordering in all the three groups of alloys is initiated by different stars. In the intermediate group it will be the (0 0 1) star. In the third group we encounter, in addition, the long-period star, usually the star (1 0 1/2M), with the variable parameter M. At the same time, the star (0 0 1) is absent in the alloys of the first group, with the exception of the alloys with the $L1_0$ superstructures.

In close relation with these findings are the data concerning the part that different coordination shells play in the stabilization of superstructures (Table 2). The three groups of alloys described above differ in magnitudes and signs of the contributions of the distant coordination shells (i.e., of the second and the third shells). Supposedly, the long-range interatomic interaction in various structures can substantially influence upon the features of the order-disorder transition, besides the nonlinearity effects.

To conclude the general survey of the behavior of the long-range order parameter during the order-disorder phase transition, we shall note that, besides the features common to all superstructures, several types of the $\eta(T/T_c)$ curves have been established as well as three groups of alloys differing in their behavior during the phase transition (Table 21). The first group of alloys showing a pure 1st-order phase transition comprises most of the low-symmetry superstructures with stoichiometry differing from A_3B. This group is characterized by a significant role of correlation and weakly pronounced effects in the order-disorder transition. In Fig. 65 the points corresponding to this group of alloys are scattered around the ascending part of the curve. Alloys of the third group are distinguished by strong nonlinear effects. This group includes low-symmetry superstructures with cubic APBs. Typically, they undergo either the 1st-order phase transition from a partly disordered phase and close to the anomalous 1st-order transition, or the phase transition which resembles the 2nd-order one. This group of alloys is represented in Fig. 65 by the initial part of the curve with a relatively shallow slope. And finally, there exists an intermediate group of alloys with a high-symmetry cubic superstructure of the $L1_2$ type, whose representatives display different types of behavior. Alloys of this group are located throughout the whole field of Fig. 65.

Size factor and specific features of the order-disorder transformation. Here we shall consider the experimental data showing finer details of the influence of the size effect upon the regularities observed in the order-disorder transformation.

In alloys with the $L1_2$ superstructure, a linear dependence a on η has been found. The equation of the tangent to the $\eta(T/T_c)$ curve has the form

$$f(a) = \frac{\partial \eta}{\partial a} \frac{\partial a}{\partial T} \Delta T. \tag{4.4.5}$$

Let us compare the values of $\partial\eta/\partial a$ for the alloys Ni_3Fe and Cu_3Au. As seen from Fig. 40, $\partial\eta/\partial a$ is negative, and $|\partial\eta/\partial a|$ for Cu_3Au is less than for Ni_3Fe. Our data show that the values of $\partial a/\partial T$ in these alloys differ insignificantly. The same conclusion is to be made from the comparison of the alloys Cu_3Au and Au_3Cu, Cu_3Au and Ni_3Mn, Cu_3Pt and Au_3Cu, Ni_3Mn, Ni_3Fe, and Ni_3Pt. Therefore, for most alloys with the superstructures, the rate at which long-range order decreases with temperature is considerably affected by the dependence of the unit cell parameter (the volume of the alloy) on the order parameter. The data for the $L1_2$ superstructure (which support this

Table 21. Three superstructure groups.

Superstructures	$D1_a$, Pt_2MoC11_b	$L1_2$, $L1_0$	$L1_2(M)$, $L1_2(MM)$, DO_{22}, DO_{23}, $L1_3$
Symmetry	mean	high;	mean and low
Star of Transition	(hk0)	(001)	$(001) + (10\frac{1}{2M})$ $(001) + (\frac{1}{2}\frac{1}{2}\frac{1}{2})$
Stoichiometry	AB_2, AB_4	AB_3, AB	AB_3
Temperature Stability	high temperature	any	low temperature
Break at Transition Point	$\Delta\eta$ large	any	$\Delta\eta$ small
Maximum Attainable η	$\eta = 1$	any event	$\eta \neq 1$
Nearness of T_k of the Temperature of the loss of rigidity T_{pu}	T_{pu} is far from T_k	average distance T_{pu} from T_k	T_{pu} near to T_k
Type of Phase Transition	clearly 1st type	any	1st type close to anomoly of the 1st type
Two-Phase	rarely	interval case	frequently
Dependence $\eta(T)$	linear	various	nonlinear
Coorrelation Effect	strong	average	weak

relation) are given in Fig. 66. The role of the size factor is evident. In essence, the relation between the $\eta(T)$ and $a(T)$ functions is due to the relation between the configurational and vibrational contributions to the entropy of the alloy in the disordering process. Alloys with low stability of a crystal lattice will undergo a slower transition to the disordered state than the alloys characterized by higher stability. Consequently, the closeness to the temperature of the absolute loss of stability of the ordered state for each particular alloy depends upon the influence of the degree of order on stability of its crystal lattice. Obviously, this temperature could coincide with the temperature of the order-disorder transformation only in the alloy in which the 1st-order transition persists and the degree of long-range order and the unit cell volume are not connected to each other. The same analysis is possible when comparing the electron contribution into the configurational energy.

For the low-symmetry superstructures, the role of the volume effect is more difficult to single out explicitly. As was pointed out earlier, the tetragonality parameter, for example, is determined not so much by the size effects as by the electronic structure, the positions of the Brillouin zones, the α resonance energy, etc. The interrelation between the electron structure and size effects does not permit one to differentiate these effects. Besides, for the low-symmetry superstructures the unit cell is specified by several parameters.

No specific trends within the individual superstructures have been revealed. But for some of them, in general, the relation between the electron structure and size effects is likely to exist. By analogy with the cubic superstructures, in the low-symmetry superstructures a greater dependence of a and c on η weakens the dependence of the order parameters on temperature. This trend can be deduced from the collected data for all superstructures or established specifically for the $D1_a$ and $L1_2(M)$ superstructures. Since, as will be explained below, the size of unit cells for low-symmetry superstructures is determined to a large extent by the band structure, such dependence characterizes the relation between the electronic structure and the configurational factor. On the other hand, Figs. 65 and 66, together with Fig. 67, demonstrate the interaction of the volume (size) and the configurational factors. The conclusion can be made that in the cubic superstructures the behavior of the order parameter depends to a greater extent on the primary atomic properties of the components of a solid solution, while in the low-symmetry ones -- on the band structure of an alloy. In these structures, variations of different order parameters, antiphase period, size of the unit cell and ratio of crystal axes are interrelated in a complex manner. The conclusion may follow that the role of the size factors can be ascertained most explicitly only for high-symmetry cubic structures.

Let us return now to the derivative $\partial a/\partial\eta$. It might be expected that this quantity depends on the size factor, i.e., on the difference of the atomic ratios of the component atoms in the alloy, Δr. The corresponding data are presented in Fig. 67. It is quite clear that the size effect is not the only factor determining the value of $\partial a/\partial\eta$. Nevertheless,

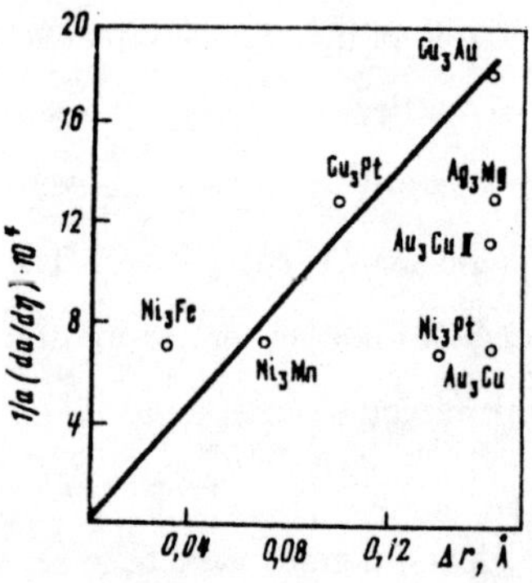

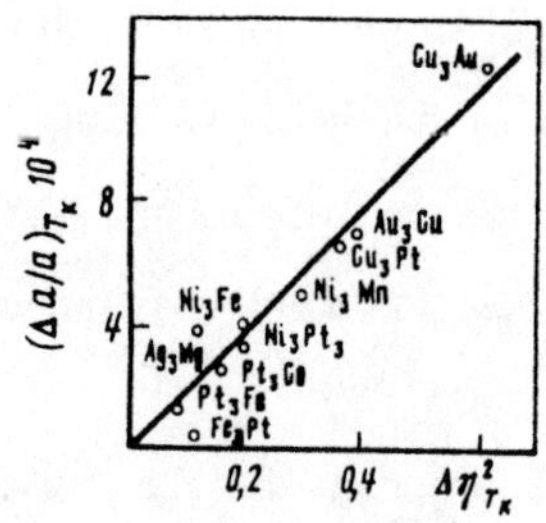

Fig. 67. Correlation between the configurational expansion coefficient and the size factor.

Fig. 68. Correlation between the relative change of the lattice parameter and the jump in the long-range order parameter for cubic and quasicubic superstructures.

the tendency to increase the component of the expansion coefficient related to disordering with increasing the size factor is distinctly traced. From comparison of the data in Fig. 67 and examples illustrating the application of formula (4.4.5), one can arrive at the conclusion that the greater is the difference in the sizes of the component atoms, the more stable is, in general, the superstructure $L1_2$ with respect to disordering near T_C.

In other words, the smaller the difference in the atomic radii of the components in the alloy, the closer can be the point of the absolute loss of stability of the ordered state to the temperature of the order-disorder 1st-order phase transition. This is indicative of a considerable contribution of the energy of lattice deformation into the ordering energy. For noncubic superstructures, we have no such well-argumented facts. In a low-symmetry superstructure, the parameter C usually changes more strongly upon ordering than the parameter a. The values of $\partial a/\partial \eta$ in cubic and noncubic superstructures are approximately of the same order, but the value of $\partial c/\partial \eta$ is larger by at least one order of magnitude. The conclusion can be drawn that in cubic superstructures the relation is more pronounced between the configurational and size factors while in low-symmetry superstructures -- between the configurational and band-structure factors. In other words, the volume alterations in noncubic superstructures are mainly determined by the band-structure effects. In low-symmetry superstructures there exists another quantity connected with the size of the unit cell, namely the tetragonality parameter |1-c/a|. It turns out to be fully independent of the size factor Δr. This fact shows that, at least in the alloys undergoing the order-disorder transformation in the solid state, tetragonality is entirely determined by the band structure of the material under consideration, i.e., by the electronic contribution to the total energy. The comparison of ∂c/∂η, ∂a/∂η and ∂(c/a)/∂η with Δr shows that the band-structure effects in noncubic superstructures are not stimulated by the size factor. For low-symmetry superstructures, the increase of the size factor either does not affect the change of the unit cell parameters at all or leads to the increase of the latter.

If the growth of the size effect decreases the disordering in the $L1_2$ superstructure, then the jump in the order parameter at T_c should increase, and the relation between Δη and the jump in the lattice constant Δa should occur at this temperature. The relation between Δη and Δa can be derived from the Clapeyon-Clausius equation: dP/dT=ΔS/ΔV. For the order-disorder transformation $L1_2 \rightarrow A1$, this equation may be written, with no account of correlations, as

$$\frac{1}{a}\left(\frac{\Delta a}{\Delta \eta^2}\right) = \frac{\kappa A}{a^3 \beta} \, \frac{V(k)}{T_\kappa}, \tag{4.4.6}$$

where β is the thermal expansion coefficient, k is the compressibility coefficient, V(k) is the ordering energy; A being a constant if the correlation is neglected. The value of $k/(a^3\beta)$ in (4.4.6) is the constant for a given class of substances, and the expression on the left side of (4.4.6) must have close values for the superstructures of the same type if the ratios $V(k)/T_c$ are not strongly different. For example, the values of $1/a(\Delta a/\Delta \eta^2)$ obtained for the alloys Cu_3Au and Ni_3Fe are equal to $2.2 \cdot 10^{-3}$ and $2.1 \cdot 10^{-3}$, respectively. The validity of the Clapeyron-Clausius equation in the form (4.4.6) for cubic and quasicubic superstructures is demonstrated by Fig. 68. Obviously, the required dependence holds with a good accuracy. The jump in the long-range order parameter squared, $\Delta\eta^2$, is proportional to the jump in the lattice parameter Δa. Hence, the application of (4.4.6) reminds one of the existence of the close relation between the configurational and volume (vibrational or electronic) contributions to the entropy of disordering.

In low-symmetry superstructures, the jump in the order parameter, in compliance with the Clapeyron-Clausius equation, has to be compared to the abrupt change in the volume at the point T_c. The relation between these quantities is virtually absent. The volume change is not connected with the the size factor as well. These findings provide evidence that in the low-symmetry superstructures the band-structure effects are more important than the initial atomic properties. Since in the low-symmetry superstructures the role of the band-structure effects is enhanced, one can expect to find the relationship between the jump in the order parameter and the change of tetragonality after the formation of the ordered structure. Analysis of the results shows undoubtedly the trend to the increase of the order parameter jump $\Delta\eta$ with an increase in the maximum tetragonality attainable in the alloy. This tendency becomes more and more pronounced in the series of superstructures: $D1_a \rightarrow L1_2(M) \rightarrow L1_2(MM) \rightarrow DO_{23} \rightarrow DO_{22}$. It provides evidence of a considerable effect of the anisotropy of bonding and the ordering energy on the order parameter in an alloy.

The jumps in the lattice constant and in the order parameter at T_c are not only closely related to each other but turn out to be dependent on the size factor Δr. The relevant data for the $L1_2$ superstructure are presented in Fig. 69. As the size factor increases, the stability of the ordered state with respect to the disordered state diminishes. The more the difference between the sizes of the component atoms forming the alloy, the higher the rate at which the free energy of the state with incomplete long-range order increases with the fall of the latter and the greater is the value of the long-range order parameter corresponding to the jumplike phase transition into a disordered state. The same drawing demonstrates that a certain part in this effect plays the volume energy. It will be shown further that in this case the increase in the free energy is due to the increase of statical displacements in a lattice.

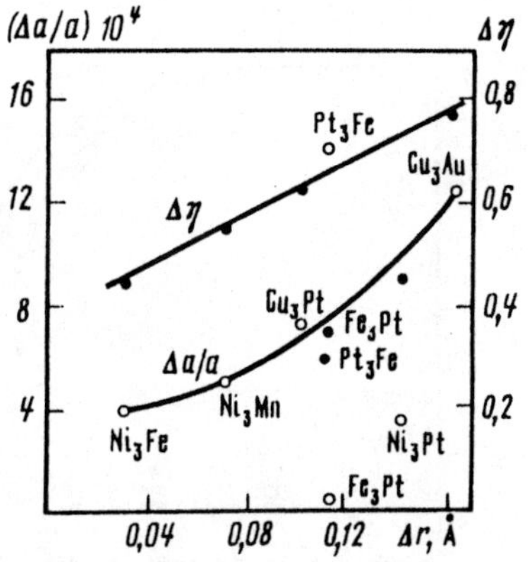

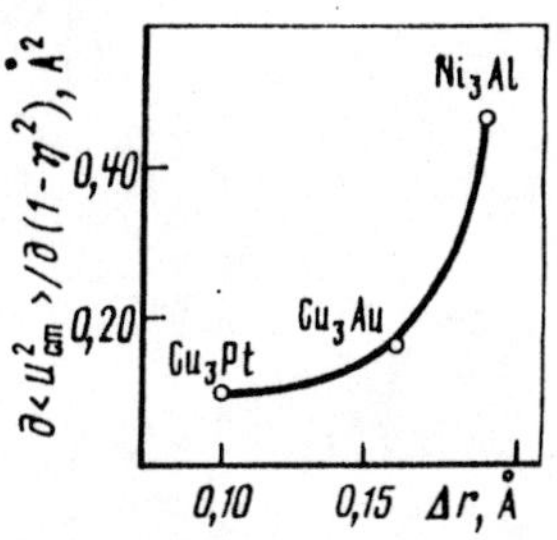

Fig. 69. Effect of the size factor on the jump in the long-range order parameter and the relative change of the lattice parameters at T_c. The $L1_2$ superstructure. The Fe_3Pt, Ni_3Pt and Pt_3Fe alloys show anomalous effect.

Fig. 70. Effect of the size factor on the rate of the growth of static displacements in disordering.

Situation is changed when we proceed to noncubic superstructures. With an increase in Δr, the $L1_2$ superstructure becomes more stable with respect to the intraphase disordering (i.e., to the decrease of the degree of order in the region where the superstructure exists), whereas the low-symmetry superstructures exhibit the opposite behavior, in spite of an obvious dependence of Δη on Δr. The character of the dependence changes owing to a greater role of the band structure factor.

The tendency toward the appearance of static displacements in an alloy as the disordering proceeds, expressed by the derivative $\partial(\sqrt{u^2_{\text{CT}}})/\partial\eta$, also depends on the size factor (Fig. 70). In the $L1_2$ superstructure, with an increase of the size factor, static displacements increase, on disordering, at a higher rate. They play the substantial part in the energy balance of an alloy. Static displacements are connected, partly, with the phenomenon of superstructure contraction. They give a marked contribution to the entropy of the disordered state.

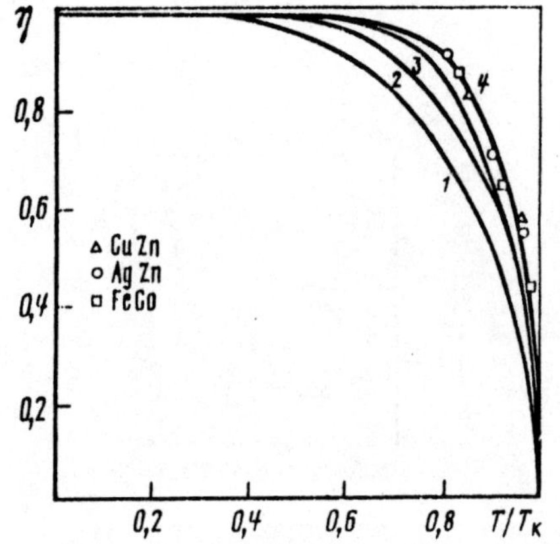

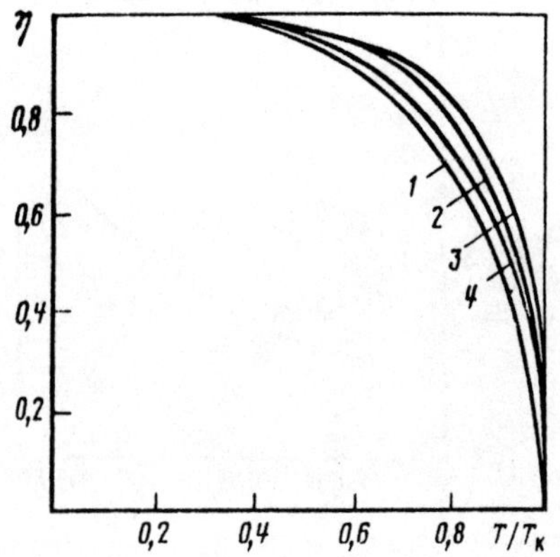

Fig. 71. Comparison of the theoretical $\eta(T/T_C)$ curves (1 - 3) with the experimental one (4). The B2 superstructure (GFM - Green function method, HTEM - high-temperature expansion method). (1) - random field method, (2) - BBGKI, (3) - GFM, (4) - HTEM.

Fig. 72. Comparison of the $\eta(T/T_C)$ curves calculated by traditional methods with different level of correlations accounted for. The B2 superstructure.

Thus, the results discussed in this section confirm once again the significant role of the volume and size effects in atomic disordering. Not only the type of the forming superstructure and its stability depend on these effects, but also fine details of the order-disorder phase transition and the structure of a solid solution in the state close to the transformation. This emphasizes a need for developing the quantum and statistical theories which would take these effects into consideration. At the same time, the role of the volume effects (in particular, the size factor) is often far from being dominant. With the reduction in the superstructure symmetry, the band structure contributions begin to play a greater part. The content of this section shows the existence of a close relation between the configurational contribution into the free energy of an alloy and the vibrational and electronic contributions. Consider now in detail the trends revealed in the transformation of particular superstructures, basically from the theoretical standpoint.

Disordering in the B2 superstructure. The equations of equilibrium are given in Section 4.3. Their numerical solutions have been obtained using the computer codes. Various entities have been evaluated with the use of different models. We shall consider first the behavior of $\eta(T/T_c)$ curves, starting from the results achieved with the aid of the Ising model in the nearest-neighbor approximation. In all these cases, the phase transition B2 $\rightarrow$ A2 is of the second order. For the B2 superstructure, the calculations were carried out by the Green function method, in the random field approximation, and by the methods based on the use of the chain of integral equations and high-temperature expansions.

The results obtained by these methods in the best approximations are given in Fig. 71 together with the experimental η values for the alloys CuZn, AgZn, and FeCo which, in accordance with the law of "corresponding states", lie on a single curve. This curve coincides virtually with the theoretical one calculated by the random field method. The four curves in Fig. 71 can be averaged according to the order of the correlations allowed for: the method of random field, the method of integral equations, the Green function method, the high-temperature expansions method.

In the first case, the pair correlations are not considered completety, the second method takes a full account of the pair correlations, in the third method the three-particle correlations are partially present, and the last one deals, in part, with the twelve-particle correlations. As the order of correlations increases, the $\eta(T/T_c)$ function changes in the following way: at low temperatures the $\eta(T/T_c)$ curve is rather flat, while at higher temperatures has a greater slope. Such change of the curve implies that the B2 $\rightarrow$ A2 phase transformation approaches the 1st-order one.

The traditional theories usually take into account one-, two-, or at most four-particle correlations. The corresponding curves are plotted in Fig. 72. The comparison of Figs. 71 and 72 confirms the above conclusion that the order of correlation accounted for by the theory affects the appearance of the $\eta(T/T_c)$ curve. It should be noted that various modifications of the methods developed here, i.e., different decouplings of the Green

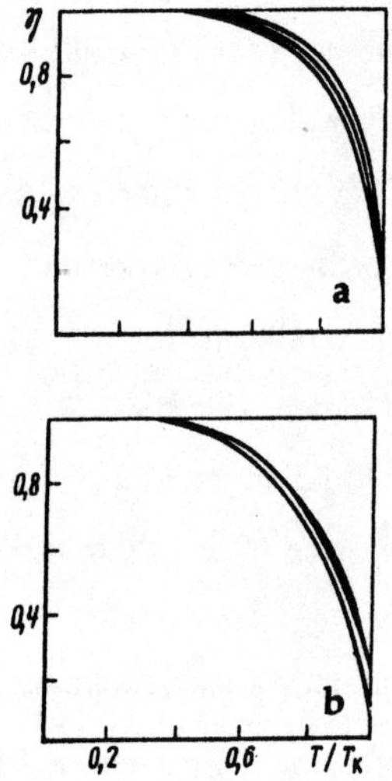

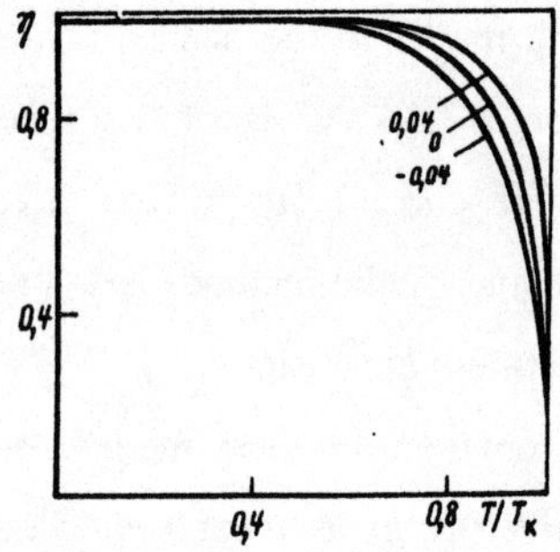

Fig.73 Theoretical $\eta(T/T_c)$ curves obtained by the Green function method (a) and by the random field method (b). The higher the degree of correlation accounted for, the steeper is the curve.

Fig. 74. Effect of the second shell contribution V_2/V_1 on the $\eta(T/T_c)$ curve. The ratio V_2/V_1 is given on the curves.

functions of the same order, or different clusters (corresponding to the same shape of the distribution functions) yield slightly different $\eta(T/T_c)$ dependences (Fig. 73). In other words, the shape of the curve is determined, in the first place, by the level of correlations accounted for and then by other details of the approximations used in the theories, such as the cluster size, the way of decoupling of the Green functions and so on. This also maintains the idea about the decisive influence of the character of the correlations taken into account in the theory on the $\eta(T/T_c)$ dependence.

The critical temperature of the order-disorder transition, T_c (or kT_c/V_1), is also a measure of the perfectness of the theory. The results obtained by different methods can be compared to the asymptotic estimate. The refinement of the theory leads, as a rule, to the decrease of kT_c/V_1. This is relevant to the increase in the number of terms in a power series used in the high-temperature expansions method, to the increase of the cluster size in the random fields method, and to the expansion of the statistical ensemble. When comparing different theories, it can be seen that the increase of the order of the correlations considered by the theory entails the decrease in the value of kT_c/V_1. The asymptotic estimate falls into the middle of the obtained kT_c/V_1 values.

A straightforward expansion of the model enables the long-range interactions and the nonlinear effects to be involved in it. The involvement of the interaction on the second coordination shell greatly extends the potentialities of the theory. If V_1 and V_2 have the same signs, the curve $\eta(T/T_c)$ shows a steeper slope at higher values of V_2/V_1 provided the way in which the correlations are taken into account is retained. If the signs of V_1 and V_2 are opposite, the rise in V_2/V_1 leads to the reverse effect (Fig.74).

The slope of the $\eta(T/T_c)$ curve near T_c is closely related to the behavior of the configurational contribution to the heat capacity. The comparison of the literary data with the experimental curve and calculations carried out in the present work is shown in Fig. 75. It is evident that the comparison of the heat capacities calculated by different methods is a stronger indicator of the quality of the theory than the comparison of the η (T/T_c) curves which are less sensitive to the level of the approximation involved. Only the theory taking the correlations into account completely is able to reach good agreement with experiment. As a matter of fact, the behavior of the real heat capacity curve is poorly reproduced by the GBW theory and much better by the method of the high-temperature expansions. The curves predicted by other theories lie much lower than those in Fig. 75. It should be reminded that the method of high-temperature expansions, along with the involvement of higher correlations, operates with the ensemble that is wider than the canonical one.

Another factor affecting the height of the C_V peak and the magnitude of the jump in the heat capacity is the involvement of the second coordination shell. With the increase in V_2/V_1 (on condition that V_2 and V_1 have the same signs), the jump in the heat capacity at T_c increases and approaches the experimental value.

As the contribution of the second shell increases, the ratio kT_c/V_1 declines. The higher the order of the correlations considered by the theory, the stronger is the influence of the second coordination shell. Finally, at a certain value of V_2/V_1, the phase transition of the 1st order changes for the 2nd-order one.

Allowing for the nonlinear effects also influences the slope of the curve and the proximity of the 2nd-order B2 → A2 phase transition to that of the 1st order. Nonlinear effects have been introduced by including in a straightforward way the dependence of the ordering energy on the order parameter:

$$V = V_0(1 + \kappa\eta^2), \tag{4.4.7}$$

Here V_0 represents the η-independent part of the ordering energy, k is the parameter describing the nonlinearity effect. A positive nonlinearity increases the slope of the curve (Fig. 76). A strong nonlinearity produces the change of the 2nd-order phase transition for the 1st-order one.

It was shown in the foregoing sections that a better account of the correlation effects makes the B2 → A2 transition closer to the 1st-order phase transition. Therefore, the theory neglecting the correlation effects needs in a considerable nonlinearity for changing the order of the phase transition, while, on a serious account of the correlation, even a minor nonlinearity leads to the 1st-order phase transition. The same can be said about the role of the long-range interactions in changing the order of the phase transition. In this case, the theory which takes better the correlation effects into account requires much smaller contribution from the second coordination shell to the ordering energy to go over to the 1st-order transition. Hence, the calculations and analysis made above showed that in the B2 superstructure the allowance for higher correlations, the positive contributions of the long-range interactions and nonlinear effects changes the shape of the $\eta(T/T_c)$ curve bringing the decribed dependences closer to the experimental data and the 2nd-order transformation to the 1st-order one.

Disordering in the $L1_2$ superstructure. The formation of the $L1_2$ superstructure represents a classical example of the order-disorder phase transition of the 1st-order. It has been most extensively studied experimentally in the present work (see Section 4.2).

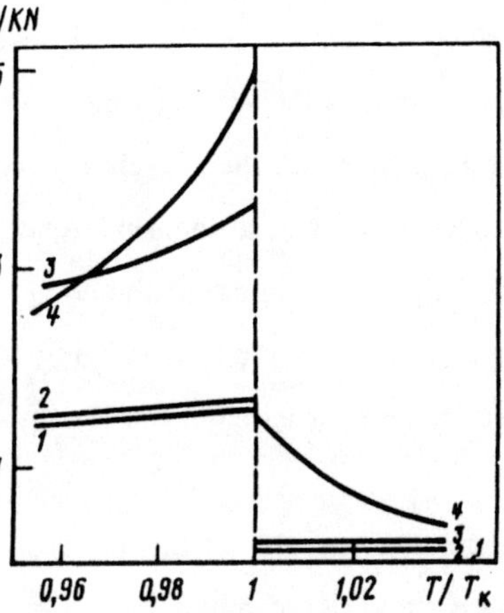

Fig. 75

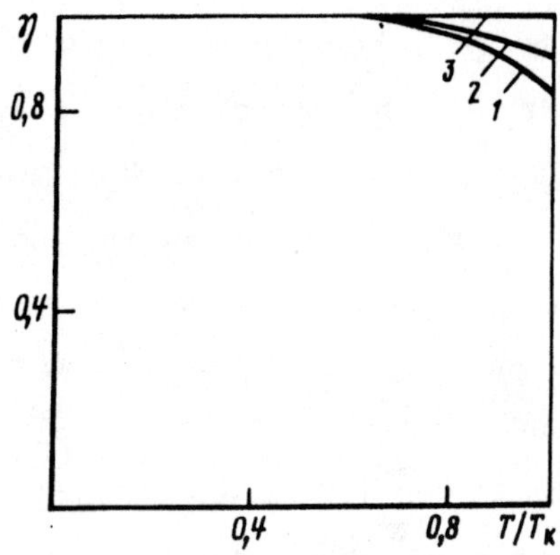

Fig. 77

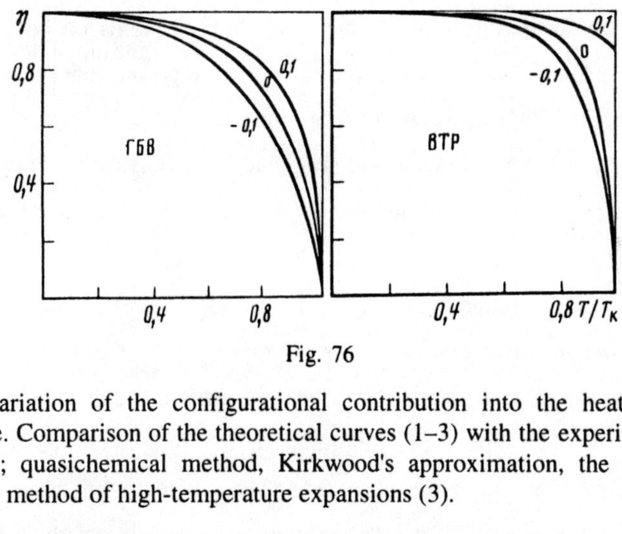

Fig. 76

Fig. 75. Variation of the configurational contribution into the heat capacity with temperature. Comparison of the theoretical curves (1–3) with the experiment (4): GBW method (1); quasichemical method, Kirkwood's approximation, the Green function method (2); method of high-temperature expansions (3).

Fig. 76. The influence of the nonlinear η-dependence of the ordering energy on the shape of the $\eta(T/T_c)$ curve. The values of the coefficient k in (4.4.7) are given next to the curves.

Fig. 77. The theoretical $\eta(T/T_c)$ curves with different level of correlations accounted for. The $L1_2$ superstructure. Methods: GFM (1); entropy expansion (2); high-temperature expansions (3).

The theoretical formulas required to calculate the long-range order parameter, correlation, free energy, entropy, transformation temperatures are presented in Section 4.3. The layout of the present subsection is similar to that of the section where the B2 superstructure was described. The difference is due to the fact that for the 1st-order transitions, the problem of the critical states is not so crucial but, at the same time, the determination of the transformation temperature becomes very important and nontrivial. For this reason, more attention is given here to the calculation of the free energies. Most of the results in the present section were obtained in the approximation of the "point" transition.

The η versus T/T_c curves calculated by the methods developed in this work are shown in Fig. 77. When using the Green function method, the calculation was carried out taking the pair correlation into account (and indirectly, the three-particle correlations). A rigorous treatment of the three-particle correlations has been performed by the method based on the expansion of the entropy into a power series in correlations. And finally, using the high-temperature expansions technique, as high as the eight-particle correlations have been accounted for, though incompletely. The higher the order of the correlations involved in the theory, the more distant from the temperature axis are the $\eta(T/T_c)$ curves exhibiting nearly flat and linear behavior (Fig. 77). It was shown earlier that such variation in the form of the $\eta(T/T_c)$ curves can be deduced from the general thermodynamical considerations. In other words, the account of higher correlations makes the 1st-order phase transition more distinct and distant from the 2nd-order transition. At the same time, the jump in the long-range order parameter increases, whereas, the intraphase disordering just before the phase transition diminishes, i.e., in the alloy with a high correlation the stability with respect to the internal disordering is lowered.

The comparison of the calculated and experimental $\eta(T/T_c)$ curves shows that the description of the order-disorder $L1_2 \rightarrow A1$ transformation in the one-shell Ising model is hardly attainable and the long-range interaction and nonlinear effects are important in different alloys. The greater the drop in η at higher temperatures in the particular versions of the theory, the faster the correlations grow up in the alloy. The phase transition is accompanied by the jump in the correlation parameters.

The $\eta(T/T_c)$ curves calculated within the framework of the traditional theories are shown in Fig. 78. Analysis of these curves and comparison with those in Fig. 77 confirm the conclusion that the shape of the $\eta(T/T_c)$ curve and the character of the $L1_2 \rightarrow A1$ transformation depend on the extent to which the correlations are taken into consideration. Another trend can also be pointed out: the increase in the jump in the η value at T_c is accompanied by the decrease in the T_{sl} value.

In the case of the 1st-order phase transition, alloys can approach the point of stability loss in different ways: the loss of stability (the internal one) may happen at different values of η. This problem has been discussed above from the experimental point of view. The results of the corresponding theoretical calculations are given in Fig. 79. This time, the entire $\eta(T/T_c)$ curve was calculated, including the region of the unstable state. The increase in the order of highest correlations considered by the theory decreases the internal stability: the degree of order in the point of instability grows up. The curves in Fig. 79 confirm the previous conclusion that the closeness of the behavior of the alloy to the linear type is determined by the level of the highest correlations in the theory. Fig. 79 demonstrates clearly that the 1st-order $L1_2 \rightarrow A1$ phase transition approaches the 2nd-order type with decreasing correlation effects. In each version of the theory, the 1st-order phase transition temperature lies close to the temperature of the stability loss ($T_c/T_{sl} \geq 0.95$). The highest T_c/T_{sl} ratio (0.99) is predicted by the high-temperature expansion method.

The regions where the $L1_2$ phase is stable, metastable or unstable are clearly seen on the free energy versus temperature curves, obtained by the method in which the entropy is expanded into a series in correlations. The appearance of the $L1_2 \rightarrow A1$ phase transition in this method is provided mainly by the different behavior of the internal energy versus temperature curves. The entropy affects only the position of the phase transition point in the temperature scale [317].

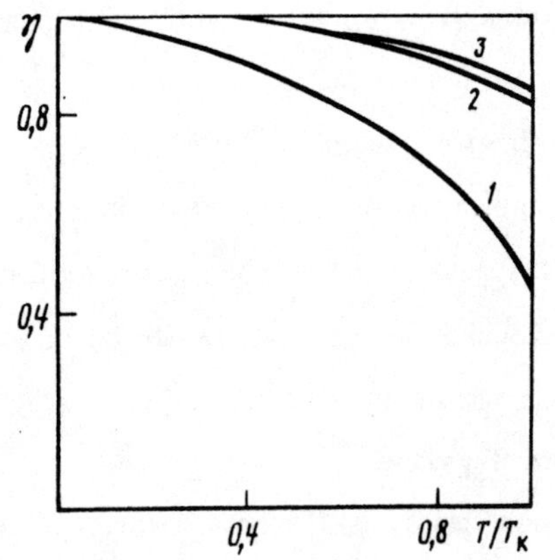

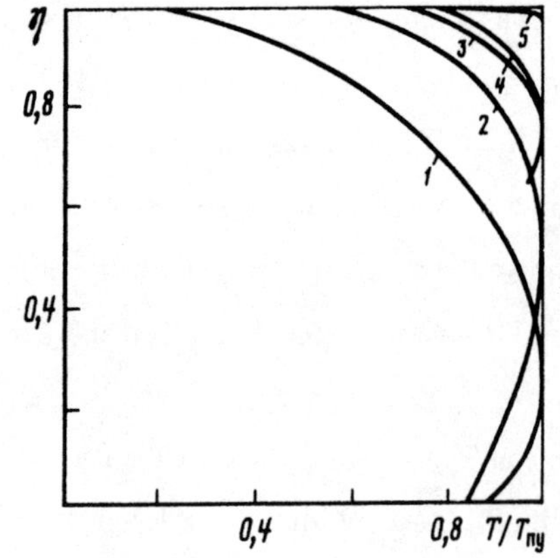

Fig. 78. Comparison of the theoretical $\eta(T/T_c)$ curves calculated with different level of correlations by traditional methods. The $L1_2$ superstructure. Methods: GBW (1); Cowley (2); cluster variation (3).

Fig. 79. Theoretical $\eta(T/T_{sl})$ curves for the $L1_2$ superstructure extended to the region of metastable and unstable configurations. Methods: GBW (1); BBGKI (2); series in correlations (3); GFM (4); cluster variation (5).

Also of interest is the structure of the entropy in the ordered state. The contributions of the long-range order and the three-particle correlations are positive and increase with temperature, while the pair correlation contribution is negative and decreases with increasing temperature. Hence, in the approximation taking into account the three-particle correlations, the order-disorder transition to the $L1_2$ phase is largely due to the stochastic contribution of the one- and three-particle correlations. The contribution of the pair correlation impedes the formation of a superstructure. This explains why the method involving only the pair correlation has failed to describe the phase transition to the $L1_2$ superstructure in a reasonable temperature range. The calculation shows that the phase in the state of long-range order is unstable at high temperatures in this approximation, and metastable at low temperatures. Only near T = 0 the contribution from the internal energy provides the possibility of the A1 $\rightarrow$ $L1_2$ transition. The partial addition of the contribution from the three-particle correlations (the Green function method) leads to a reasonable description of this transition.

The effect of "switching on" the interaction in the second coordination shell on the shape of the $\eta(T/T_c)$ curve for the $L1_2$ superstructure is similar in many respects to that described for the B2 superstructure. If V_2 and V_1 have the opposite signs, the $\eta(T/T_c)$ curves move downward to the horizontal axis with increasing V_2, and vice versa. Note that, at the opposite signs of V_1 and V_2, an increase in the ratio V_2/V_1 decreases the jump in the η value and shifts the transformation toward the 2nd-order one, and T_c to T_{sl}.

Nonlinear effects were introduced into the ordering energy in the same way as for the B2 superstructure (see formula (4.4.7)). A positive nonlinearity makes the curve not so steep and increases the jump in the order parameter at T_c. A negative nonlinearity causes an adverse effect and shifts T_c to T_{sl} and the $L1_2 \rightarrow A1$ phase transition to the 2nd-order one. In other words, only the negative nonlinearity in the ordering energy gives rise to the nonlinear behavior of the $\eta(T/T_c)$ curve. Thus, the theoretical investigation of the role of correlation effects, long-range interaction and nonlinearity shows that these factors exert a considerable effect both on the temperature dependence of the order parameters and the behavior of a solid solution during the phase transformation $L1_2 \rightarrow A1$. The enhancement of the correlation effects in an alloy, the positive nonlinearity in the ordering energy, and the long-range interaction (the positive contribution of the second coordination shell) reduces the stability of the $L1_2$ phase to disordering, shifting the transformation toward the greater η values (the external loss in stability). The behavior of an alloy becomes more linear. The negative nonlinearity and long-range interaction in the ordering energy as well as the neglect of the correlation effects make the phase transition more similar to the 2nd-order one or, at least, to an anomalous 1st-order transition and enhances the nonlinear dependence of η on T/T_c.

Disordering in low-symmetry superstructures. Among these superstructures, we chose those having not been studied earlier: the superstructure $D1_a$ and that of the Pt_2Mo type based on an f.c.c. lattice, and the $C11_b$ superstructure based on a b.c.c. lattice. Only for the superstructure of the Pt_2Mo type, the calculation carried out in the GBW approximation is now available [321]. To the aforementioned structures we added that of the $L1_0$ type because it was relatively well studied previously [189, 322, 323]. Significant correlation effects were found to exist in the low-symmetry superstructures [320].

4.5 Volume effects in the theory of ordering

The consistent theory of the volume effects in alloys should be based on the methods of quantum theory. The foundations for such theory and all the formulas to be used in the pseudopotential formalism are developed and reported in Chapter 5. Everything that is needed for the calculation of the pair and many-particle interatomic potentials and the nonfactorized contributions into the internal energy will be also given there. In the present section, however, we present the calculations carried out with the use of the model theory to demonstrate explicitly the role of the volume effects.

The point is that the model theory is, in principle, diagonal. In the more consistent quantum theory, the volume effects will interfere with the purely quantum effects, such as the variations of the electronic state parameter, of the occupation of different electronic configurations and of the band structure itself. The nondiagonal character of quantum theory, which reduces its practical merits, and the need in a large amount of calculations to redetermine all the entities involved in the pseudopotential formalism (form-factors, orthogonalization holes, effective valencies, etc.) for each degree of order and the volume of the alloy motivated us to employ the model approach in the present section. It should be noted that the application of the interatomic potentials used here is justified, to some extent, by the quantum theory.

Allowance for the volume effects in the theory of long-range and short-range atomic order in metallic solid solutions has been taken by several authors [326]. Among them, we want to draw a special attention to the pioneering investigations of Pines [304], Aptekar' and Finkel'stein [306] and Rudman [327]. Most intensively has been working in this field the group of scientists under the leadership of Academician Smirnov. A series of tasks in this line is also solved in this section of the monograph. The novelty of the material is determined by several factors: the more realistic approach to some problems in the theory, a wider range of the interatomic potentials used and, particularly, a wider set of physical quantities involved which have been given insufficient attention in previous works. These quantities define the structure of a solid solution and its thermodynamical behavior.

Interatomic potentials and atomic radii. We used here several interatomic potentials: the one-shell harmonic potential, the model of "elastic spheres" and the three-sphere potentials of Morse and Rydberg. These are considered in the stated succession.

The pair harmonic interatomic interaction. The method was put into practice by Aptekar' and Finkel'stein and used in several works. Smirnov and his succesors have most intensively applied it to the problem of ordering alloys. The pair interaction energy V_{xy} can be expanded into a series in r-r_{xy}, i.e., in the deviations of the mean interatomic spacing r from the distance r_{xy} corresponding to the minimum of the interatomic potential. Retaining only the first and the second, quadratic in (r-r_{xy}), terms in such expansion, we obtain the harmonic interatomic potential in the form

$$V_{xy} = V^0_{xy} + \gamma_{xy}(r - r_{xy})^2, \tag{4.5.1}$$

which is most convenient if we work with the only (first) coordination shell. The calculations described in this section have been carried out in this approximation. The harmonic potential does not take into account the thermal expansion effect but describes the phenomenon of the configurational expansion because the equilibrium interatomic spacings and the potential parameters depend upon the kinds of atoms x and y participating in the interatomic bonding x–y. The first term V^0_{xy} in (4.5.1) describes the constant contribution into the interaction energy; the "rigidity" of the potential is determined by the coefficient γ_{xy}=D r_{xy} K_{xy}, where D is the structure-dependent constant (D=$\sqrt{3}/2$ for a b.c.c. lattice and D=$3\sqrt{2}/8$ for an f.c.c. lattice), and K_{xy} is the bulk compressibility modulus for the perfectly ordered alloy. If the atoms at both ends of the r_{xy} vector are identical, we use the properties of pure components to evaluate the potential parameters. In fact, we have changed here the way in which γ_{xy} is determined for the alloys of nonequilibrium composition and took into account the fact that in some alloys in the state of perfect order the sites in the first coordination shell of a given atom may be partly occupied by atoms of the same kind.

Method of "elastic spheres" has been suggested by Pines. An impurity atom is replaced by an elastically isotropic sphere inserted into a spherical cavity within the elastic continuum representing the region where a uniform solid solution exists. Both spheres should coincide at an intermediate value of the radius. The work extended for deformation of the impurity atom W^x_{AT}, and of the surrounding medium W^x_{cp}, are determined in accord with the classical theory of elasticity:

$$W^x_{\text{AT}} = \sigma\pi K_x R^0_x (R_x - R^0_x)^2,$$
$$W^x_{\text{cp}} = 8\pi G \rho_x (R_x - \rho_x)^2. \tag{4.5.2}$$

Here, Kx and R^0_x are, respectively, the bulk modulus and the atomic radius of the pure component x, ρ is the cavity radius in a solid solution, G is the shear modulus for a solid solution, and Rx is the final radius of the impurity atom x in a solid solution. The value of Rx is determined by minimizing the work $(W^x_{\text{AT}} + W^x_{\text{cp}})$. It should be noted that the substitution of the impurity atom for the spherical inclusion is valid only when this atom occupies a high-symmetry position in the lattice (the site in a b.c.c. or f.c.c. lattice, the octahedral interstice in an f.c.c. lattice). In the present work we have modified the Rudman method to apply it to the alloys with long-range order. In this case, the mean radius of the cavity is equal to $R = \sum_{x=A,B} cR_x$, where C_x is the atomic fraction of the component x. The mean distance from a given atom to its second coordination shell measured in the direction of close packing will be 3R. The cavity radius ρ_x determined before the impurity atom is placed into it depends on the real distribution of the surrounding atoms

$$\rho_x = 3R - 2[P_{xy} R_y + P_{xx} R_x]. \tag{4.5.3}$$

Here, P_{xy} and P_{xx} are the pair probabilities. In using the method of elastic spheres, it is exceptionally important to take into account the short-range order in an alloy. In the selected model, the major part of the elastic distortion energy is due to the displacements of atoms in the first coordination shell; in the case of imperfect long-range order this energy depends strongly on correlation. It follows from this fact that the formation of short-range order may result from the tendency of a system to reach a minimum of the elastic energy. It may happen in such situation that the structure of short-range order will not comply with that of long-range order, as in the case of the Ni_4Mo alloy.

The method of elastic spheres and the method using the pair harmonic potential have much in common. The energy of elastic distortions in a crystal lattice per one atomic bond, as determined by the second method, is equal to $\omega = Dk_x r_x (r - r_x)^2$, where D = 0.53 for an f.c.c. lattice and D = 0.86 for a b.c.c. one. The first method gives

$$\omega = \frac{6\pi K_x R_x^0 (R_x - R_x^0)}{Z} + \frac{8\pi G \rho_x (R_x - \rho_x)^2}{Z} =$$

$$= \frac{\pi}{4z} \; [3K_x r_x (r - r_x)^2 + 4G(2\rho_x)(r - 2\rho_x)^2].$$

Putting $r_x(r-r_x)2 \approx 2\rho_x(r-2\rho_x)2$ and $G = 3/8\ K_x$ into the last expression, we obtain

$$\omega = 9\pi/8Z; \quad K_x r_x (r - r_x)^2 \approx D_1 K_x r_x (r - r_x)^2.$$

Here D_1 = 0.3 for an f.c.c. lattice, and D_1 = 0.44 for a b.c.c. one. Hence, the specific energies of elastic distortions calculated by the two methods are expressed by the same functional relationships and have the same order of magnitude. Thus, the method of elastic spheres can be classsified in accordance with the type of the pair interaction and must be referred to the harmonic potential approximation.

Anharmonic potentials. The calculations have also been carried out with the anharmonic potentials of Born-Mayer, Rydberg and Morse. The latter, like Johnson's potential, has an oscillatory behavior. Account of different number of coordination shells in the interaction has been taken in these calculations. We present here the results obtained with the Morse potential, allowing for the interaction on the three nearest coordination spheres. As Leibfrid showed first [328], this is only a different way to take the long-range interaction into account.

The pair anharmonic Morse potential is expressed by the formula:

$$\varphi(r_{ij}) = D_0 \left[e^{-2\alpha(r_{ij}-r_0)} - 2e^{-\alpha(r_{ij}-r_0)}\right], \tag{4.5.4}$$

where D_0 is the parameter that is numerically equal to the energy of the bond rupture (i.e., the dissociation energy), r_0 is the equilibrium distance between the two atoms, the indices i and j denote the kinds of atoms, and α is the parameter defining the "rigidity" of the potential. Thus, the energy per one atom in a crystal is expressed by

$$E = \frac{1}{2N} \Sigma\Sigma\, \varphi(r_{ij}), \tag{4.5.5}$$

where the summation is over all Z_l atoms in the l-th coordination shell.

As in the case of the model potentials, the problem arises here of finding the potential parameters. Let us consider it, for example, for the Morse potential, which depends on the three parameters: D_0, α and r_0. Since the alloy consists of two components, the nine parameters should be determined as a whole. In principle, we can make up the system of the nine equations to find the necessary parameters for each concentration, temperature and the degree of order. Of course, D_0, α and r_0 are the functions of temperature, concentration and the degree of order. In reality, there will be a shortage of experimental data for any alloy, including the well-studied alloy Cu_3Au, to set up, under fixed conditions, the system of nine equations. In our work we have tested several versions of such system [329], but could not find the experimental data required for each of them. As a consequence, the potentials describing the interaction between atoms of the same kind in the Cu_3Au alloy were derived from the experimental data available for the pure components using the following system of equations:

$$E(a_0) = u_0(a_0); \quad \left(\frac{\partial E(a)}{\partial a}\right)_{a = a_0} = 0; \quad \left(\frac{\partial^2 E(a)}{\partial a^2}\right)_{a = a_0} = \frac{18Na_0}{K_{00}}. \tag{4.5.6}$$

Here, a_0 is the equilibrium lattice constant, K_{00} is the coefficient of bulk compressibility, $U_0(a_0)$ is the sublimation energy. All the quantities are taken for a b.c.c. lattice and at pressure P = 0. After the simple transformations, the system takes the following form:

$$\begin{aligned}
&Z\beta^2 \sum_j Z_j \exp(-2\alpha a\mu_j) - 2Z\beta \sum_j Z_j \exp(-\alpha a\mu_j) = u_0(a_0),\\
&Z\beta^2 \alpha a \sum_j Z_j \exp(-2\alpha a\mu_j) - Z\beta\alpha a \sum_j Z_j \mu_j \exp(-\alpha a\mu_j) = 0,\\
&2Z\beta^2 \alpha^2 a^2 \sum_j Z_j \exp(-2\alpha a\mu_j) - Z\beta\alpha^2 a^2 \sum_j \mu_j^2 Z_j \exp(-\alpha a\mu_j) =\\
&= 9NV_0/2K_{00}.
\end{aligned} \tag{4.5.7}$$

Here

$$Z = \frac{1}{2} ND; \quad \beta = e^{\alpha r_0}; \quad r_j = a_0\mu_j; \quad \mu_j = (h_j^2 + k_j^2 + l_j^2)^{\frac{1}{2}}.$$

This system is easy to solve with respect to β and Z, and the problem of the determination of the potential parameters reduces virtually to the subsequent solution of each of the equations. In contrast to the preceding calculations, we did not simulate here a limited block of a crystal but, instead, carried out the computations for an indefinitely large crystal involving one gram-atom of the substance. Such procedure simplifies the calculations and eliminates the boundary effects. The available experimental data which are needed for inserting them into the equation are contradictory. We chose, naturally, the most reliable of them. For Cu, for example, two values of the sublimation energy, 92 and 83.2 kkal/mole, were used.

To find the parameters describing the interaction between unlike atoms, the following system was used:

$$2D_{AB}\nu_{AB}^{(i)}(\nu_{AB}^{(i)}-2)=D_{AA}\nu_{AA}^{(i)}(\nu_{AA}^{(i)}-2)+ \\ +D_{BB}\nu_{BB}^{(i)}(\nu_{BB}^{(i)}-2)+V_i, \quad i=1,2,3, \tag{4.5.8}$$

Here $V_{\mu\nu}^{(i)}=\beta_{\mu\nu}\ e^{-\alpha_{\mu\nu} a_{\mu_i}}$, μ, ν=A,B; a_0 is the lattice constant for the alloy at 390°C. The ordering energies V_1, V_2, and V_3 were obtained by processing the experimental data on diffuse X-ray scattering measured on the Cu_3Au alloy at 390°C [330]. The corresponding potentials have been given in [341]. The lattice stability for the potentials was checked using the Born criterion [305].

Harmonic approximation. B2 superstructure. The calculation has been carried out for the CuZn alloy [332]. Since the size effects are sensitive not only to long-range order but also to correlation, we took into account in the expression for the free energy the pair correlation in a manner analogous to the quasichemical theory. In this approximation, the free energy is expressed by

$$\begin{aligned} F\ \frac{Nz}{2}\Bigg[&\left(c+\frac{1}{2}\eta-P_{\alpha\beta}^{AB}\right)V_{AA}+\left(1-c+\frac{1}{2}\eta-P_{\alpha\beta}^{AB}\right)V_{BB}+ \\ &+2\left(P_{\alpha\beta}^{AB}-\frac{1}{2}\eta\right)V_{AB}\Bigg]-\frac{kTN}{2}\Bigg\{\left(Z-1\right)\Bigg[\left(c+\frac{1}{2}\eta\right)\ln\left(c+\frac{1}{2}\eta\right)+ \\ &+\left(c-\frac{1}{2}\eta\right)\ln\left(c-\frac{1}{2}\eta\right)+\left(1-c-\frac{1}{2}\eta\right)\ln\left(1-c-\frac{1}{2}\eta\right)+ \\ &+\left(1-c+\frac{1}{2}\eta\right)\ln\left(1-c+\frac{1}{2}\eta\right)\Bigg]-Z\Bigg[P_{\alpha\beta}^{AB}\ln P_{\alpha\beta}^{AB}+ \\ &+\left(c+\frac{1}{2}\eta-P_{\alpha\beta}^{AB}\right)\ln\left(c+\frac{1}{2}\eta-P_{\alpha\beta}^{AB}\right)+\left(P_{\alpha\beta}^{AB}-\eta\right)\ln\left(P_{\alpha\beta}^{AB}-\eta\right)+ \\ &+\left(1-c+\frac{1}{2}\eta-P_{\alpha\beta}^{AB}\right).\ \ln\left(1-c+\frac{1}{2}\eta-P_{\alpha\beta}^{AB}\right)\Bigg]\Bigg\}. \end{aligned} \tag{4.59}$$

Thus, the calculations in this section are carried out in the statistical ensemble which is wider than that used by Smirnov. In the ensemble classification, our ensemble takes a higher place than Smirnov's ensemble; its level corresponds to the quasichemical approximation. The condition for equilibrium is determined, as usual, by minimizing the free energy in η, $P_{\alpha\beta}^{AB}$ and r. As a result, we come to the following system:

$$\frac{\left(\frac{1}{2}\eta + c\right)\left(1 - c + \frac{1}{2}\eta\right)}{\left(1 - c - \frac{1}{2}\eta\right)\left(c - \frac{1}{2}\eta\right)} = \left[\frac{P^{AB}_{\alpha\beta}}{P^{AB}_{\alpha\beta} - \eta}\right]^{8/7},$$

$$\frac{V}{kT} = \ln \frac{(P^{AB}_{\alpha\beta} - \eta)P^{AB}_{\alpha\beta}}{\left(c + \frac{1}{2}\eta - P^{AB}_{\alpha\beta}\right)\left(1 - c + \frac{1}{2}\eta - P^{AB}_{\alpha\beta}\right)},$$

$$r = \frac{c\gamma_A r_A + (1-c)\gamma_B r_B + \left(P^{AB}_{\alpha\beta} - \frac{1}{2}\eta\right)\gamma_0}{c\gamma_A + (1-c)\gamma_B + \left(P^{AB}_{\alpha\beta} - \frac{1}{2}\eta\right)\gamma}. \tag{4.5.10}$$

where $\gamma_0 = 2\gamma_{AB} r_{AB} - \gamma_A r_A - \gamma_B r_B$; $\gamma = 2\gamma_{AB} - \gamma_A - \gamma_B$. The ordering energy in the final form is

$$V = V_0 - [2\gamma_{AB}(r - r_{AB})^2 - \gamma_A(r - r_A)^2 - \gamma_B(r - r_B)^2], \tag{4.5.11}$$

where the constant portion of the ordering energy is equal to $V_0 = 2V^0_{AB} - V^0_{AA} - V^0_{BB}$. System (4.5.10) can be converted to the system used in the quasichemical method if one disregards the dependence of the interatomic potential on the distance. Another restriction of the ensemble which excludes correlation, converts (4.5.10) into the expression obtained by Smirnov [188, 331].

The numerical solution of the above system enables one to find a wider set of quantities than that obtained in Section 4.4. In the first place, it will be the dependence of η on T. It is clear from Fig. 80 that the account of the lattice dilatation caused by the ordering improves the agreement between the theoretical and experimental curves. This result might indicate the existence of nonlinear effects in the CuZn alloy. It is possible, however, that the account of correlation in a quasichemical approximation is insufficient (see Section 4.4). The correlation effects play an important role in the B2 superstructure. In addition, this kind of calculations seems to underestimate the contribution of the distant coordination shells. Most likely, the three aforementioned effects take place here: nonlinearity, correlation, and the long-range interaction.

Second, the variation of the interatomic distance with temperature can also be found by the numerical solution of equations (4.5.10). The comparison of the theoretical predictions with the experimental observations was made for the CuZn alloy. The effect of thermal expansion has been excluded from the database because the theory does not predict such effect. The results show that the theory is capable of reflecting correctly the main trend in the change of the interatomic interaction, but at the same time the one-shell harmonic approximation may hardly pretend to a higher accuracy.

Third, the configurational heat capacity C_p can be evaluated from (4.5.10). Referred to one atom, it is written as

$$\frac{C_p}{Nk} = \frac{1}{Nk}\frac{\partial E}{\partial T} = \frac{4}{k}\left\{\left(\frac{1}{2}\frac{\partial \eta}{\partial T} - \frac{\partial P_{\alpha\beta}^{AB}}{\partial T}\right)V + \left(\frac{1}{2}\eta - P_{\alpha\beta}^{AB}\right)\frac{\partial V}{\partial T} + \right.$$

$$\left. + 4[r(\gamma_A + \gamma_B) - \gamma_A r_A - \gamma_B r_B]\frac{\partial r}{\partial T}\right\}. \quad (4.5.12)$$

This expression takes into account both the indirect (through the V value in the first two terms) and the direct (the third term) contributions of the size effect into the configurational heat capacity. Figure 81 displays the importance of these contributions, especially near T_c. The inclusion of the elastic effects seriously improves the agreement with the experiment both in the general behavior of the heat capacity curve and in the jump in this quantity at T_c. The type of transition remains unchanged. Nevertheless, the reader should be aware of the substantial role of higher correlations and the long-range order interaction near the temperature T_c.

Fourth, the ordering energy V in the considered model turns out to be a function of η. Its constant part is to be evaluated at T_c where $\eta = 0$. Theoretical processing of the experimental data of Chipman and Worren [273] reveals the occurrence of strong nonlinear effects in this alloy. The account of the volume effects also leads to nonlinear effects but the latter are less pronounced. In both cases, the ordering energy increases with an increase in the degree of order [Fig. 82).

Harmonic approximation. $L1_2$ superstructure [333]. The principal effects introduced by the harmonic approximation have already been illustrated by the calculation of the B2 structure. So, we chose for the Cu_3Au alloy a noncorrelation

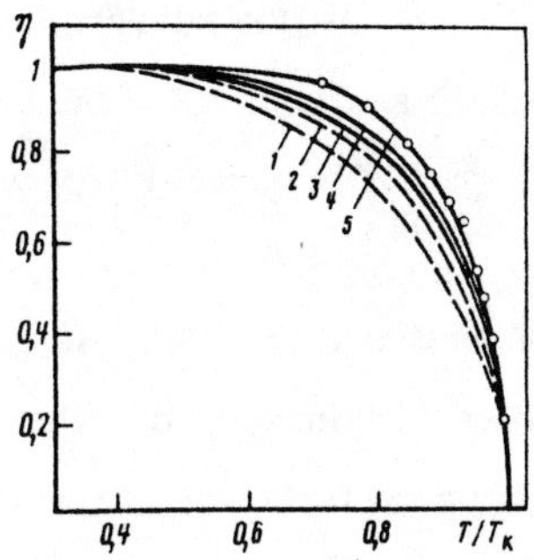

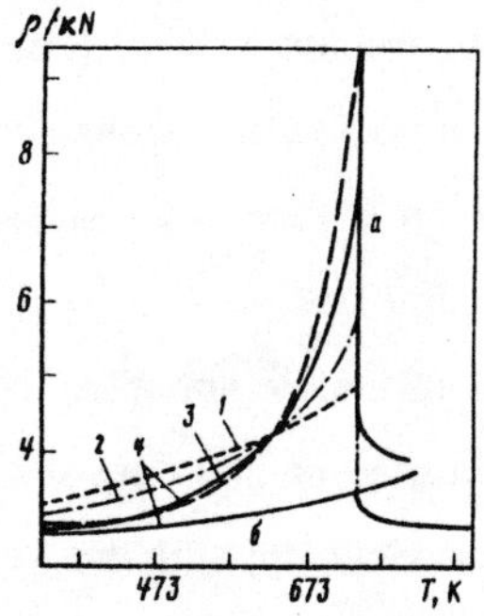

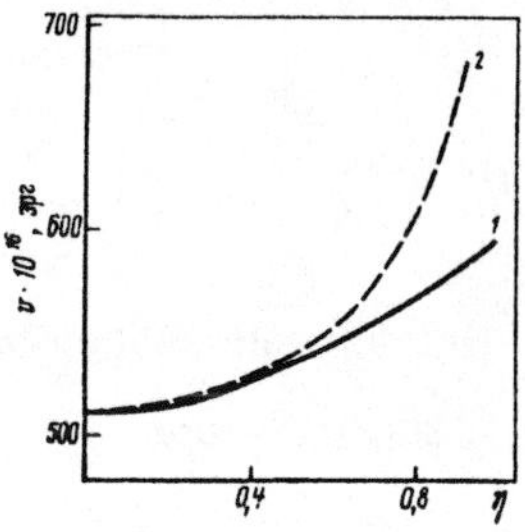

Fig. 80. The variation with temperature of the long-range order parameter for the CuZn alloy. Methods: GBW and quasichemical (1, 2); GBW and quasichemical, with account of variation in size effects (3, 4); experiment [273] (5).

Fig. 81. Configurational heat capacity (per atom) as a function of temperature for the CuZn alloy. Methods: GBW (1); quasichemical (2); quasichemical with the volume effects taken into account (3); experimental (4). Phases: ordered (a), disordered (b).

Fig. 82. Dependence of the ordering energy on the long-range order parameter for CuZn. On account of the volume effects (1); on account of the experimental $\eta(T)$ dependence [273] (2).

approach, or, in other words, carried out the calculation using Smirnov's ensemble. In contrast to the known works, the potential parameters were determined not for the AuCu alloy but directly for the Cu_3Au alloy. The theory predicts, in full agreement with the experimental observation, a linear dependence of a on η. The description of the nonlinear effect, i.e., the $V(\eta)$ function, turns out to be successful as well [333].

Therefore, the account of the volume effects within the framework of a simple one-shell harmonic approximation showed that such an expansion of the ensemble is quite efficient. At the same time, a considerable part of nonlinear effect was shown to be related to the volume one.

The method of elastic spheres. B2 superstructure [334]. Although the level of approximation in this method does not differ, in principle, from that in the previous method, it offers wider opportunities, enabling, in particular, the atomic sizes in this alloy to be evaluated. The radius of the cavity for an atom x on the site j may be written, for the B2 superstructure, as

$$\rho_x^j = 3R - \frac{2}{P_j^x}\left(R_x^k P_{kj}^{xx} + R_y^k P_{kj}^{xy}\right). \tag{4.5.13}$$

As a result, we obtain the following ultimate relation between the radius of the cavity and the order parameter for the B2 superstructure

$$\begin{aligned} \rho_x^\alpha &= \left(\frac{3}{2}c + \frac{3}{4}\eta\right)R_A^\alpha + \frac{3}{2}\left(1 - c - \frac{1}{2}\eta\right)R_B^\alpha + \left(\frac{3}{2}c - \frac{3}{4}\eta - D1\right)R_A^\beta + \\ &+ \left(\frac{3}{2} - \frac{3}{2}c + \frac{3}{4}\eta - D_2\right)R_B^\beta; \\ \rho_x^\beta &= \left(\frac{3}{2}c + \frac{3}{4}\eta - D_3\right)R_A^\alpha + \left(\frac{3}{2} - \frac{3}{2}c - \frac{3}{4}\eta - D_4\right)R_B^\alpha + \\ &+ \frac{3}{2}\left(c - \frac{1}{2}\eta\right)R_A^\alpha + \frac{3}{2}\left(1 - c + \frac{1}{2}\eta\right)R_B^\beta; \end{aligned} \tag{4.5.14}$$

where, for x=A,

$$D_1 = \frac{2\left(c + \frac{1}{2}\eta\right) - P_{\alpha\beta}^{AB}}{c + \frac{1}{2}\eta}; \qquad D_2 = \frac{2P_{\alpha\beta}^{AB}}{c + \frac{1}{2}\eta};$$

$$D_3 = \frac{2\left(c + \frac{1}{2}\eta - P_{\alpha\beta}^{AB}\right)}{c - \frac{1}{2}\eta}; \qquad D_4 = \frac{2(P_{\alpha\beta}^{AB} - \eta)}{c - \frac{1}{2}\eta},$$

and, for x=B,

$$D_1 = \frac{2(P^{AB}_{\alpha\beta} - \eta)}{1 - c - \frac{1}{2}\eta}; \qquad D_2 = \frac{2\left(1 - c + \frac{1}{2}\eta - P^{AB}_{\alpha\beta}\right)}{1 - c - \frac{1}{2}\eta};$$

$$D_3 = \frac{2P^{AB}_{\alpha\beta}}{1 - c + \frac{1}{2}\eta}; \qquad D_4 = 2 - \frac{2P^{AB}_{\alpha\beta}}{1 - c + \frac{1}{2}\eta}.$$

The atomic radii R^j_x can be found by minimizing the total deformation energy for the atom x placed at the site j, W^{xj}_{AT}, and the energy of the deformation of the medium surrounding the cavity near this site, $W^{xj}_{c,p}$:

$$\frac{\nu(W^{Aj}_{\mathrm{AT}} + W^{Aj}_{\mathrm{cp}})}{\partial R^j_A} = 3K_A R^0_A (R^j_A - R^0_A) + 2G(R^j_A - \rho^j_A)\left[\frac{3}{2}Z_1 R^j_A + \right.$$
$$\left. + \left(2 - \frac{9}{2}Z_1\right)\rho^j_A\right] = 0;$$
$$\frac{\nu(W^{\beta j}_{\mathrm{AT}} + W^{\beta j}_{\mathrm{cp}})}{\partial R^j_\beta} = 3K_B R^0_B (R^j_B - R^0_B) + 2G(R^j_B - \rho^j_B) \times$$
$$\times \left[\frac{3}{2} Z_2 R^j_B + \left(\frac{9}{2}Z_2 - \frac{5}{2}\right)\rho^j_B\right] = 0, \tag{4.5.15}$$

Here for j = α: $Z_1 = c + 0.5\eta$, $Z_2 = 1 - c - 0.5\eta$; for j = β: $Z_1 = c - 0.5\eta$, $Z_2 = 1 - c + 0.5\eta$. The calculation of the mean interatomic spacing as a function of η for the CuZn alloy gave the result similar to that obtained in the experiment, with some discrepancy in the absolute values of the lattice parameter [334].

The contribution of the energy of elastic distortions into W/V is estimated to be of the order of 650 kkal/cm^3. The dependence of the elastic distortion energy on the degree of order shows that in the CuZn alloy this contribution promotes the ordering. Thus, in the case of CuZn, the theory predicts a significant contribution of the volume effects into the ordering energy.

The method of elastic spheres. $L1_2$ superstructure. While for the B2 superstructure the correlation effects were taken into account within the quasichemical approximation, for the $L1_2$ superstructure we used the approach similar to that of Cowley. The mean interatomic halfdistance R in a solid solution is determined as:

$$R = \frac{1}{N} \sum_{\substack{x = A, B \\ j = \alpha, \beta}} N_x^j R_x^j = \frac{1}{4} \sum_{x = A, B} (P_\alpha^x R_x^\alpha + 3P_\beta^x R_x^\beta). \tag{4.5.16}$$

For the radii of cavity we have in this case:

$$\begin{aligned} \rho_x^\alpha &= 3R - 2[P_{xx}^{(1)} R_x^\beta + P_{xy}^{(1)} R_y^\beta]; \\ \rho_x^\beta &= 3R - \frac{2}{3} [P_{xx}^{(1)}(R_x^\alpha + 2R_x^\beta) + P_{xy}^{(1)} (R_y^\alpha + 2R_y^\beta)]; \\ &(x = A, B, \quad x \neq y). \end{aligned} \tag{4.5.17}$$

The atomic radii can be derived, correspondingly, from the equations:

$$\begin{aligned} &\frac{3K_x R_x^0 (R_x^\alpha - R_x^0)}{2G} + (R_x^\alpha - \rho_x^\alpha)\left[\left(2 - \frac{9}{2} P_x^\alpha\right)\right) \rho_x^\alpha + \frac{3}{4} P_x^\alpha R_x^\alpha \Big] = 0, \\ &\frac{3K_x R_x^0 (R_x^\beta - R_x^0)}{2G} + (R_x^\beta - \rho_x^\beta)\left[\left(2 - \frac{27}{4} P_x^\beta\right) + 4P_{xx}^{(1)} \rho_x^\beta + \right. \\ &\left. + \left(\frac{9}{4} P_x^\beta - \frac{4}{3} P_{xx}^{(1)}\right) R_x^\beta \right] = 0. \end{aligned} \tag{4.5.18}$$

In the B2 superstructure the dependence of the mean interatomic distance on the long-range order parameter resembles the experimental one; the difference will be in the absolute values.

The difference of the atomic sizes in the alloy was measured in two independent experiments [335 - 337]. The results of the present calculation are compared to experimental data in Tables 22 and 23. It is evident that the model calculation described here reproduces well the difference in the atomic radii and the change in going from a pure metal to an alloy. Simultaneously, it follows from this result that the method of elastic spheres can be used to calculate statistical displacements in an alloy.

Table 22. Atomic sizes in Cu_3Au alloy.

Measured values	η	Numerical values	
		Experiment	Theory
$\epsilon = \frac{1}{4\pi}(\omega_{Cu} - \omega_{Au})$	0,8	$4{,}3 \cdot 10^{-25}$ см^3	$4{,}00 \cdot 10^{-25}$ см^3

Table 23. Variation of atomic sizes in Cu_3Au alloy.

Measured values	T, K	Numerical values	
		Experiment	Theory
$(\Delta r/r)$ Cu	873	−0,031	−0,031
$(\Delta r/r)$ Au	873	0,095	0,083

Note: r is the mean interatomic spacing; Δr is the difference between r and the distance between like atoms.

Table 24. Thermal effects in Cu_3Au alloy, kal/mole.

Experiment	Present calculation	GBW method.
505–525	395	180

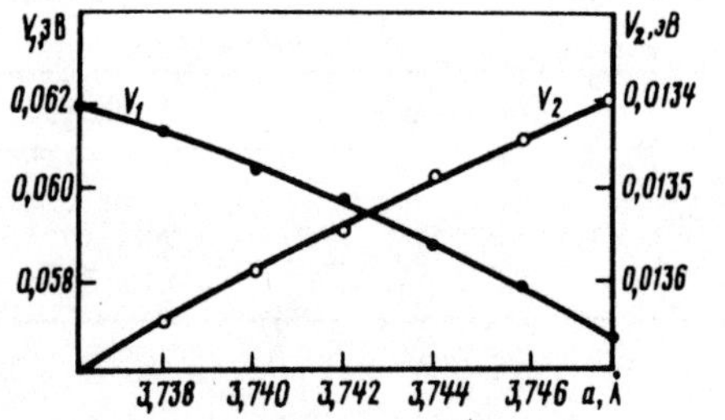

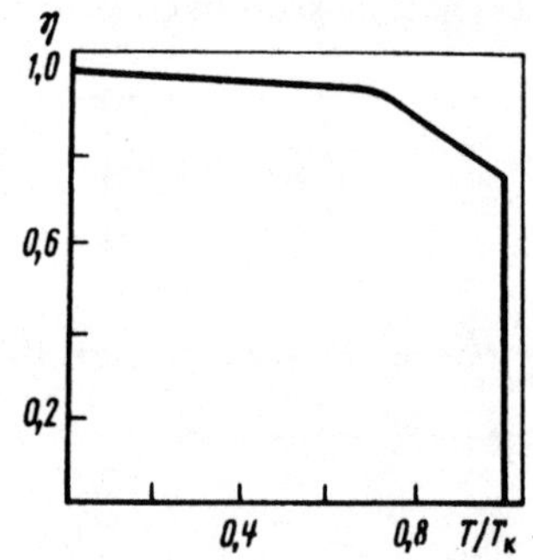

Fig. 83. The ordering energy for two nearest coordination shells as a function of the lattice parameter. Anharmonic potential.

Fig. 84. Variation with temperature of the long-range order parameter for Cu_3Au. Anharmonic potential.

And finally, we estimated the energy of lattice distortions. Its contribution to the transition latent heat was shown to be about 100 kkal/mole. This value is slightly smaller than one-fifth of the total transformation latent heat. Elastic distortions promote the long-range ordering in the Cu_3Au alloy (Table 24). In closing this subsection, it is worth noting that the refinement of the method of elastic spheres in the present work which has been written about in a series of publications, has proved to be sufficiently effective.

Anharmonic potential. $L1_2$ superstructure. The calculation has been carried out for the Cu_3Au alloy without taking the correlation into account. The corresponding free energy has the form analogous to that in the GBW method when the latter accounts for the interactions on the three coordination shells. However, the interatomic distances are now dependent on the degree of order, and the ordering energy -- on the interatomic distances. This is, in fact, Smirnov's ensemble because in this case the choice of the potential different from that used by Smirnov does not affect the nature of the statistical ensemble.

The ordering energies plotted against the lattice constant are presented in Fig. 83. With an increase in the degree of order, the ordering energy for the first coordination shell decreases, for the second -- decreases, and for the third -- remains constant (the latter is not shown in Fig. 83). The equations of equilibrium have been obtained in a usual way. The order parameter has been determined from the equation:

$$\ln \frac{\left(C_A + \frac{3}{4}\eta\right)\left(C_B + \frac{1}{4}\eta\right)}{\left(C_A - \frac{1}{4}\eta\right)\left(C_B - \frac{3}{4}\eta\right)} = \frac{2\eta}{kT}\left[2V_1(\eta) - 3V_2(\eta) + 4V_3(\eta)\right] . \tag{4.5.19}$$

The transformation temperature and the jump in the long-range order parameter at T_c were found by an ordinary procedure. The jump $\Delta\eta$ was estimated to be 0.77. This value is in reasonable agreement with the corresponding experimental values obtained for the Cu_3Au alloy (see Section 4.2). The same may be said about the theoretical dependence of η on T which displays a marked nonlinear behavior (Fig. 84). It is such behavior that the experimental $\eta(T/T_c)$ curve shows for Cu_3Au. Here we can draw some conclusions. The use of the anharmonic potential leads to a noticeable nonlinear effect. The 1st-order phase transition becomes more pronounced even if the correlation is disregarded. In contrast to the experimental a s(η) dependence, the theoretical curve is somewhat nonlinear, although the absolute @a values are close to each other. Thus, the account of the volume effects has significantly improved the agreement between the theory and experiment not only in the shape of the $\eta(T/T_c)$ curve but also in the heat of the phase transition in comparison to the GBW method (see Table 24). The same can be asserted in relation to the heat capacity jump at T_c. We have $\Delta C_p = 4.2$ kN, while in the GBW approximation $\Delta C_p = 2.5$ kN.

In closing this section, it is worthwhile noting that the expansion of the statistical ensemble toward the involvement of size effects enhances the potentialities of the theory and improves its agreement with experiment. The volume effects, though depending on the order parameter, contribute markedly to the energy balance in the order-disorder transformation. In addition, the inclusion of such effects into the theory changes the critical parameters of an order-disorder transformation and even the form of the $\eta(T/T_c)$ curve causing a tendency for the model alloy to exhibit a nonlinear behavior.

4.6 Thermal effects in order-disorder transformations.

So far special attention has been focused on the two quantities specifying the order-disorder transformation: the long-range order parameter and the unit cell parameter. These characteristics are of great importance, because they provide the most general description of the phase transformation. We could make sure of this by considering the foregoing sections of the present chapter. Nevertheless, a comprehensive treatment of the order-disorder transformation with the aid of these two quantities alone cannot be attained. There exist two other important thermodynamic characteristics, namely, the configurational heat capacity (in this case, at constant pressure, i.e., C_p) and the heat of the order-disorder transition, Q. From these quantities one can go over directly to the free energy and the entropy of the ordered alloy which, along with the long-range order parameter, are the main characteristics in the quantum and statistical theories of ordered solid solutions. The measurements of the configurational heat capacity and latent heat of transition require fine and laborious experiments with help of a specialized equipment and, probably for this reason, have been carried out so far on a small number of alloys. The corresponding data are available for the alloys CuZn, Cu_3Au, CuAu, Au_3Cu, Ag_3Mg, Au_3Cd, Cu_3Pd, FeCo, Fe_3Al, Mg_3Cd, Cd_3Mg, MgCd, Ni_3Fe, and Ni_2Cr.

The data concerning the transformation heat, the ordering energy and the configurational part of heat capacity can be used to gain a deeper insight into the nature of ordered solid solutions in several aspects. One of them consists in unusual testing of the adequacy of the statistical theory of atomic ordering with the use of the experimental data concerning not the long-range order parameter but directly the temperature dependence of the free energy and entropy. Even in the cases where the

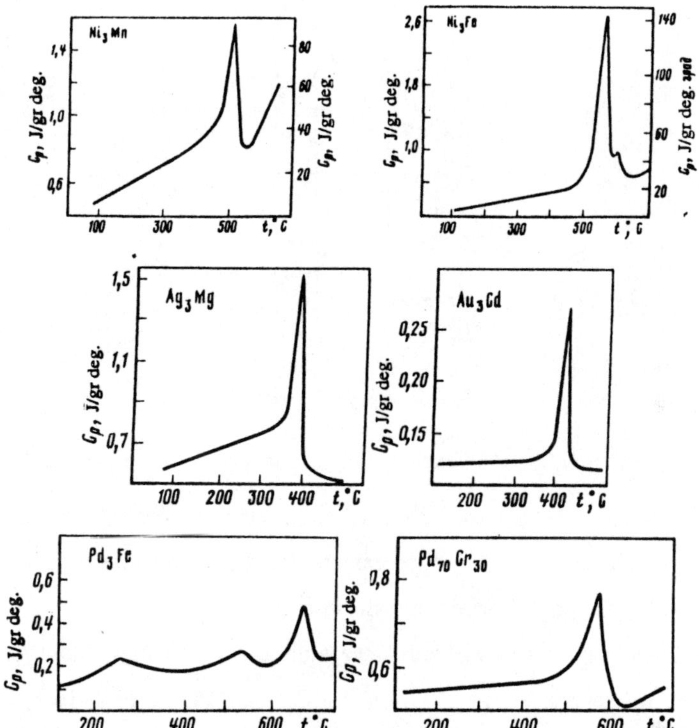

Fig. 85. Heat capacity vs temperature plots for different alloys (author's results).

theoretical and experimental temperature dependences of the long-range order parameter are coincident, it may happen that the calculated temperature dependences of the free energy and entropy differ considerably from the experimental curves.

Such a comparison for some of the known theories has been made for the Cu_3Au alloy. The results show that the energy and entropy calculated in the GBW approximation do not differ to a greater extent from experimental data than those obtained in the quasichemical approximation or by the Cowley's or the cluster variation methods. At the same time, the results of all these theories are sufficiently far from experimental values. It should be emphasized here that a complete check on the

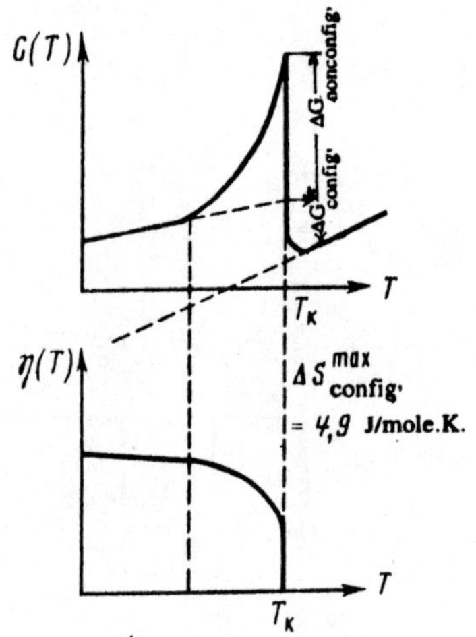

Fig. 86. Schematic temperature behavior of heat capacity for ordering alloy.

adequacy of theory and experiment must involve necessarily not only the comparison of the temperature behavior of the order parameter but also of the internal energy, entropy and latent heat of the order-disorder transformation.

The authors and their coworkers measured heat capacities and transformation temperatures for several alloys: Ag_3Mg, Au_3Cd, Ni_3Mn, Ni_3Fe, Pd_3Cr, and Pd_3Fe. The experiments have been carried out on a specially designed calorimeter operating on the triple heat bridge principle. The sensitivity of this device is at least not worse than 0.05 kal/gr. The experiment consisted in measuring C_p versus T curves in the ordered and disordered states (Fig. 85). Extrapolation of the linear segment of the heat capacity curve for the ordered state, i.e., of the part where the long-range order parameter remains constant, allowed one to cut off the peak the area under which gives the difference in heats between the state with a perfect long-range order and that with a short-range order (Fig. 86). Then, by using the well-known formulas, the configurational and nonconfigurational contributions into internal energy and entropy of the alloy were calculated.

Table 25. Thermal characteristics of alloys studied.

Alloy;	Superstructure;	T_c, K;	Q, kJ/gr.-atom;	ΔS, J/mole.K.
Cu_3Au	$L1_2$	678 ± 3	2,13	12,50
Ag_3Mg	DO_{23}	660 ± 3	1 ,81	9,60
Au_3Cd	DO_{23}	668 ± 3	2,37	2,59
Ni_3Mn	$L1_2$	778 ± 3	4,73	16,64
Ni_3Fe	$L1_2$	807 ± 3	4,26	24,36
Pd_3Fe	$L1_2$	910 ± 3	1,12	4,09
Pd_3Cr	$L1_2$	873 ± 3	0,78	3,89

The experimental data combined with the literary data for other alloys [338] are presented in Table 25. All the alloys in this table display the 1st-order phase transition. Their latent heats of transformation lie in the range 0.5 to 2.0 kJ/gr.-at., except for the Ni_3Fe and Ni_3Mn alloys which have markedly higher heats. The order-disorder transformation occurring in these two alloys is accompanied by a transition from the ferromagnetic to paramagnetic state, and the origin of the ordering energy is largely due to magnetism. The different values of the latent heat per atom in various alloys show that the dependence of the ordering energy upon the interatomic distance as well as the role of many-particle contributions and the volume-dependent terms are different. Evidently, the Cu_3Au alloy differs strongly from the others both in the behavior of the long-range order parameter and in the thermal effects connected with the transformation. The comparison of theory and experiment should be made, in principle, for each alloy individually.

Next column of Table 25 shows the values of the transformation entropy $\Delta S = S(\eta=0) - S(\eta=1)$. If the order-disorder transformation had been of a purely configurational nature, the values of ΔS would have not exceeded 4.9 J/mole K, on condition that the correlation effects are negligible. In reality, the measured values of ΔS in the alloys Cu_3Au, Ni_3Fe and Ni_3Mn exceed the configurational contribution. In the remaining alloys, the transformation entropy is much less than 4.9 J/mole K. These data give evidence of the important role of the nonconfigurational contributions into the order-disorder transformation. The lattice contribution in the Cu_3Au alloy and the magnetic contributions in Ni_3Mn and Ni_3Fe promote the ordering. In the two last alloys the order-disorder transformation superimposes on the transition from the ferromagnetic to paramagnetic state increasing the jump in entropy. Since the volume effects in the Ni_3Fe and Ni_3Mn alloys are quite small, it may be thought that the magnetic transformation gives a significant contribution to the configurational part of entropy. That is what is assumed in the theories describing simultaneously these two transformations [310]. In other alloys studied, the lattice contributions reduce the transformation entropy. The possibility of such behavior has been demonstrated in our theoretical calculations for Cu-Au alloys (see Section 4.5). As stressed there, the configurational contribution may be as great as to supress the ordering itself. It did not happen in the alloys Ag_3Mn, Au_3Cd, Pd_3Fe and Pd_3Cr but the entropy in these alloys turned out to be strongly lowered.

The role of the configurational contributions can be demonstrated with the aid of the temperature dependence of the heat capacity (Figs. 85 and 86). First of all, we must pay attention to the fact that in several alloys the heat capacity at temperatures above T_c is lower than the value obtained by extrapolation to these temperatures. This difference determines the nonconfigurational jump in the heat capacity. Comparison of Figs. 85, 86 and Table 25 shows that in the alloys where the nonconfigurational contribution into the transformation entropy is opposite to the configurational one, the crystal lattice of the disordered alloy is more stable, i.e., it has a lower heat capacity compared to the ordered alloy. On the other hand, the occurrence of the jump in the configurational part of heat capacity gives evidence that the 1st-order phase order-disorder transformation in the investigated alloys is accompanied by the structural 2nd-order nonconfigurational transition. Interestingly, this effect is more pronounced for the DO_{23} superstructure than for $L1_2$; in the latter, the crystal lattice undergoes merely the volume changes, whereas in the DO_{23} superstructure, the order-disorder transition is accompanied by the change from cubic to tetragonal lattice.

Another interesting effect seen in Fig. 85 is an anomalous behavior of the heat capacity just above T_c. It gives evidence of a considerable role of heterophase fluctuations prior to the transformations and enables one to estimate the relevant heat effect. It is noteworthy that the thermal effect of heterophase fluctuations is pronounced only in those alloys where the nonconfigurational contribution to the entropy impedes the ordering. Heterophase fluctuations increasing as the temperature approaches T_c from above are now poorly investigated. In the alloys under consideration, they can manifest themselves in the growth of a short-range order or in the formation of microdomain structure. In any case, the up-to-date statistical theories do not describe this fact, although the related heat effects can be rather significant. It is possible that the variations in the band structure related to the short-range order before the transformation play a certain role for explaining these effects.

The change in the configurational energy with disordering must be proportional to $(1 - \eta^2)$; also consistent must be the jumps in the energy and the order parameter at the transition point. Such behavior is not always observed experimentally. Inconsistency in the heat capacity jumps is connected with the different nature of energy changes occurring in the course of formation of long-range order in various alloys. Near T_c the internal energy is not proportional to $(1-\eta^2)$ varying with η somewhat more rapidly. Along with the influence of nonconfigurational contributions, this phenomenon may be related to the nonlinear effects: many-particle interactions, the dependence of the ordering energy upon η and the correlation effects.

Therefore, the study and analysis of the behavior of the heat capacity, enthalpy, internal energy, entropy and latent heat of transformation, in conjunction with the material discussed in the foregoing sections of this book, show that, in addition to the underlying configurational effect which is determined by the long-range order and correlations, the important part in order-disorder transformations also play the nonconfigurational and nonlinear effects.

5. THE NATURE OF BONDING AND STABILITY OF ORDERED PHASES

5.1 The binding forces in metals and alloys

In the preceding chapters we stressed the important role of the contributions into the ordering energy from different coordination shells in relation to the problem of the superstructure stability and the problems connected with order-disorder phase transitions. Many details of the structure of ordered solid solutions were shown to be determined decisively by the nature of the ordering energy and the relative part of different contributions to it. Attempts have been made to obtain some information on the relative contributions into the ordering energy either directly from experimental data or by applying statistical theories of atomic ordering to the experimental results. This approach enables one to obtain information on some properties of the ordering energy.

Nevertheless, in many cases the treatment of experimental findings turns out to be ambiguous. In our opinion, it is hardly possible to describe completely the ordering energy relying solely upon the experimental data. The point is that the ordering energy and various contributions into it do not manifest themselves in experiments in a straightforward manner but enter the complicated functional relationships which are not known to the end and, even if they had been, the solution of the inverse problem could not have been obtained in many cases. For this reason, for the evaluatioin of the role of different contributions into the ordering energy and for better understanding of its nature, the theory is needed which would permit one to deduce these values in a certain way from the "first principles".

Since the ordering alloys are chiefly the alloys of transition and noble metals (at the same time, the phases based on simple and polyvalent metals occur also), this theory must cover, from a single point of view, a wide class of metallic materials. Another condition is that the theory must be capable of describing the structures with different degree of short-range and long-range order and with different concentrations of the components. And finally, the theory must be able to explain a variety of the alloy properties (electronic and atomic ones, transport phenomena, the vibrational spectrum of atoms). In particular, it must desribe the energy of an alloy not only in terms of electronic states but also in terms of interionic interactions. It is clear that such theory must be quantum theory. The only theory satisfying nowadays all above-listed requirements is the pseudopotential theory (PP).

In this chapter we consider the results of the successive development and application of the PP method to the problem of atomic ordering. The underlying theory is based on the one-electron and nonrelativistic approximation. However, we believe that the first limitation is not important for the atoms in the considered problem. The exchange and correlation effects are taken properly into account. The second limitation should be overcome in future. We cannot exclude the possibility of the existence of alloys in which the ordering is conditioned by the relativistic effects.

Applicability of the PP method to the theory of metals is based on the possibility to replace formally a strong crystal potential acting upon the collectivized electronic states by a weak pseudopotential. If we add to the first the repulsive potential arising from the inner d shells, the wave function ψ_k of the conduction band state will be replaced by the pseudowave function φ_k . Such transition is possible in either fundamental method of the calculation of the electronic band structure such as the augmented plane-wave (APW) method, the orthogonalized plane-wave method (OPW) method, the method developed by Korringa-Kohn-Rostocker (KKR or the Green function method) and others [339 - 341]. The possibility of going over to the PP method has been substantiated in [32, 339]. The transition to the equation of the pseudo-potential theory makes it possible to solve the task of finding the wave functions and the dispersion law, $E(\mathbf{k})$, for the collectivized electrons using the perturbation theory similar in many respects to the nearly-free electron approximation [56]. The most developed schemes of the PP theory are those based on the OPW [33] and the quantum defect [57] methods. In the present chapter, we employ, with rare exceptions, the PP method in Harrison's formulation [342, 343] because it does not involve any fitted parameters and is conceivable to be the most consistent and physically plausible.

If a metallic crystal, along with the electronic states originating from the outer s states, contains the outer d-like states, the scheme of the PP method becomes more complicated. The d states do not represent the core or the collectivized states being, in fact, quasilocalized (overlap within the nearest coordination shells and the formation of the collectivized orbitals are possible). In this case, the Harrison method based on the OPW scheme involves the so-called model Hamiltonian, with the matrix H in the form [34]

$$H = \left(\begin{array}{c|c} s-s & s-d \\ \hline s-d & d-d \end{array} \right). \tag{5.1.1}$$

The diagonal terms describe the collectivized s states and the quasilocalized d states, while the nondiagonal terms take hybridization into account. The d states are described by one of the variants of the tight-binding method [340]. The inclusion of the model Hamiltonian makes it possible to take account of the s-d hybridization and the resonance character of the d states. The dispersion law $E(\mathbf{k})$ is modified in such a way as to replace an idealized scheme of the nearly-free s states and the "rigid" d level for the real pattern of the dispersion curves. The states with $k=\sqrt{E_d}$ are resonant. As a matter of fact, this version of the Harrison-Moriarty PP method [344 - 347] represents the generalization of the nearly-free electron and tight-binding methods.

Without going into details of the pseudopotential method in the form developed by Harrison (see, for example, [56, 344]), we shall outline in brief the main approximations adopted in this theory. All electronic states are subdivided into three categories: the core, quasilocalized and collectivized. The core states do not change their wave functions but the transition from a free atom to metal may alter the energies of the corresponding levels. The quasilocalized states are described in the tight-binding scheme, usually in the one-center approximation. The wave functions corresponding to these states are expanded into a series, using a base system of atomic wave functions. This approach is clearly a simplification compared to the symmetrized expansion

in spherical harmonics [30, 31]. On the other hand, the system of basic functions in the OPW method is overdefined since it includes a complete system of the plane waves and, additionally, the orthogonalized core states and the atomic d states, on account of their overlap. The result of this is ambiguity of the pseudopotential. In this representation, the d states fall into a separate group beforehand. This is substantiated by the results of the KKR-Z (Ziman) and APW band structure calculations for transition metals in which such approximations have not been adopted in advance [340, 341].

So far we have been concerned with traditional approximations of the PP method. The points should be noted, however, where, following the logic of the investigation, we were forced to go beyond the limits of this method. First, the PP scheme developed for noble and transition metals has been generalized so as to apply it to binary alloys. As a matter of fact, in some of the works in which the method has been applied to the theory of alloys in general, a formal transfer of the results obtained for pure metals has led to the loss of the features substantial for describing the properties and the behavior of alloys [348 - 351]. Second, the ordering should have been included in the PP procedure devised for transition and noble metal alloys. In several works devoted to the theory of ordering alloys of simple and polyvalent metals [352 - 361] and accomplished simultaneously or before the present work, the substantial results have been obtained which we employed to construct the theory of ordering in alloys of noble and transition metals. Third, the contact interaction and overlap of the d shells of the neighboring ions take place for transition elements. As a consequence, we inserted into the electrostatic energy the terms allowing for the deviation of the ionic potential from the point one.

Some other alterations have been made in the general formalism, perhaps not so noticeable but critically important for constructing the theory of atomic ordering and understanding of the nature of the ordering energy. Many approximations in the Harrison-Moriarty scheme applied to the problem of pure metals were determined by the required accuracy of calculations. Since the characteristic energies in the problem of atomic ordering are of the same order, sensitivity of the theory should be enhanced in comparison to the solution of problems connected with pure metals. It has been done in the following way: (1) The higher orders of the perturbation theory were taken into account in the most straightforward and understandable manner; (2) Since the order-disorder transformation is due to the behavior of small groups of electrons near the Fermi surface, the nonspherical character of the latter has been taken into account in the second order of the perturbation theory. As a result, some peculiarities in the electron spectrum of the collectivized and hybridized states have become more distinct, and their role in the ordering energy has been cleared up; (3) The calculation of the polarization operator of the electron subsystem beyond the limits of the random phase approximation (RPA) has been required to take into account the shape of the Fermi surface.

Allowing for the characteristic features of the method and the assumptions made in it, the total energy E of a binary solid solution based on transition metals can be expressed as the sum of the following contributions (per ion) [35, 362, 363]:

$$E = E_0 + E_{b.s.} + E_{ov} + E_{el.-st.} \tag{5.1.2}$$

where E_0 is the portion of the electron energy independent of the structure, $E_{b.s.}$ is the band-structure energy, E_{ov} is the energy of the overlap of the electron states, and $E_{el.-st.}$ is the electrostatic energy. The electron energy in the PP method is represented as a series of the perturbation theory in the pseudo-potential and hybridization parameters. The structure-independent portion of the energy is determined solely by the alloy volume:

$$E_0 = \frac{1}{N} \sum_{k<k_F} (k^2 + \langle \mathbf{k} | \tilde{W} | \mathbf{k} \rangle + \sum_{\mu, d\mu} \langle \mathbf{k} | \tilde{W} | \mathbf{k} \rangle \langle \mathbf{k} | d_\mu \rangle \langle d_\mu | \mathbf{k} \rangle) +$$
$$+ \frac{1}{\pi N} \sum_{\mu, d\mu} E_d^{(\mu)} \left(\operatorname{arctg} \frac{E_F - E_d^{(\mu)}}{\omega_d^{(\mu)}} + \frac{\pi}{2} \right) +$$
$$+ \frac{1}{N} \sum_{\mathbf{k}} \sum_{\mu, d_\mu} \frac{\langle \mathbf{k} | \Delta_\mu | d_\mu \rangle \langle d_\mu | \Delta_\mu | \mathbf{k} \rangle}{(k^2 - E_d^{(\mu)})^2 + (\omega_d^{(\mu)})^2} \Bigg\{ k^2 - E_d^{(\mu)} - \langle \mathbf{k} | \tilde{W} | \mathbf{k} \rangle +$$
$$+ \sum_{\nu, d'_\nu} \Bigg[\langle \mathbf{k} | d'\nu | \rangle \langle d'\nu | \mathbf{k} \rangle (k^2 - E_{d'}^{(\nu)} +$$
$$- \langle \mathbf{k} | \Delta_\nu | d'_\nu \rangle \langle d'_\nu | \Delta_\nu | \mathbf{k} \rangle \frac{k^2 - E_{d'}^{(\nu)}}{(k^2 - E_{d'}^{(\nu)})^2 + (\omega_{d'}^{(\nu)})^2} \Bigg] \Bigg\} \vartheta(k^2 - E_F). \tag{5.1.3}$$

Here the first term determines an ordinary "free-electron" energy, the second gives the contribution into E_0 from the d states, and the third -- the contribution into E_0 due to the hybridized s and d states.

The band-structure energy $E_{b.s.}$ depends on the atomic order in an alloy, i.e., on the arrangement of ions:

$$E_{b.s.} = \frac{1}{N} \sum_{\mathbf{q}} \Bigg\{ \sum_{k<k_F} \frac{\langle \mathbf{k}+\mathbf{q} | \tilde{W} | \mathbf{k} \rangle \langle \mathbf{k} | \tilde{W} | \mathbf{k}+\mathbf{q} \rangle}{k^2 - |\mathbf{k}+\mathbf{q}|^2} \theta(E_F - k^2) +$$
$$+ \sum_{\mathbf{k}} \Bigg[\sum_{\mu, d_\mu} \frac{\langle \mathbf{k}+\mathbf{q} | \tilde{W} | \mathbf{k} \rangle \langle \mathbf{k} | \Delta_\mu | d_\mu \rangle \langle d_\mu | \Delta_\mu | \mathbf{k}+\mathbf{q} | \rangle + \mathrm{KC}}{k^2 - |\mathbf{k}+\mathbf{q}|^2} \times$$
$$\times \frac{k^2 - E_d^{(\mu)}}{(k^2 - E_d^{(\mu)})^2 + (\omega_d^{(\mu)})^2} +$$
$$+ \sum_{\mu, \nu} \sum_{d_\mu, d_\nu} \frac{\langle \mathbf{k}+\mathbf{q} | \Delta_\mu | d_\mu \rangle \langle d_\mu | \Delta_\mu | \mathbf{k} \rangle \langle \mathbf{k} | \Delta_\nu | d'_\nu \rangle \langle d'_\nu | \Delta_\nu | \mathbf{k}+\mathbf{q} \rangle}{k^2 - |\mathbf{k}+\mathbf{q}|^2} \times$$
$$\times \frac{(k^2 - E_d^{(\mu)})(k^2 - E_{d'}^{(\nu)})}{[(k^2 - E_d^{(\mu)})^2 + (\omega_d^{(\mu)})^2][(k^2 - E_{d'}^{(\nu)})^2 + (\omega_{d'}^{(\nu)})^2} \Bigg] \vartheta(k^2 - E_F) \Bigg\}, \tag{5.1.4}$$

where the first term is the structure-dependent electron energy associated with the renormalized electron-ion interaction. The contribution to the second term is originated from the hybridized s and d states. It should be noted that the summation of $\bar{\succcurlyeq}$ in (5.1.4) is over the volume occupied by the real Fermi surface.

Overlap of the states originating ,from the outer d states gives the following contribution to the total energy [364]:

$$E_{\text{пер}} = \sum_{\mu,\nu}\ \sum_{d_\mu, d'_\nu} \Big\{ \langle d_\mu | d'_\nu \rangle \langle d'_\nu | d_\mu \rangle \theta (E_K - E_d^{(\mu)}) -$$
$$- \langle d_\mu | d'_\nu \rangle \Bigg[\sum_{k < k_F} ((K^2 - E_d^{(\mu)}) \langle d'_\nu | \mathbf{k} \rangle \langle \mathbf{k} | d_\mu \rangle +$$
$$+ \langle d'_\nu | \mathbf{k} \rangle \langle \mathbf{k} | \Delta_\mu | d_\mu \rangle + \langle d'_\nu | \Delta_\nu | \mathbf{k} \rangle \langle \mathbf{k} | d_\mu \rangle) +$$
$$+ \sum_{\mathbf{k}} \frac{\langle d'_\nu | \Delta\nu | \mathbf{k} \rangle \langle \mathbf{k} | \Delta_\mu | d_\mu \rangle}{(k^2 - E_{d'}^{(\nu)}) + (\omega_{d'}^{(\nu)})^2} \vartheta (k^2 - E_F) \Bigg] -$$
$$- \langle d_\mu | \Delta_\mu | d'_\nu \rangle \Bigg[\sum_{k < k_F} \langle d'_\nu | \mathbf{k} \rangle \langle \mathbf{k} | d_\mu \rangle +$$
$$+ \sum_{\mathbf{k}} \Bigg(\frac{2 \langle d'_\nu | \mathbf{k} \rangle \langle \mathbf{k} | \Delta_\mu | d_\mu \rangle}{(k^2 - E_d^{(\mu)})^2 + (\omega_d^{(\mu)})^2} +$$
$$+ \frac{\langle d'_\nu | \Delta_\nu | \mathbf{k} \rangle \langle \mathbf{k} | \Delta_\mu | d_\mu \rangle}{[(k^2 - E_d^{(\mu)})^2 + (\omega_d^{(\mu)})^2][(k^2 - E_{d'}^{(\nu)})^2 + (\omega_{d'}^{(\nu)})^2]} \Bigg) \vartheta (k^2 - E_F) \Bigg] \Big\} . \tag{5.1.5}$$

This contribution into the electron energy is proportional to the overlap integrals

$$\langle d_\mu | d'_\nu \rangle = \int \psi_\lambda^{(\mu)*} (\mathbf{r} - \mathbf{R}_i) \psi_{\lambda'}^{(\nu)} (\mathbf{r} - \mathbf{R}_j) \, d\mathbf{r} \tag{5.1.6}$$

and to the matrix elements of the hybridization potential

$$\langle d_\mu | \Delta_\nu | d'_\nu \rangle = \int \psi_\lambda^{(\mu)*} (\mathbf{r} - \mathbf{R}_i) \Delta_\nu (\mathbf{r} - \mathbf{R}_j) \psi_{\lambda'}^{(\nu)} (\mathbf{r} - \mathbf{R}_j) \, d\mathbf{r} \tag{5.1.7}$$

i.e., the overlap energy includes that of the outer d states and that of the s and d hybridized states (here λ is a set of the quantum numbers specifying the d state). Let us pay attention to the structure of the energy-dependent denominators in (5.1.4) and (5.1.5) which are of two kinds:

$$(k^2 - |\mathbf{k} + \mathbf{q}|^2)^{-1} \quad \text{and} \quad [(k^2 - E_d^{(\mu)})^2 + (\omega_d^{(\mu)})^2]^{-1} .$$

If the denominators are small, the so-called resonant states appear. As we shall see further, these states play the important role in the stabilization of particular superstructures.

Taking into account the non-point form of the ionic potential, the interaction between the core states and the nuclear charges may be expressed by the sum:

$$E_{\text{эл.ст}} = \frac{1}{N} \sum_{\mu,\nu} \sum_{\substack{s,\gamma \\ s'\gamma'}} \sum_{n_i n_j} \sum_{m=-n_L}^{+n_L} C^{\mu}_{s\gamma} C^{\nu}_{s'\gamma'} \iint B^{|m|}_{n_i n_j} \times$$

$$\times \rho_{n_i m}(\mathbf{r}_i) \rho_{n_j m}(\mathbf{r}_2) r_1^2 \, d\mathbf{r}_1 \, r_2^2 \, d\mathbf{r}_2. \tag{5.1.8}$$

Here $B^{|m|}_{n_i n_j}$ are the coefficients in the two-centered expansion depending on the degree of the overlap of the charge density $\rho_{m,m}$ [365, 366]. In the case of the non-point ions, the overall potential of all charges is divided by the contributions from the charge concentrated at one point and the distribution represented by a multipole series. The corresponding expressions depend on whether the ions overlap or not. In the simplest case of the point ions, expression (5.1.8) transforms to the ordinary expression for the electrostatic energy [367]

$$E_{\text{эон.стр}} = \frac{1}{N} \sum_{\mu,\nu} \sum_{i,j} \sum_{\mathbf{q}} \frac{4\pi z^*_\mu z^*_\nu e^2}{\Omega_0 q^2} e^{-q^2/4\xi} e^{-i\mathbf{q}(\mathbf{R}_i - \mathbf{R}_j)}. \tag{5.1.9}$$

Here z^*_μ and z^*_ν are the effective valencies of the atoms of μ and ν kind, respectively, ξ is the Evald parameter. It should be noted that the interionic interaction in (5.1.8) is noncentral: the more the overlap of the neighboring ions, the greater the departure of the interaction from the central one.

The values {k + q/w/k} in (5.1.4) and (5.1.5) are the matrix elements of the pseudopotential of the transition metal alloy:

$$\widetilde{W} = V + \sum_{\mu, d\mu} (k^2 - E^{(\mu)}_\alpha) | \alpha_\mu \rangle \langle \alpha_\mu | + \sum_{\mu, d\mu} [(k^2 - E^{(\mu)}_d) | d_\mu \rangle \langle d_\mu | +$$

$$+ | d_\mu \rangle \langle d_\mu | \Delta_\mu + \Delta_\mu | d_\mu \rangle \langle d_\mu |], \tag{5.1.10}$$

where $\alpha\mu\rangle$ and $d\mu\rangle$ are the core and d states of the ion of the μ kind (μ = A, B); $|\mathbf{k}\rangle$ is the plane wave vector; Δμ is the potential related to hybridization: $E^{(\mu)}_d$ and $\omega^{(\mu)}_d$ are the position and the width of the d resonance, respectively [344]. The factor υ in (5.1.4) and (5.1.5) is given by the expression

$$\vartheta(k^2 - E_F) = \theta(E_F - \mathbf{k}^2) - \frac{1}{\pi}\left(\mathrm{arctg}\,\frac{E_F - E_d^{(\mu)}}{\omega_d^{(\mu)}} + \frac{\pi}{2}\right), \tag{5.1.11}$$

where $\Theta(X)$ is the Heaviside function, E_F is the Fermi energy.

Expressions (5.1.2)–(5.1.5) are simplified in the case of the noble metal alloys. In particular, the conventional approximations for noble metals and their alloys involve the neglect of the overlap of the d shells and the change of the finite width of the d resonance for the zero one: $\omega_\alpha^{|\mu|} = 0$ [368]. In this case, the expression for the total electron energy takes the form

$$E_{\text{эл}} = E_0 + E_{\text{зон.стр}}. \tag{5.1.12}$$

where
$$E_0 = \frac{1}{N}\sum_{k<k_F}(\mathbf{k}^2 + \langle\mathbf{k}|\widetilde{W}|\mathbf{k}\rangle + \sum_{\mu, d_\mu}\langle\mathbf{k}|\widetilde{W}|\mathbf{k}\rangle\langle\mathbf{k}|d_\mu\rangle\langle d_\mu|\mathbf{k}\rangle) +$$

$$+ \cdot\sum_{\mu, d\mu} E_d^{(\mu)} - \frac{1}{N}\sum_{\mu, d\mu}\sum_{k>k_F}\frac{\langle\mathbf{k}|\Delta_\mu|d_\mu\rangle\langle d_\mu|\Delta_\mu|\mathbf{k}\rangle}{K^2 - E_d^{(\mu)}}\Bigg\{k^2 - E_d^{(\mu)} -$$

$$-\langle\mathbf{k}|\widetilde{W}|\mathbf{k}\rangle + \sum_{\nu, d'_\nu}\Bigg[\langle\mathbf{k}|d'_\nu\rangle\langle d'_\nu|\mathbf{k}\rangle(k^2 - E_{d'}^{(\nu)}) -$$

$$-\frac{\langle\mathbf{k}|\Delta_\nu|d'_\nu\rangle\langle d'_\nu|\Delta_\nu|\mathbf{k}\rangle}{k^2 - E_{d'}^{(\nu)}}\Bigg]\Bigg\}. \tag{5.1.13}$$

and

$$E_{\text{зон.стр}} = \frac{1}{N}\sum_{\mathbf{q}}{}'\Bigg\{\sum_{k<k_F}\frac{\langle\mathbf{k}+\mathbf{q}|\widetilde{W}|\mathbf{k}\rangle\langle\mathbf{k}|\widetilde{W}|\mathbf{k}+\mathbf{q}\rangle}{k^2 - |\mathbf{k}+\mathbf{q}|^2} -$$

$$-\sum_{k>k_F}\Bigg[\sum_{\mu, d\mu}\frac{\langle\mathbf{k}+\mathbf{q}|\widetilde{W}|\mathbf{k}\rangle\langle\mathbf{k}|\Delta_\mu|d_\mu\rangle\langle d_\mu|\Delta_\mu|\mathbf{k}+\mathbf{q}\rangle + \text{KC}}{(k^2 - |\mathbf{k}+\mathbf{q}|^2)(k^2 - E_d^{(\mu)})} +$$

$$+\sum_{\mu,\nu}\sum_{d\mu, d'_\nu}\frac{\langle\mathbf{k}+\mathbf{q}|\dot{\Delta}_\mu|d_\mu\rangle\langle d_\mu|\Delta_\mu|\mathbf{k}\rangle\langle\mathbf{k}|\Delta_\nu|d'_\nu\rangle\langle d'_\nu|\Delta_\nu|\mathbf{k}+\mathbf{q})}{(k^2 - |\mathbf{k}+\mathbf{q}|^2(k^2 - E_d^{(\mu)})(k^2 - E_{d'}^{(\nu)})}\Bigg]\Bigg\}. \tag{5.1.14}$$

Note that expressions (5.1.3), (5.1.4), (5.1.5), (5.1.13), and (5.1.14) have been obtained in the second order in the pseudopotential W and in the fourth order in the hybridization potentials Δμ.

Transition to the conventional formulas for the pseudopotential of simple and polyvalent metals [352 - 358] is accomplished if we omit in (5.1.13) and (5.1.14) the terms related to the d states and set Δμ = 0. The total energy calculated in the third order for the pseudopotential is written as

$$E_{эл} = \frac{1}{N}\sum_{k<k_F}(\mathbf{k}^2 + \langle \mathbf{k}|\widetilde{W}|\mathbf{k}\rangle) - \frac{\Omega_0}{2N}\sum_{\mathbf{q}\neq 0}|W(\mathbf{q})|^2\pi(\mathbf{q}) + \\ + \Omega_0 \sum_{\mathbf{q}_1\mathbf{q}_2\mathbf{q}_3} W(\mathbf{q}_1)W(\mathbf{q}_2)W(\mathbf{q}_3)\Lambda^{(3)}(\mathbf{q}_1\mathbf{q}_2\mathbf{q}_3)\Delta(\mathbf{q}_1\mathbf{q}_2\mathbf{q}_3), \tag{5.1.15}$$

where

$$W = V + \sum_{\mu,\alpha_\mu}(k^2 - E_\alpha^{(\mu)})|\alpha_\mu\rangle\langle\alpha_\mu| \tag{5.1.16}$$

is the pseudopotential of the alloy made up of simple or polyvalent metals, $\pi(\mathbf{q})$ is the polarization operator, and $\Lambda^{(3)}$ is the irreducible three-vertex network [371]. The last expression will become equivalent to the formulas derived in [352 - 358] if one uses the approximation of the point ions in the electrostatic energy, retains only the second order perturbation terms and replaces the integration over the real Fermi surface by the integration over the Fermi sphere. The latter assumption will always be made with a special reservation because, as we shall see in the following, there is a risk to lose the details of the theory directly responsible for the formation of the superstructure.

After the physical meaning of different contributions into the total energy is ascertained, we must proceed to an explicit form of the dependence of the energy on the order parameters. It will be achieved by the factorization of the matrix elements of the pseudopotential and hybridization potentials:

$$\langle \mathbf{k}+\mathbf{q}|W|\mathbf{k} = \frac{1}{N}\sum_{\mathbf{R}_i} e^{-i\mathbf{q}\mathbf{R}_i}\left\{C_i\,\omega_A(\mathbf{q}) + (1-C_i)\,\omega_B(\mathbf{q})\right\}, \tag{5.1.17}$$

where

$$\omega_\mu(\mathbf{q}) = \frac{1}{\Omega_0} \int \omega_\mu(\mathbf{r}) e^{-i\mathbf{q}\mathbf{r}} d\mathbf{r} \tag{5.1.18}$$

is the pseudopotential form-factor of the ion of the μ kind and Ω_0 is the volume of the alloy per ion. We obtain next:

$$\sum_{\mu, d\mu} \langle \mathbf{k}+\mathbf{q} \mid \Delta_\mu \mid d_\mu \rangle \langle d_\mu \mid \Delta_\mu \mid \mathbf{k} \rangle = \tag{5.1.19}$$

$$= \frac{1}{N} \sum_{\mathbf{R}_i} e^{-i\mathbf{q}\mathbf{R}_i} \left\{ C_i \sum_{\lambda_A} q_\lambda^{(A)}(\mathbf{q}) + (1 - C_i) \sum_{\lambda'_B} q_{\lambda'}^{(B)}(\mathbf{q}) \right\},$$

$$g_\lambda^{(\mu)}(\mathbf{q}) = \frac{1}{\Omega_0} \int e^{-i(\mathbf{k}+\mathbf{q})\mathbf{r}} \Delta_\mu(\mathbf{r}) \psi_\lambda^{(\mu)}(\mathbf{r})\, d\mathbf{r} \int \psi_\lambda^{(\mu)*}(\mathbf{r}') \Delta_\mu(\mathbf{r}') e^{i\mathbf{k}\mathbf{r}'} d\mathbf{r}'. \tag{5.1.20}$$

where λ_μ is a set of the quantum numbers that specifies the d state of the ion centered at the site $\mathbf{R}_i$; $\mid d_\mu \rangle = \psi_\lambda^{(\mu)}(\mathbf{r} - \mathbf{R}_i)$. The factorization is described in detail for a one-component material by Harrison [56] and for ordering alloys based on simple metals in [358]. The lattice potentials can be expressed by the sum of the one-center potentials with the random occupation numbers:

$$V(\mathbf{r}) = \sum_{\mathbf{R}_i} \left\{ C_i V_A(\mathbf{r} - \mathbf{R}_i) + (1 - C_i) V_B(\mathbf{r} - \mathbf{R}_i) \right\},$$

and the relevant matrix elements are found by integration.

Various contributions into the total electron energy depend on atomic order in different ways. The structure-independent part of the energy E_0 is related to the degree of long-range order implicitly (through Ω_0 and k_{F_0}). As a result, we obtain the following dependence of the band structure energy $E_{b.s.}$ on the order parameters $\{C_i\}$ and correlation parameters $\{C_i C_j\}$:

$$E_{b.s.} = \frac{1}{N} \sum_{\mathbf{q}}' \left\{ \frac{1}{N^2} \sum_{i,j} e^{-i\mathbf{q}(\mathbf{R}_i - \mathbf{R}_j)} [\langle C_i C_j \rangle F^{(2)}(\mathbf{q}) + \right.$$
$$\left. + 2\langle C_i \rangle (F_{AB}^{(2)}(\mathbf{q}) - F_{BB}^{(2)}(\mathbf{q})) + F_{BB}^{(2)}(\mathbf{q})] \right\}, \tag{5.1.21}$$

where the characteristic functions $F^{(2)}_{\mu\nu}(\mathbf{q})$ for the transition metal alloys have the form

$$F^{(2)}_{\mu\nu}(\mathbf{q}) = \frac{4\pi z^*_\mu z^*_\nu e^2}{\Omega_0 q^2} e^{-q^2/4\xi} + \tilde{w}_\mu(\mathbf{q})\tilde{w}_\nu(\mathbf{q})\chi(\mathbf{q})\epsilon(\mathbf{q}) +$$
$$+ \tilde{W}_\mu(\mathbf{q}) \sum_{\lambda'_\nu} q^{(\nu)}_{\lambda'}(\mathbf{q})\pi^{(\nu)}_{\lambda'}(\mathbf{q}) + \tilde{w}_\nu(\mathbf{q}) \sum_{\lambda\mu} q^{(\mu)}_{\lambda}(\mathbf{q})\pi^{(\mu)}_{\lambda}(\mathbf{q}) +$$
$$+ \sum_{\lambda_\mu\lambda'_\nu} q^{(\mu)}_{\lambda}(\mathbf{q}) q^{(\nu)}_{\lambda'}(\mathbf{q})\pi^{(\mu\nu)}_{\lambda\lambda'}(\mathbf{q}). \tag{5.1.22}$$

The function

$$\chi(\mathbf{q}) = \frac{1}{N} \sum_{K < K_F} (\mathbf{k}^2 - |\mathbf{k}+\mathbf{q}|^2)^{-1} \tag{5.1.23}$$

is the susceptibility of the collectivized electron states;

$$\epsilon(\mathbf{q}) = 1 - (1 - G(\mathbf{q}))\frac{8\pi e^2}{\Omega_0 q^2}\chi(\mathbf{q}) \tag{5.1.24}$$

is the dielectric function which takes into account the contribution (through the $G(\mathbf{q})$ function) of the exchange and correlation effects;

$$\pi^{(\mu)}_\lambda(\mathbf{q}) = \frac{1}{N}\sum_{\mathbf{k}} \frac{(K^2 - E^{(\mu)}_\lambda)\vartheta(K^2 - E_F)}{(K^2 - |\mathbf{k}-\mathbf{q}|^2)[(K^2 - E^{(\mu)}_\lambda)^2 + (\omega^{(\mu)}_\lambda)^2]}, \tag{5.1.25}$$

and

$$\pi^{(\mu\nu)}_{\lambda\lambda'}(\mathbf{q}) =$$
$$= \frac{1}{N}\sum_{\mathbf{k}} \frac{(K^2 - E^{(\mu)}_\lambda)(K^2 - E^{(\nu)}_\lambda)\vartheta(K^2 - E_F)}{(K^2 - |\mathbf{k}+\mathbf{q}|^2)[(K^2 - E^{(\mu)}_\lambda + \omega^{(\mu)}_\lambda)^2][(K^2 - E^{(\nu)}_{\lambda'})^2 + (\omega^{(\nu)}_{\lambda'})^2]} \tag{5.1.26}$$

are the "susceptibilities" connected with the hybridized s and d states, respectively; z^*_μ is the effective valency of the ion of the μ kind; ξ is the Evald's parameter.

$$F^{(2)}(\mathbf{q}) = F^{(2)}_{AA}(\mathbf{q}) + F^{(2)}_{BB}(\mathbf{q}) - 2F^{(2)}_{AB}(\mathbf{q}). \tag{5.1.27}$$

The characteristic functions for noble metal alloys are written in the form

$$F^{(2)}_{\mu\nu}(\mathbf{q}) = \frac{4\pi z^*_\mu z^*_\nu e^2}{\Omega_0 q^2} e^{-q^2/4\xi} + \tilde{W}_\mu(\mathbf{q})\,\tilde{W}(\mathbf{q})\,\chi(\mathbf{q})\,\epsilon(\mathbf{q}) + $$
$$+ \tilde{W}_\mu(\mathbf{q}) \sum_{\lambda_\nu} g^{(\nu)}_{\lambda'}(\mathbf{q})\,\chi^{(\nu)}_{\lambda'}(\mathbf{q}) + \tilde{W}(\mathbf{q}) \sum_{\lambda_\mu} g^{(\mu)}_{\lambda}(\mathbf{q})\,\chi^{(\mu)}_{\lambda}(\mathbf{q}) +$$
$$+ \sum_{\lambda_\mu\lambda_\nu} g^{(\mu)}_{\lambda}(\mathbf{q})\,g^{(\nu)}_{\lambda'}(\mathbf{q})\,\chi^{(\mu\nu)}_{\lambda\lambda'}(\mathbf{q}), \tag{5.1.28}$$

where

$$\chi^{\mu}_{\lambda}(\mathbf{q}) = \frac{1}{N} \sum_{K > K_F} [(\mathbf{k}^2 - |\mathbf{k}+\mathbf{q}|^2)(K^2 - E^{(\mu)}_{\lambda})]^{-1}, \tag{5.1.29}$$

$$\chi^{(\mu\nu)}_{\lambda\lambda'}(\mathbf{q}) = \frac{1}{N} \sum_{K > K_F} [(\mathbf{k}^2 - |\mathbf{k}+\mathbf{q}|^2)(K^2 - E^{(\mu)}_{\lambda})(K^2 - E^{(\nu)}_{\lambda'})]^{-1}. \tag{5.1.30}$$

For the alloys of simple and polyvalent metals

$$F^{(2)}_{\mu\nu}(\mathbf{q}) = \frac{4\pi z^*_\mu z^*_\nu l^2}{\Omega_0 q^2} e^{-q^2/4\xi} + W_\mu(\mathbf{q})\,W_\nu(\mathbf{q})\,\chi(\mathbf{q})\,\epsilon(\mathbf{q}). \tag{5.1.31}$$

Given the probabilities $\langle C_i C_j \rangle$ are known, it is easy to write down the formulas for the electron energy for particular crystal structures. For example, in the case of an f.c.c. lattice, we have

$$E_{\text{b.s.}} = \frac{1}{N}\left\{ \sum_{\mathbf{q}=\mathbf{k}'_n} |W(\mathbf{q})|^2 + \frac{1}{16}\eta^2 \sum_{\mathbf{q}=\mathbf{k}''_n} |\Delta W(\mathbf{q})|^2 + \right.$$
$$\left. + \frac{1}{N} \sum_{\mathbf{q}}{}' \left[\left(c_A c_B - \frac{3}{16}\eta^2 \right) + \sum_{\rho \neq 0} \epsilon(\rho)\cos(\mathbf{q}\rho) \right] |\Delta W(\mathbf{q})|^2 \right\} \chi(\mathbf{q})\,\epsilon(\mathbf{q}). \tag{5.1.32}$$

The summation over $\mathbf{q} = \mathbf{k}'_n$ corresponds to that over the principal vectors of the reciprocal lattice while that over $\mathbf{q} = \mathbf{k}''_n$ is equivalent to the summation over the superstructure vectors. The first term is responsible for the energy of mixing which, as formula (5.1.32) shows, is determined by the average form-factor of the alloy $W(\mathbf{q}) = c_A W_A(\mathbf{q}) + c_B W_B(\mathbf{q})$; the second term responsible for the ordering or decomposition of the alloy is determined by the degree of the component form-factors, $\Delta W(\mathbf{q}) = W_A(\mathbf{q}) - W_B(\mathbf{q})$. The last term in (5.1.32) corresponds to the fluctuation energy.

The electron energy of the alloys of simple and polyvalent metals calculated in the third order in pseudopotential also depends on the three-particle correlation function [372, 373]

$$\langle E^{(3)} \rangle = \frac{1}{N^3} \sum_{q_1, q_2, q_3} \Big\{ N [c_A F^{(3)}_{AAA}(q_1, q_2, q_3) + c_B F^{(3)}_{BBB}(q_1, q_2, q_3) +$$

$$+ \sum_{i \neq j} e^{-iq_1(\mathbf{R}_i - \mathbf{R}_j)} [F^{(3)}_{BBB}(q_1, q_2, q_3) + 2 \langle C_i \rangle (F^{(3)}_{ABB}(q_1, q_2, q_3) -$$

$$- F^{(3)}_{BBB}(q_1, q_2, q_3)) + \langle C_i C_j \rangle F_2^{(3)}(q_1, q_2, q_3)] +$$

$$+ \sum_{i \neq j = k} e^{-iq_1(\mathbf{R}_i - \mathbf{R}_k)} e^{-iq_2(\mathbf{R}_j - \mathbf{R}_k)} [F^{(3)}_{BBB}(q_1, q_2, q_3) +$$

$$+ 3 \langle C_i \rangle (F^{(3)}_{ABB}(q_1, q_2, q_3) - F^{(3)}_{BBB}(q_1, q_2, q_3) +$$

$$+ 3 \langle C_i C_j \rangle F_2^{(3)}(q_1, q_2, q_3) + \langle C_i C_j C_k \rangle F_3^{(3)}(q_1, q_2, q_3)] \Big\} \Delta(q_1^+, q_2^+, q_3), \tag{5.1.33}$$

where

$$F^{(3)}_{\mu\nu\lambda}(q_1, q_2, q_3) = W_\mu(q_1) W_\nu(q_2) W_\lambda(q_3) \Lambda^{(3)}(q_1, q_2, q_3),$$
$$F_2^{(3)}(q_1, q_2, q_3) = \Delta W(q_1)[W_A(q_2) W_A(q_3) - W_B(q_2) W_B(q_3)] \times$$
$$\times \Lambda^{(3)}(q_1, q_2, q_3),$$
$$F_3^{(3)}(q_1, q_2, q_3) = \Delta W(q_1) \Delta W(q_2) \Delta W(q_3) \Lambda^{(3)}(q_1, q_2, q_3). \tag{5.1.34}$$

Some specific features of these expressions should be noted for the alloys of transition and noble metals which distinguish the factorization performed here from the results obtained in [374, 375] for the alloys of simple metals. The pseudopotential form-factors $W_\mu(\mathbf{q})$ and the matrix elements of the hybridization potentials $g_\lambda^{(\mu)}(\mathbf{q})$ prove to be dependent on the parameters of order and correlation, etc. Owing to this dependence, the behavior of transition metal alloys differs principally from that of the alloys of simple and polyvalent metals. Taking the charge distribution of the neighboring ions into account, the hybridization potential $\delta V_\mu(\mathbf{r} - \mathbf{R}_i)$ can be written in the form

$$\delta V_\mu(\mathbf{r}') = \delta V_\mu^{(0)}(\mathbf{r}') + \delta V_\mu^{(c)}(\mathbf{r}'), \tag{5.1.35}$$

where $\delta V_\mu^{(0)}$ is the potential of the valence electrons of the free atom minus the potential of free electrons in the alloy plus the potential produced by the orthogonalization hole centered at the given ion and by the difference in the exchange potentials of the free atom and the crystal. The second term , $\delta V_{\mu}^{(c)}$ defines the potential created by the charge distribution of the neighboring ions including the fraction originated from the orthogonalization holes centered on the neighboring ions:

$\mathbf{r}' = \mathbf{r} - \mathbf{R}_j$;

$$\delta V_{\mu}^{(c)}(\mathbf{r}') = \sum_{j \neq i} \left\{ C_i V_A^{(c)}(\mathbf{r}' - \mathbf{R}_j) + (1 - C_j) V_B^{(c)}(\mathbf{r}' - \mathbf{R}_j) \right\}. \tag{5.1.36}$$

This term gives the following contribution into the matrix element of the hybridization parameter

$$\sum_{\mu, d\mu} \langle \mathbf{k} + \mathbf{q} \mid \Delta_\mu \mid d_\mu \rangle \langle d_\mu \mid \Delta_\mu \mid \mathbf{k} \rangle = \frac{1}{N} \sum_{i,j}{}' e^{-i\mathbf{q}\mathbf{R}_i} \left\{ C_i C_j V_{AA}^{(c)}(\mathbf{R}_j) + \right.$$

$$\left. + C_i(1 - C_j) V_{AB}^{(c)}(\mathbf{R}_j) + (1 - C_i) C_j V_{BA}^{(c)}(\mathbf{R}_j) + (1 - C_i)(1 - C_j) V_{BB}^{(c)}(\mathbf{R}_j) \right\}, \tag{5.1.37}$$

where

$$V_{\mu\nu}^{(c)}(\mathbf{R}_j) = \frac{1}{\Omega_0} \sum_{\lambda_\mu, \lambda'_\nu} \left\{ \int e^{-i(\mathbf{k}+\mathbf{q})\mathbf{r}} v_\mu^{(c)}(\mathbf{r} - \mathbf{R}_j) \psi_\lambda^{(\mu)}(\mathbf{r}) \, d\mathbf{r} \int \psi_{\lambda'}^{(\nu)*}(\mathbf{r}') e^{i\mathbf{k}\mathbf{r}'} d\mathbf{r}' + \right.$$

$$\left. + \int e^{-i(\mathbf{k}+\mathbf{q})\mathbf{r}} \psi_\lambda^{(\mu)}(\mathbf{r}) \, d\mathbf{r} \int \psi_{\lambda'}^{(\nu)*}(\mathbf{r}') v_\nu^{(c)}(\mathbf{r}' - \mathbf{R}_j) e^{i\mathbf{k}\mathbf{r}'} d\mathbf{r}' \right\}. \tag{5.1.38}$$

Thus, the matrix element $\langle k + q / \tilde{W} / k \rangle$ turns out to be dependent on the pair correlation function, i.e., on η^2 and on the correlation parameter ε [363]. It means that the band structure energy calculated in the second order of the perturbation theory will depend not only on the pair correlation function but also on the three- and four-particle correlation functions. The same may be said about the overlap terms in the energy. Besides, the term in the structure-independent part of the electron energy E' will now be dependent on the structure since it is affected by the magnitude of the pair correlation function.

In other words, the factorization of the pseudopotential for transition and noble metal alloys would be meaningful only for some particular values of the order and correlation parameters. How strongly the pseudopotential form-factors and other matrix elements depend on the degree of order is to be solved in each individual case after the numerical calculations are carried out. However, the possibility of such dependence should be always kept in mind. The implications of this, important for the interpretation of experimental data and constructing statistical theories, will be discussed further.

Upon factorization, expressions (5.1.4) and (5.1.5) in which the summation over the lattice is singled out in an explicit form, are written as follows

$$E_{tot} = E_k^{(0)} + \frac{1}{N} \sum_{\mu = A} \sum_i C_i^\mu V_\mu(\mathbf{R}_i) + \frac{1}{2N} \sum_{\mu, \nu = A, B} \sum_{i \neq j} C_i^\mu C_j^\nu V_{\mu\nu}(\mathbf{R}_i + \mathbf{R}_j) +$$

$$+ \frac{1}{3!N} \sum_{\mu, \vartheta, \lambda = A, B} \sum_{i \neq j \neq k} C_i^\mu C_j^\nu C_k^\lambda V_{\mu\nu\lambda}(\mathbf{R}_i - \mathbf{R}_k, \mathbf{R}_j - \mathbf{R}_k), \quad (5.1.39)$$

Here, $V_{\mu\nu}(\mathbf{R}_i - \mathbf{R}_j)$ is the potential of the indirect pair interaction of the ions of the μ and ν kind located at the lattice sites $\mathbf{R}_i$ and $\mathbf{R}_j$ respectively; $V_{\mu\nu\lambda}(\mathbf{R}_i - \mathbf{R}_k, \mathbf{R}_j - \mathbf{R}_k)$ is the potential of the indirect three-particle interaction of the ions of the μ, ν, and λ kind at the sites $\mathbf{R}_i$, $\mathbf{R}_j$ and $\mathbf{R}_k$ respectively. The rearrangement of the terms in this expression has been made in such a way as to explicitly separate the energy contributions depending upon the coordinates of one, two, three, ... ions [72, 373]. Thus, we determine the contributions into the total electron energy which describe the pair, three-particle, ... effective interatomic interactions.

Attention should be given to the fact that the higher order terms of the perturbation theory will contribute to the lower orders of the interionic interactions. For example, the pair interaction incorporates the contributions related to the terms of the second, third and higher orders of the perturbation theory and the contribution from the overalp energy,

$$V_{\mu\nu}(\mathbf{R}_i - \mathbf{R}_j) = V_{\mu\nu}^{(2)}(\mathbf{R}_i - \mathbf{R}_j) + V_{\mu\nu}^{(3)}(\mathbf{R}_i - \mathbf{R}_j) + \ldots + V_{\mu\nu}^{ov}(\mathbf{R}_i - \mathbf{R}_j), \quad (5.1.40)$$

where

$$V_{\mu\nu}^{(2)}(\mathbf{R}_i - \mathbf{R}_j) = \frac{2}{N} \sum_{\mathbf{q}}{}' F_{\mu\nu}^{(2)}(\mathbf{q}) e^{-i\mathbf{q}(\mathbf{R}_i - \mathbf{R}_j)} \quad (5.1.41)$$

represents the contribution into the pair interaction corresponding to the second order of the perturbation energy;

$$V_{\mu\nu}^{(3)}(\mathbf{R}_i - \mathbf{R}_j) = \frac{6}{N^2} \sum_{\mathbf{q},\, \mathbf{p} \neq 0} F_{\mu\nu\lambda}^{(3)}(\mathbf{q}, \mathbf{p}, -\mathbf{q} - \mathbf{p}) e^{-i\mathbf{q}(\mathbf{R}_i - \mathbf{R}_j)} \tag{5.1.42}$$

gives the contribution into the pair interaction from the terms of the third order. The expressions for the characteristic functions $F_{\mu\nu}^{(2)}(\mathbf{q})$ and $F_{\mu\nu\lambda}^{(3)}(\mathbf{q}_1\, \mathbf{q}_2\, \mathbf{q}_3)$ are determined by the formulas (5.1.22) and (5.1.34). The physical meaning of the second term in (5.1.40) is determined by the fact that the energy $E^{(3)}$ comprises, in addition, the indirect pair interionic interaction resulting from the double scattering of an ion and single scattering of another ion.

The three-particle interaction has the form

$$V_{\mu\nu\lambda}^{(3)}(\mathbf{R}_i - \mathbf{R}_k, \mathbf{R}_j - \mathbf{R}_k) =$$

$$\frac{6}{N^2} \sum_{\mathbf{q},\, \mathbf{p} \neq 0} F_{\mu\nu\lambda}^{(3)}(\mathbf{q}, \mathbf{p}, -\mathbf{q} - \mathbf{p}) e^{-i\mathbf{q}(\mathbf{R}_i - \mathbf{R}_k)} e^{-i\mathbf{p}(\mathbf{R}_j - \mathbf{R}_k)}. \tag{5.1.43}$$

Thus, in alloys, as in pure metals, the interionic interaction via the conduction electrons exists that can be divided into the pair and many-particle interactions.

A certain part of the band-structure energy which cannot be expressed as the contributions to the interionic interactions entered the first term, $E_k^{(0)}$, in (5.1.39) together with other structure-independent contributions:

$$E_k^{(0)} = E_{f.e.} + \sum_{l \geq 1} E_0^{(l)}, \tag{5.1.44}$$

where

$$E_0^{(1)} = Z(c_A W_A(0) + c_B W_B(0)), \tag{5.1.45}$$

$$E_0^{(2)} = \frac{1}{N} \sum_{\mathbf{q} \neq 0} (c_A F_{AA}^{(2)}(\mathbf{q}) + c_B F_{BB}^{(2)}(\mathbf{q})), \tag{5.1.46}$$

$$E_0^{(3)} = \frac{1}{N^2} \sum_{\mathbf{q},\,\mathbf{p} \neq 0} (c_A F^{(3)}_{AAA}(\mathbf{q}, \mathbf{p}, -\mathbf{q}-\mathbf{p}) + c_B F^{(3)}_{BBB}(\mathbf{q}, \mathbf{p}, -\mathbf{q}-\mathbf{p}). \tag{5.1.47}$$

On the whole, the contributions into the interionic interactions are due to the direct electrostatic interaction (the first term in (5.1.22)) and to the indirect interaction via the conduction band electrons, d electrons and hybridized states. Besides, the interaction due to the overlap of d states and to the appearance of collectivized orbitals takes place, according to [362] (In this work we restricted ourselves to the consideration of the one-center orbitals, see (5.1.5))

$$\begin{aligned} V^{ov}_{\mu\nu}(\mathbf{R}_l - \mathbf{R}_j) = \sum_{\lambda_\mu \lambda'_\nu} \Bigg\{ & S^{(\mu\nu)}_{\lambda\lambda'}(\mathbf{R}_l - \mathbf{R}_j)\Delta^{(\mu\nu)}_{\lambda\lambda'}(\mathbf{R}_l - \mathbf{R}_j)\theta(E_F - E^{(\mu)}_\lambda) - \\ & - S^{(\mu\nu)}_{\lambda\lambda'}(\mathbf{R}_l - \mathbf{R}_j)\Bigg[\sum_{k<k_F} e^{-i\mathbf{k}(\mathbf{R}_l-\mathbf{R}_j)}((k^2 - E^{(\mu)}_\lambda)S^{(\nu)}_{k\lambda} + S^{(\nu)}_{\lambda'k}\Delta^{(\mu)}_{k\lambda} + \\ & + \Delta^{(\nu)}_{\lambda'k}S^{(\mu)}_{k\lambda}) + \sum_{\mathbf{k}} e^{-i\mathbf{k}(\mathbf{R}_l-\mathbf{R}_j)} \frac{\Delta^{(\nu)}_{\lambda'k}\Delta^{(\mu)}_{k\lambda}}{(k^2 - E^{(\mu)}_\lambda)^2 + (\omega^{(\mu)}_\lambda)^2}\nu(k^2 - E_F) - \\ & - \Delta^{(\mu\nu)}_{\lambda\lambda'}(\mathbf{R}_l - \mathbf{R}_j)\Bigg[\sum_{k<k_F} e^{-i\mathbf{k}(\mathbf{R}_l-\mathbf{R}_j)} S^{(\nu)}_{\lambda'k}S^{(\mu)}_{k\lambda} + \\ & + \sum_{\mathbf{k}} \frac{e^{-i\mathbf{k}(\mathbf{R}_l-\mathbf{R}_j)}}{(k^2 - E^{(\mu)}_\lambda)^2 + (\omega^{(\mu)}_\lambda)^2}\left(2S^{(\nu)}_{\lambda'k}S^{(\mu)}_{k\lambda}\frac{\Delta^{(\nu)}_{k\lambda'}\Delta^{(\mu)}_{k\lambda}}{(k^2 - E^{(\nu)}_{\lambda'})^2 + (\omega^{(\nu)}_{\lambda'})^2}\right)\vartheta(k^2 - E_F)\Bigg]\Bigg\}. \end{aligned} \tag{5.1.48}$$

Here,

$$\begin{aligned} S^{(\mu\nu)}_{\lambda\lambda'}(\mathbf{R}_l - \mathbf{R}_j) &= \int \psi^{(\mu)*}_\lambda(\mathbf{r})\,\psi^{(\nu)}_{\lambda'}(\mathbf{r} - (\mathbf{R}_l - \mathbf{R}_j))\,d\mathbf{r}, \\ \Delta^{(\mu\nu)}_{\lambda\lambda'}(\mathbf{R}_l - \mathbf{R}_j) &= \int \psi^{(\mu)*}_\lambda(\mathbf{r})\Delta_\mu(\mathbf{r})\,\psi^{(\nu)}_{\lambda'}(\mathbf{r} - (\mathbf{R}_l - \mathbf{R}_j))\,d\mathbf{r}, \\ S^{(\mu)}_{k\lambda} &= \frac{1}{\sqrt{\Omega_0}}\int e^{-i\mathbf{k}\mathbf{r}}\psi^{(\mu)}_\lambda(\mathbf{r})\,d\mathbf{r}, \\ \Delta^{(\mu)}_{k\lambda} &= \frac{1}{\sqrt{\Omega_0}}\int e^{-i\mathbf{k}\mathbf{r}}\psi^{(\mu)}_\lambda(\mathbf{r})\Delta_\mu(\mathbf{r})\,d\mathbf{r}, \end{aligned} \tag{5.1.49}$$

and $\psi^{(\mu)}_\lambda(\mathbf{r})$ is the d-electron wave function specified by the set of the quantum numbers λ_μ of the ion of the μ kind.

The derivation of formulas (5.1.41) and (5.1.48) underlines one of the merits of the PP method, namely, the possibility of going over to the description of the alloy energy in terms of the pair and many-particle interactions.

In general, the indirect interionic interaction would be of the long-range oscillatory character. As may be seen from formulas (5.1.39) - (5.1.48), (5.1.22), and (5.1.34), the amplitude of the interaction depends upon the matrix elements of the pseudopotential $W_{\mu}(\mathbf{q})$ as expressed by (5.1.18), of the hybridization potential $g_{\lambda}^{(\mu)}(\mathbf{q})$ defined by (5.1.20) and of the overlap integrals (5.1.49). The phase of the interaction is determined by the geometry and the parameters of the Fermi surface (through the susceptibilities $\chi(q)$, $\pi_{\lambda}^{(\mu)}(\mathbf{q})$ and $\pi_{\lambda\lambda'}^{(\mu\nu)}(\mathbf{q})$ given by (5.1.23), (5.1.25) and (5.1.26), respectively, and the three-vertex network $\Lambda^{(3)}(\mathbf{q}_1, \mathbf{q}_2, \mathbf{q}_3)$ expressed by formula (5.1.15)) as well as by the position $E_d^{(\mu)}$ and the width $W_d^{(\mu)}$ of the resonances. In other words, the amplitude of the interaction is determined, to a greater extent, by the atomic properties of the components of the alloy rather than the electronic structure. The variation of the potentials with distance depends chiefly on the subtle features of the electron energy spectrum of the alloy. Expressions (5.1.39) - (5.1.42) are written in the most general form. They are valid both for transition metal alloys and the alloys of noble, polyvalent and simple metals. The distinction consists only in the way of writing down the characteristic functions $F_{\mu\nu}^{(2)}(\mathbf{q})$ (the three-particle energy (5.1.43) is written only for the alloys of simple and polyvalent metals): for transition metal alloys expression (5.1.22) must be used, for noble metal alloys -- (5.1.28), and for the alloy of simple and polyvalent metals -- (5.1.31). The form of the potentials of the pair and many-particle interactions, as seen from (5.1.40) - (5.1.44), depends on a variety of factors that are determined by the electronic structure of the alloys. Along with the major factors considered in the present section which affect the form of the theoretically calculated potentials $V_{\mu\nu}(\mathbf{R}_i - \mathbf{R}_j)$, we obtain the means of taking into account the interelectron interaction in the one-electron approximation.

5.2 The energy of atomic ordering

The ordering energy is the fundamental entity in the problem of atomic ordering. The treatment in Section 5.1 of the structure of the contributions into the total energy of a solid solution allows one to proceed to analysis of the ordering energy.

The pair ordering energy can be obtained by comparing the expressions for the potentials of the pair interaction in Section 5.1 with the conventional definition of the pair ordering energy (see Section 3.2):

$$V(\mathbf{R}) = V_{AA}(\mathbf{R}) + V_{BB}(\mathbf{R}) - 2V_{AB}(\mathbf{R})$$

As a result, we arrive at the formula of quantum theory for $V(\mathbf{R})$. The expressions for the pair ordering energy (in the second order of the perturbation theory in the pseudopotential) are the same for all classes of metallic alloys [376]:

$$V_2^{(2)}(\mathbf{R}_i - \mathbf{R}_j) = \frac{2}{N} \sum_{\mathbf{q}}{}' F_2^{(2)}(\mathbf{q}) e^{-i\mathbf{q}(\mathbf{R}_i - \mathbf{R}_j)}. \tag{5.2.1}$$

The structure of the formula (5.2.2) for the alloys of transition, noble and simple metals differs by the form of the characteristic function $F_2^{(2)}(\mathbf{q})$. For transition metal alloys,

$$F_2^{(2)}(\mathbf{q}) = \frac{4\pi e^2(\Delta Z^*)^2}{\Omega_0 q^2} e^{-q^2/4\xi} + |\Delta \tilde{W}(\mathbf{q})|^2 \chi(\mathbf{q})\epsilon(\mathbf{q}) +$$
$$+ 2\Delta\tilde{W}(\mathbf{q}) \left[\sum_{\lambda_A} g_\lambda^{(A)}(\mathbf{q}) \pi_\lambda^{(A)}(\mathbf{q}) - \sum_{\lambda'_B} g_{\lambda'}^{(B)}(\mathbf{q}) \pi_{\lambda'}^{(B)}(\mathbf{q}) \right] +$$
$$+ \sum_{\mu,\nu = A,B} \sum_{\lambda_\mu} g_\lambda^{(\mu)}(\mathbf{q}) \left[g_\lambda^{(\mu)}(\mathbf{q}) \pi_{\lambda\lambda}^{(\mu\mu)}(\mathbf{q}) - (1-\delta_{\mu\nu}) \sum_{\lambda'_\nu} g_{\lambda'}^{(\nu)}(\mathbf{q}) \pi_{\lambda\lambda'}^{(\mu\nu)}(\mathbf{q}) \right]. \tag{5.2.2}$$

For noble metal alloys, $\omega_d^{(\mu)} = 0$ and

$$F_2^{(2)}(q) = \frac{4\pi e^2(\Delta Z^*)^2}{\Omega_0 q^2} e^{-q^2/4\xi} + |\Delta \tilde{W}(\mathbf{q})|^2 \chi(\mathbf{q})\epsilon(\mathbf{q}) +$$
$$+ 2\Delta\tilde{W}(\mathbf{q}) \left[\sum_{\lambda_A} g_\lambda^{(A)}(\mathbf{q}) \chi_\lambda^{(A)}(\mathbf{q}) - \sum_{\lambda'_B} g_{\lambda'}^{(B)}(\mathbf{q}) \chi_{\lambda'}^{(B)}(\mathbf{q}) \right] +$$
$$+ \sum_{\mu,\nu = A,B} \sum_{\lambda_\mu} g_\lambda^{(\mu)}(\mathbf{q}) \left[g_\lambda^{(\mu)}(\mathbf{q}) \chi_{\lambda\lambda}^{(\mu\mu)}(\mathbf{q}) - (1-\delta_{\mu\nu}) \sum_{\lambda'_\nu} g_{\lambda'}^{(\nu)}(\mathbf{q}) \chi_{\lambda\lambda'}^{(\mu\nu)}(\mathbf{q}) \right]. \tag{5.2.3}$$

For the alloys of simple metals [348 - 360],

$$F_2^{(2)}(q) = \frac{4\pi e^2(\Delta Z^*)^2}{\Omega_0 q^2} e^{-q^2/4\xi} + |\Delta\widetilde{W}(\mathbf{q})|^2 \chi(\mathbf{q})\epsilon(\mathbf{q}). \tag{5.2.4}$$

These expressions show that the pair ordering energy is determined by the square of the difference of the form-factors $\Delta W(\mathbf{q}) = W_A(q) - W_B(q)$ and of the effective valencies $\Delta Z^* = Z_A^* - Z_B^*$, and, in the case of the alloys of noble and transition metals, also by the difference of the matrix elements of the hybridization potential multiplied by the appropriate susceptibilities. The corresponding quantities have been defined in Section 5.1. The long-range character of the ordering energy is worth to be noted. The interaction ensuring the ordering extends over long distances. Another specific feature of the actual pair ordering energy is, in general, the lack of the spherical symmetry, i.e., V turns out to be the function of $V(\mathbf{R}_i - \mathbf{R}_j)$, but not the modulus $|\mathbf{R}_i - \mathbf{R}_j|$. Just for few alloys with the nearly spherical Fermi surface having the pseudopotential of almost local type, the ordering energy will be approximately spherical.

Simple theories of atomic ordering have usually restricted themselves by the use of the pair ordering energy. Quantum theory of solids gives evidence of the important part of the many-particle interionic interactions. The PP method enables the higher orders of the corresponding quantities to be calculated. In the third order of the perturbation theory in the pseudopotential one can find from (5.1.39) or (5.1.43) the following expression for the three-particle ordering energy (which is valid for the alloys of simple and polyvalent metals)

$$V_3^{(3)}(\mathbf{R}_i - \mathbf{R}_k, \mathbf{R}_j - \mathbf{R}_k) = \\ = \frac{6}{N^2} \sum_{\mathbf{q},\mathbf{p}\neq 0} F_3^{(3)}(\mathbf{q},\mathbf{p},-\mathbf{q}-\mathbf{p}) e^{-i\mathbf{q}(\mathbf{R}_i - \mathbf{R}_k)} e^{-i\mathbf{p}(\mathbf{R}_j - \mathbf{R}_k)}, \tag{5.2.5}$$

where

$$F_3^{(3)}(\mathbf{q}_1,\mathbf{q}_2,\mathbf{q}_3) = \Delta W(\mathbf{q}_1)\Delta W(\mathbf{q}_2)\Delta W(\mathbf{q}_3)\Lambda^{(3)}(\mathbf{q}_1,\mathbf{q}_2,\mathbf{q}_3). \tag{5.2.6}$$

The corresponding expressions for the alloys of transition and noble metals are much more cumbersome and will not be given here. It should be noted only that in transition metal alloys the many-particle interactions are substantially affected by the many-centered orbitals of d electrons. In other words, the role of many-particle interactions rises as the delocalization of the d states increases and diminishes again as these states become completely collectivized.

In addition, the third-order term of the perturbation theory, $E^{(3)}$, defined by (5.1.39), contributes to the pair ordering energy due to the double scattering on one ion and the single scattering on the other $(\mathbf{R}_i \neq \mathbf{R}_j = \mathbf{R}_k)$:

$$V_2^{(3)}(\mathbf{R}_i - \mathbf{R}_j) = = \frac{6}{N^2} \sum_{\mathbf{q},\, \mathbf{p} \neq 0} F_2^{(3)}(\mathbf{q}, \mathbf{p}, -\mathbf{q} - \mathbf{p}) e^{-i\mathbf{q}(\mathbf{R}_i - \mathbf{R}_j)}, \tag{5.2.7}$$

where

$$F_2^{(3)}(\mathbf{q}_1, \mathbf{q}_2, \mathbf{q}_3) = = \Delta W(\mathbf{q}_1)[W_A(\mathbf{q}_2) W_A(\mathbf{q}_3) - W_B(\mathbf{q}_2) W_B(\mathbf{q}_3)] \Lambda^{(3)}(\mathbf{q}_1, \mathbf{q}_2, \mathbf{q}_3). \tag{5.2.8}$$

This leads to the renormalization of the pair ordering energy. Taking the three-particle ordering energy into account proves to be principally important because it implies the nonpair, essentially noncentral, three-particle interaction. The presence of the many-particle contributions into the ordering energy may significantly complicate the structure of alloys and the general appearance of the corresponding phase diagram. This question has not been given the proper consideration in literature.

Along with the ordering energy, the important entity in the theory of atomic ordering is the Fourier transform. By the Fourier transformation of (5.2.1), (5.2.5) and (5.2.7), one can obtain

$$V_2^{(2)}(\mathbf{k}) = 2 \sum_{\mathbf{K}_n} F_2^{(2)}(\mathbf{k} + \mathbf{K}_n) - \frac{2}{N} \sum_{\mathbf{q} \neq 0} F_2^{(2)}(\mathbf{q}), \tag{5.2.9}$$

$$V_2^{(3)}(\mathbf{k}) = \frac{6}{N} \sum_{\mathbf{q} \neq 0} \sum_{\mathbf{K}_\Pi} F_2^{(3)}(\mathbf{k} + \mathbf{K}_\Pi, \mathbf{q}, -\mathbf{k} - \mathbf{K}_\Pi - \mathbf{q}) -$$
$$- \frac{6}{N^2} \sum_{\mathbf{q}, \mathbf{p} \neq 0} F_2^{(3)}(\mathbf{q}, \mathbf{p} - \mathbf{q} - \mathbf{p}), \tag{5.2.10}$$

$$V_3^{(3)}(\mathbf{k}, \mathbf{k}') = 6 \sum_{\mathbf{K}_\Pi, \mathbf{K}_l} F_3^{(3)}(\mathbf{k} + \mathbf{K}_\Pi, \mathbf{k}' + \mathbf{K}_l, -\mathbf{k} - \mathbf{K}_\Pi - \mathbf{k}' - \mathbf{K}_p) -$$
$$- \frac{6}{N} \sum_{\mathbf{q} \neq 0} \sum_{\mathbf{K}_\Pi} \Big[F_3^{(3)}(\mathbf{q}, \mathbf{k}' + \mathbf{K}_\Pi, -\mathbf{q} - \mathbf{k}' - \mathbf{K}_\Pi) + F_3^{(3)}(\mathbf{k} + \tag{5.2.11}$$

Here, $\mathbf{K}_\Pi$, and $\mathbf{K}_l$ are the vectors in the reciprocal lattice. For the alloys of transition, noble, polyvalent and simple metals, the Fourier transforms expressed in terms of the characteristic functions have the same form. The distinction is determined by the different form of the characteristic functions themselves.

The structure of short-range order at high temperatures is directly determined by the coordinates of the minima of the Fourier transform of the pair and many-particle ordering energy (see Section 3.2). Such connection is not quite evident near the transition temperature, but takes place even in the lowest approximation. The state with a long-range order is not always determined by the minimum in $V(\mathbf{k})$, $V(\mathbf{k}, \mathbf{k}')$, . . . Nevertheless, in some important cases the structure of long-range order is connected in a straightforward way with the minimum of the ordering energy corresponding to the wave vector of the low-symmetry star. Because all these factors determine the significance of the concrete position of the minima of $V(\mathbf{k})$, $V(\mathbf{k}, \mathbf{k}')$, the origin of these minima will be considered in detail.

For the alloys of simple and polyvalent metals, the Fourier transform of the pair ordering energy is determined by the characteristic functions $F_2^{(2)}(\mathbf{q})$ and $F_2^{(3)}(\mathbf{q}, \mathbf{p}, -\mathbf{q} - \mathbf{p})$ defined by (5.2.4) and (5.2.8) respectively. The ordering energy in the second-order, $V_2^{(2)}(\mathbf{k})$, depends on the direct electrostatic and indirect interactions (the first and second terms in (5.2.4)); the energy in the third and higher orders -- only on the indirect interaction, according to (5.2.8). The minima of $V(\mathbf{k})$, $V(\mathbf{k}, \mathbf{k}')$, etc. may be due either to the indirect interaction or to the combination of the latter with the direct one. The minima may be of two types of origin.

The first type appears accidentally. Since the direct and indirect interactions have opposite signs, their combination may give occasionally the minimum of $V_2^{(2)}(\mathbf{k})$. The same may occur if the second and higher orders in $V(\mathbf{k})$ compensate for each other. The positions of such minima cannot be predicted from the general standpoint. They can be revealed just after the numerical computations.

The second type is determined by the atomic properties of the components. Its origin is connected with the actual behavior of the pseudopotential form-factors of the elements making up the alloy. This is the term $|\Delta W(\mathbf{q})|^2$ in (5.2.4) and (5.2.8). The structure of the order may be predicted, in its main features, without additional computations by comparing the $\dot{W}_\mu(\mathbf{q})$ dependences for different components of the alloy. Assuming the Fermi surface to be spherical, the remaining factors involved in the expressions for the characteristic functions will be smooth functions of $|\mathbf{q}|$. Most of the quantum theories of ordering have been devised and all the calculations within the PP method performed just within such approximation [348 - 360]. Meanwhile, there exists the third cause for the appearance of the minimum of $V(\mathbf{k})$ related to the contribution into the characteristic function from $\chi(\mathbf{q})$ and $\epsilon(\mathbf{q})$, expressed by (5.1.23) and (5.1.24), respectively, and from $\Lambda^{(3)}(\mathbf{q}_1, \mathbf{q}_2, \mathbf{q}_3)$. These three functions describe the properties of the electronic subsystem of the alloy. In other words, atomic ordering may arise due to the peculiarities in the electronic structure of the disordered alloy; as a result, the structure of atomic order will be determined by the electronic structure. For the first time, his fact has received attention, from different positions in the works of Sato and Toth [52], Teramoto and Tachiki [377], Krivoglaz [378, 379], Moss [380, 381] and others. In the present work this question is considered in detail with the aid of the PP method (see Sections 5.3 and 5.4). For this aim, we have developed a variant of the method taking into account the departure of the Fermi surface from the sphere. The point is that the singularities in the electronic subsystem due to the specific topological features and local geometry of the Fermi surface entail the appearance of the minimum of $V(\mathbf{k})$ and yield a certain class of structures of short-range and long-range order.

Such approach provides genetically relation of the order-disorder phase transitions with other transformations in solids, such as the transition to the superconducting state, martensitic transformations, magnetic and segnetoelectric transitions, the formation of the charge density and spin density waves and so on.

In the alloys of noble (and, all the more, of transition) metals, as is evident from (5.2.2) and (5.2.3), the role of the electronic subsystem in the appearance of the minima of $V(\mathbf{k})$ is more pronounced. Here, besides the reasons listed for simple metals, the minimum of $V(\mathbf{k})$ may be connected with the resonance character of the d states, and with the overlap of the d shells of the neighboring ions.

It should be noted here that all the treatment in this section is restricted to the local form of the pseudopotential $\langle \mathbf{k} + \mathbf{q} \mid W \mid \mathbf{k} \rangle = W(\mathbf{q})$. In the case of the substantially nonlocal pseudopotential, the form-factor cannot be separated from the functions characterizing the electronic subsystem and, accordingly, one cannot distinguish the cases of ordering due to the atomic properties and the electronic structure of the alloy.

Interatomic interaction in alloys is essentially of a long-range nature. This is very important for the problem of atomic ordering. The way in which the ordering energy varies with the distance determines the relative stability of different superstructures (see Section 3.2), the temperature dependence of long-range order parameters (Section 4.4), the orientation of antiphase boundaries, and some other phenomena. It is customary for the interaction to decline as R^{-3}, but specific features of the electronic subsystem of the alloy stimulating phase transitions and, particularly, the order-disorder transitions, may cause a weaker damping ($\sim R^{-2}$ or even R^{-1}).

Consider this problem in some more detail. It is known that in the alloys of simple metals which possess the nearly spherical Fermi surface,

$$V(R) \sim \frac{\cos(2K_{F_0}R)}{R^3}. \tag{5.2.12}$$

These are the well-known Friedel oscillations [25] arising due to the logarithmic singularity of the susceptibility $\chi(q)$ at $q = 2K_{F_0}$. The above dependence of oscillations on R takes place only at T = 0 K. At non-zero temperatures, however, the singularities responsible for the long-range interaction will be spread out and the interatomic potential falls off more rapidly with distance. In the case of the spherical Fermi surface, the amplitude of oscillations should be multiplied by the factor $\exp\left(-\frac{\pi RT}{K_{F_0}}\right)$, according to [382]. All other variations, statistical displacements, imperfect long-range order, or many-electron effects also produce the enhancement of damping of the interatomic interaction with distance. Analysis of such damping has been carried out in several works; among them the investigation by Krivoglaz [383] is to be noted. In spite of damping, the long-range interaction in alloys remains significant also at non-zero temperature; this has been established even for liquid metals [384].

For transition metal alloys, the contribution of the indirect interatomic interaction into the ordering energy due to the band structure effects (at T = 0 K) may be written in the form [363]

$$\begin{aligned} V(R) &\sim A\cos(2K_F R)/R^n + (C_{AB} - C_{AA}) \times \\ &\times \cos[\sqrt{2E_d^{(A)} \pm 2\sqrt{(E_d^A)^2 + (W_d^A)^2}}\, R]/R^n + (C_{BA} - C_{BB}) \times \\ &\times \cos[\sqrt{2E_d^B \pm 2\sqrt{(E_d^{(B)})^2 + (W_d^B)^2}}\, R]/R^n. \end{aligned} \tag{5.2.13}$$

Here A, C_{AA}, C_{BB} and C_{BA} are the nonsingular factors in $F(q)$, dF/dq and $d^2F(q)/dq^2$ calculated at $q = 2K_F(A)$, $q = 2\sqrt{E_d^A}$ and $q = 2\sqrt{E_d^B}\,(C_{\mu\nu})$. They can be derived from (5.2) by the conventional procedure used in calculating asymptotic values: the singular terms should be kept under the sign of the integral over q, while the remaining terms are taken out of the integral. These terms depend on the long-range order, including the term proportional to η^4. It may be seen from (5.2.13) that the oscillations do not obey the Friedel law. The exponent n determining the decrease of $V(R)$ c with distance depends upon geometry of the Fermi surface. For the Fermi sphere of radius $K_F = K_{F_0}$ the exponent n = 3; for the cylindrical portion of the Fermi surface with the base radius $K_F = \rho$, n=2; and for the plane portion of the surface with the radius vector $K_F = K_0$, n = 1. Thus, the local geometry of Fermi surface is an important factor which characterizes the tendency of the alloy to the long-range interactions. These interactions increase abruptly if the Fermi surface contains the plane portions [378], i.e., just before the structure phase transitions. This is connected with the poor screening properties of electrons near these regions. This effect also causes the anisotropy of the interatomic interaction in alloys. The second and the third terms in (5.2.13) are due, by the origin, to the singularity in the polarizability of s and d hybridized states $\pi_\lambda^{(\mu)}(q)$. It is evident that each element may contribute to the long-range interaction, and the contribution may be different for various groups of the d states. To conclude the consideration of transition metal alloys, the contribution of the overlap energy into the interaction should be pointed out. It will depend on the actual shape of the cluster's collectivized orbitals on the nearest coordination shells and seems to be insignificant for distant shells.

For noble metal alloys, formula (5.2.13) must be changed by equating the width of the d resonance to zero ($\omega_\Lambda^{|\mu|}$=0). In this case, the period of oscillations depends on the energy of the resonant states while the amplitude -- on the long-range order parameter. In the alloys of simple metals just the first term in (5.2.13) remains. This case has been considered by Krivoglaz and Tju Hao [378].

The many-particle interactions in alloys also have the long-range character. For example, for the three-particle interaction, one can derive the analogous expressions describing the asymptotic behavior of the ordering energy which is determined by the band structure. If the Fermi surface contains the portions with a small curvature, the three-particle interaction will be damped as $(\sim R^{-1})$ i.e., it will contribute both to the long-range interaction and its anisotropy.

5.3 Driving forces of atomic ordering

Generally, the conditions for stability of the ordered state must be determined by the thermodynamical criterion of stability of a certain phase or mixture of phases. To formulate such a criterion for the case of the ordered phase, it is necessary to calculate thermodynamical potentials of existing phases, i.e., their internal energy and entropy as functions of temperature and concentration. With the aid of the formulas given in Chapters 4 and 5, such criterions can be formulated mathematically, but they will prove to be unwieldy and unsuitable to use. It has been shown in Section 3.2 that the comparatively simple absolute criteria of stability of ordered phases do not exist.

The simpler case can be considered, however, namely, the condition for the formation of an ordered phase from a solid solution near the absolute zero of temperature where the entropy terms are small. In this case the difference of the total energies must be negative:

$$\Delta E = E(\eta = 1) - E(\eta = 0) < 0. \tag{5.3.1}$$

In the energy of a solid solution (see Section 5.1), one can single out two sorts of contributions, the configurational ones, i.e., dependent explicitly upon the parameters of long-range order and correlation, and the volume contributions depending upon the order parameters indirectly. The former determine the driving force of ordering and the fulfillment of condition (5.3.1) is necessary. The second contributions can either promote or impede the formation of the ordered state; hence the fulfillment of criterion (5.3.1) is necessary with respect to the total internal energy.

Let us consider in more detail the **necessary conditions** for the formation of the ordered state as determined by the configurational contributions. If the volume alterations on ordering are small, we have

$$\Delta E' = \frac{1}{N}\left\{ \sum_{\mathbf{q}=\mathbf{K}''_n} F_2^{(2)}(\mathbf{q}) - \frac{1}{N}\sum_q{}' [C_A C_B + \sum_{\rho\neq 0} \epsilon(\rho)\cos(\mathbf{q}\rho)]\right\} F_2^{(2)}(\mathbf{q}). \tag{5.3.2}$$

This expression holds for the superstructure $L1_2$ of the stoichiometric composition and for the point ions. The first contribution is originated from the energy of the perfectly ordered alloy and is due to the change in the electron and electrostatic energy at the formation of the superstructure Brillouin zones. The second contribution stems from the energy of the alloy with a short-range order and is connected with the fluctuations (the first term in the square brackets) and correlation in the positions of atoms in the solid solution. The characteristic function $F_2^{(2)}(\mathbf{q})$ is determined in Section 5.1.

Consequently, the occurrence of the driving force of ordering owes to the competition of the configurational contributions into the energy of solid solutions with long- and short-range order. It follows from (5.3.2) that the tendency for the alloy to form long-range order increases, firstly, with the increase in the difference in valencies or effective valencies of the components of the alloy. In other words, the effects such as those of the orthogonalization hole or the charge transfer may contribute to atomic ordering. Secondly, such a tendency is determined by the properties of the collectivized and hybridized s and d states. The latter will contribute mostly into the value of ΔE'.

And finally, the conditions for the formation of an ordered state will depend upon the values of the overlap integrals for the quasilocalized states. We have presented in Chapters 1 and 4 a lot of experimental evidences confirming these conclusions. Nevertheless, the facts were also given there showing a great significance of the nonconfigurational contributions. Thus we will emphasize that the necessary condition for the formation of a superstructure is $\Delta E' < 0$, but the ordering will not take place unless the obligatory condition $\Delta E < 0$ is simultaneously verified.

In the foregoing section we have paid attention to the important role in the origin of the driving forces of atomic ordering of the singularities in $\chi(\mathbf{q})$ caused by the d resonances and the deviation of the Fermi surface from the sphere. Let us analyze this effect in more detail. At first we consider the simplest case of the alloys based on simple metals, then the shape of the Fermi surface is made gradually more complicated and we will pass on to the alloys of transition and noble metals and, finally, to the many-particle effects.

Simple metal alloys. In the case of the spherical Fermi surface, the susceptibility $\chi(\mathbf{q})$ defined by (5.1.23) has the form

$$\chi(\mathbf{q}) = -\frac{3\bar{Z}}{4k_{F_0}}\left(\frac{4k_{F_0}^2 - q^2}{8qk_{F_0}}\ln\left|\frac{q+2k_{F_0}}{q-2k_{F_0}}\right| + \frac{1}{2}\right), \tag{5.3.3}$$

where $\bar{Z} = C_A Z_A + C_B Z_B$ is the mean valency of the alloy, k_{F_0} is the radius of the Fermi surface. At $q = 2k_{F_0}$, the Kohn singularity takes place [32]:

$$\chi(\mathbf{q}) \sim (q - 2k_{F_0})\ln|q - 2k_{F_0}|. \tag{5.3.4}$$

The respective behavior of $\chi(q \approx 2k_{F_0})$ is shown in Fig. 87. Evidently, in this case the ordered state related to the electronic subsystem does not form. For alloys of this kind, the structure of the ordered state is determined by the atomic properties of the components.

If the cylindrical portion is present on the Fermi surface with the length L and the base radius ρ_0, then, at $q_1 = 2\rho_0$, the susceptibility $\chi(q_1)$ has the root singularity [385, 386]; see Fig. 87.

$$\chi(q_1) \sim L(q_1^2 - 4\rho_0^2)^{1/2}\theta(q_1 - 2\rho_0). \tag{5.3.5}$$

The arising singularity is not sufficient for the appearance of the long-range order. However, it may appear when a certain combination of atomic and electron properties exists in the alloy. If the Fermi surface comprises a nearly flat portion of the area S at the distance k_0 in the direction q_Z c with the main radii of curvature R_1 and R_2, then $\chi(q_Z)$ will have a strong logarithmic singularity [385, 386]

$$\chi(q_Z) \sim S\frac{q_Z - 2k_0}{\sqrt{R_1 R_2}}\ln|q_Z - 2k_0|. \tag{5.3.6}$$

The latter case may obviously entail the formation of a minimum of the Fourier transform of the ordering energy near the points determined by the condition (see (5.2.9) and (5.2.4))

$$|\mathbf{k} + \mathbf{K}_{n}| = 2k_0. \tag{5.3.7}$$

The physical sense of the considered singularity can be elucidated within the scheme of contiguous sections of the Fermi surface [32]. The singularities arise at the values of the $\mathbf{q}$, vectors joining the points on the Fermi surface in which the energies are the same but the velocities of electrons have the opposite signs. The results of the next section show that it is the electronic structure of the alloy that determines frequently the structure of the ordered state.

Noble and transition metal alloys. In the case of noble metal alloys, the screening conditions are more complicated, and in the adopted pseudopotential scheme there appears, in addition to the susceptibility of the collectivized s states $\chi(\mathbf{q})$ defined by (5.1.23), also the susceptibility $\chi_{\lambda\lambda'}^{(\mu\nu)}(\mathbf{q})$ connected with the hybridized s and d states (see formulas (5.1.30) and (5.1.29), respectively). In the alloys of transition metals, we have an analogous scheme of susceptibilities, but with the different form due to the non-zero width of the d resonances (see (5.1.26) for $\pi_{\lambda\lambda'}^{(\mu\nu)}(\mathbf{q})$ and (5.1.25) for $\pi_{\lambda}^{(\mu)}(\mathbf{q})$. Consider first the role of s and d hybridized states. The appearance of the essential singularity in the case of noble metal alloys is due to the occurrence of the cylindrical part on the Fermi surface [312, 363]. In this case,

$$\chi_{\lambda}^{(\mu)} \sim \begin{cases} \dfrac{L}{\sqrt{q_1^2 - 4E_{\lambda}^{(\mu)}}} \ln \left| \dfrac{\sqrt{q_1^2 - 4E_{\lambda}^{(\mu)}} + \sqrt{q_1^2 - 4\rho_0^2}}{\sqrt{q_1^2 - 4E_{\lambda}^{(\mu)}} - \sqrt{q_1^2 - 4\rho_0^2}} \right| \theta(q_1 - 2\rho_0) \\ \rho_0 > \beta_\mu \\ \rho_0 < \beta_\mu;\ q_1 > 2\beta_\mu \\ -\dfrac{L}{\sqrt{4E_{\lambda}^{(\mu)} - q_1^2}} \operatorname{arctg} \sqrt{\dfrac{q_1^2 - 4\rho_0^2}{4E_{\lambda}^{(\mu)} - q_1^2}}\ \theta(q_1 - 2\rho_0) \\ \rho_0 < \beta_\mu \\ q_1 < 2\rho_0 \end{cases} \tag{5.3.8}$$

where $\beta_\mu^2 = E_\lambda^{(\mu)}$. Here we have two cases: at $\rho_0 > \beta_\mu$ there exists the square root singularity at $q_1 = 2\rho_0$, while at $\rho_0 < \beta_\mu$ the discontinuity point of the second type appears. Therefore, the hybridized s and d states may cause the atomic ordering. If the Fermi surface contains the flat part [312, 363], we have

$$\chi_\lambda^{(\mu)}(q_Z) \sim \frac{S}{2\sqrt{E_\lambda^{(\mu)}}}\left(\frac{8\sqrt{E_\lambda^{(\mu)}}}{q_Z^2 - 4E_\lambda^{(\mu)}} \ln\left|\frac{q_Z - 2k_0}{q_Z + 2k_0}\right| + \right.$$
$$\left. + \frac{4q_Z}{q_Z^2 - 4E_\lambda^{(\mu)}} \ln\left|\frac{k_0 + \sqrt{E_\lambda^{(\mu)}}}{k_0 - \sqrt{E_\lambda^{(\mu)}}}\right|\right), \tag{5.3.9}$$

i.e., at $q_Z = 2k_0$, the logarithmic singularity occurs, while at $q_Z = 2\sqrt{E_\lambda^{(\mu)}}$ the discontinuity point of the second kind is present (Fig. 88). Evidently, the s and d hybridized states are the strong reason for the formation of atomic order in the alloy provided the flat part is available on the Fermi surface.

The quasilocalized d states may also be responsible for the stabilization of the ordered state. The necessary singularities are present on the cylindrical and flat parts of the Fermi surface.

$$\chi_{\lambda\lambda}^{(\mu\mu)}(q_1) \sim \begin{cases} \dfrac{L}{q_1^2 - 4E_\lambda^{(\mu)}}\left(\dfrac{2\sqrt{q_1^2 - 4\rho_0^2}}{\rho_0^2 - E_\lambda^{(\mu)}} + \right. \\ \left. + \dfrac{1}{\sqrt{q_1^2 - 4E_\lambda^{(\mu)}}} \ln\left|\dfrac{\sqrt{q_1^2 - 4E_\lambda^{(\mu)}} + \sqrt{q_1^2 - 4\rho_1^2}}{\sqrt{q_1^2 - 4E_\lambda^{(\mu)}} - \sqrt{q_1^2 - 4\rho_0^2}}\right|\right)\theta(q_1 - 2\rho_0) \\ \rho_0 < \beta_\mu;\ \rho_0 > \beta_\mu;\ q_1 > 2\beta_\mu \\ L\left(\dfrac{1}{4E_\lambda^{(\mu)} - q_1^2} \dfrac{\sqrt{q_1^2 - 4\rho_0^2}}{E_\lambda^{(\mu)} - \rho_0^2} + \right. \\ \left. + \dfrac{\sqrt{2}}{(4E_\lambda^{(\mu)} - q_1^2)^{3/2}} \operatorname{arctg}\sqrt{\dfrac{q_1^2 - 4\rho_0^2}{4E_\lambda^{(\mu)} - q_1^2}}\right)\theta(q_1 - 2\rho_0) \\ \rho_0 < \beta_\mu;\ q_1 < 2\beta_\mu. \end{cases} \tag{5.3.10}$$

The consideration of susceptibilities in Fig. 89 allows one to make the important conclusion that the quasilocalized d states take part in forming the structure of the ordered state. As in the case of the s and d hybridized states, pure d states can give rise to the two types of the minima of the Fourier transform of the ordering energy. The coordinates of the first minimum are connected with those of the singular points on the Fermi surface, the coordinates of the second minimum -- with the energy positions of the d resonances, $E_{\lambda}^{(\mu)}$.

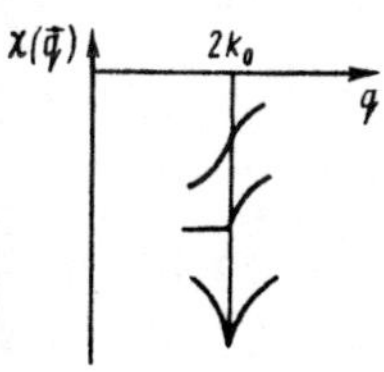

Fig. 87.

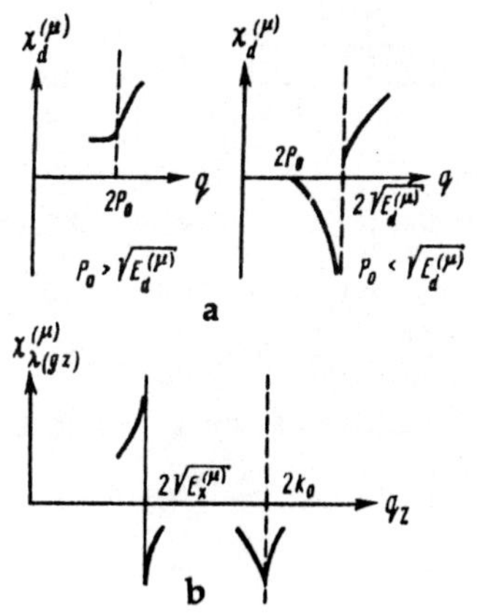

Fig. 88.

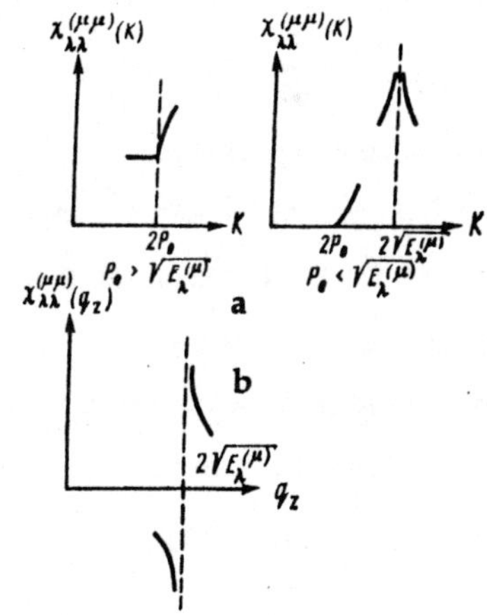

Fig. 89.

Fig. 87. Singularities in electronic susceptibility. Upper curve -- spherical Fermi surface, middle curve -- cylindrical Fermi surface; lower curve -- plane section on the Fermi surface.

Fig. 88. Singularities in electronic susceptibility of s and d hybridized states of noble metals: cylindrical Fermi surface (a), Fermi surface with plane section (b).

Fig. 89. Singularities in electronic susceptibility of quasilocalized d states for alloys of noble metals: cylindrical Fermi surface (a), Fermi surface with plane section (b).

In the alloys of transition metals the pattern described here for noble metal alloys becomes more involved. Due to the finite width of the d resonances, the singularities spread out and the coordinates of minima shift by the value proportional to the d resonance width. For example, in the case of the flat section on the Fermi surface, the discontinuity point of the second kind in the susceptibility $\pi_\lambda^{(\mu)}(\mathbf{q})$ disappears and we obtain

$$\pi_\lambda^{(\mu)}(q_Z) \sim S \, \frac{q_Z - 4E_\lambda^{(\mu)}}{(q_Z^2 - 4E_\lambda^{(\mu)})^2 + 16(\omega_\lambda^{(\mu)})^2}, \tag{5.3.11}$$

i.e., the minimum appears at $q_Z = 2\sqrt{E_\lambda^{(\mu)} + \omega_\lambda^{(\mu)}}$. Analogous changes would take place also in other cases.

The split of the d levels by the crystal field in the alloys of transition metals leads to the increase in the number of resonant levels. In these alloys the overlap of the d states also contributes to the ordering energy. The corresponding terms were not introduced into the formulas of Section 5.2 due to their complexity (see Section 5.1) and specific structure of the energy denominators which do not change the coordinates of the minima of $V(\mathbf{k})$.

To summarize the consideration of transition metal alloys, we note that, in comparison to the alloys of noble metals, the former offer a large number of ways in which the electronic structure may affect the atomic arrangement in the alloy.

Many-particle interionic interaction. Analysis of the third-order terms of the perturbation theory and the related three-particle interionic interaction shows that the latter contains new reasons that may lead to the minima of the Fourier transform of both the pair $V_2^{(3)}(\mathbf{k})$ and the three-particle $V_3^{(3)}(\mathbf{k}, \mathbf{k}')$ ordering energy. In the particular case of the alloys of simple and polyvalent metals, the appearance of the minima is connected with the singularities in the three-vertex network [387] :

$$\Lambda^{(3)}(\mathbf{q}_1,\mathbf{q}_2,\mathbf{q}_3)=\frac{1}{N}\sum_{k<k_F}\left\{\frac{1}{(k^2-|\mathbf{k}+\mathbf{q}_1|^2)(k^2-|\mathbf{k}-\mathbf{q}_2|^2)}+\right.$$
$$\left.+\frac{1}{(k^2-|\mathbf{k}+\mathbf{q}_3|^2)(k^2-|\mathbf{k}-\mathbf{q}_1|^2)}+\frac{1}{(k^2-|\mathbf{k}+\mathbf{q}_2|^2)(k^2-|\mathbf{k}-\mathbf{q}_3|^2)}\right\}. \qquad (5.3.12)$$

If the Fermi surface is spherical, the structure of singularities in $\Lambda^{(3)}(\mathbf{q}_1,\mathbf{q}_2,\mathbf{q}_3)$ is generally of the square root type [371]:

$$\Lambda^{(3)}(\mathbf{q}_1,\mathbf{q}_2,\mathbf{q}_3)\sim(k_{F_0}^2-q_R^2)^{1/2}\theta(k_{F_0}-q_R), \qquad (5.3.13)$$

(q_R is the radius of the circle circumscribed about the acute-angle triangle that is built on the vectors $\mathbf{q}_1$, $\mathbf{q}_2$, $\mathbf{q}_3$). The partial case of degeneracy $\mathbf{q}_2=-\mathbf{q}_1$, $\mathbf{q}_3=0$) corresponds to the logarithmic singularity [388]

$$\Lambda^{(3)}(\mathbf{q}_1,\mathbf{q}_2,\mathbf{q}_3)\sim\ln|q_S-2k_{F_0}|. \qquad (5.3.14)$$

There also remain the Kohn singularities:

$$\Lambda^{(3)}(\mathbf{q}_1,\mathbf{q}_2,\mathbf{q}_3)\sim(4k_{F_0}^2-q_i^2)\ln|q_i-2k_{F_0}|. \qquad (5.3.15)$$

The four-particle interaction is associated with the "tetrahedron" singularity in the four-vertex network $\Lambda^{(4)}$ [388]. For the Fermi sphere, the logarithmic singularity takes place:

$$\Lambda^{(4)}(\mathbf{q}_1,\mathbf{q}_2,\mathbf{q}_3,\mathbf{q}_4)\sim\ln|k_{F_0}^2-\kappa_R^2|, \qquad (5.3.16)$$

Thus, the many-particle interaction in the alloys of simple and polyvalent metals includes, even for the spherical Fermi surface, the driving forces of atomic ordering related to the electronic subsystem of the disordered state. The pair interaction does not exhibit such effect in the spherical case. This fact gives evidence of a nontrivial role of many-particle interactions in the problem of atomic ordering and of the need to take them into account in constructing the consistent theory of this phenomenon. Attention must be paid to to the fact that the many-angle singularities are important for polyvalent metal alloys when the first reciprocal vectors satisfy the inequality $|\mathbf{K}_{II}|<2k_{F_0}$.

$$\Lambda^{(4)}(\mathbf{q}_1,\mathbf{q}_2,\mathbf{q}_3,\mathbf{q}_4)\sim(\kappa_{F_0}^2-\kappa)^{1/2} \qquad \kappa<k_F. \qquad (5.3.17)$$

Analogous effects, complicated by the occurrence of d resonances will take place in the alloys of noble and transition metals. The formation of atomic configurations in solid solutions is also affected by the many-particle interactions.

Other features of the electronic subsystem determining the driving forces of atomic ordering. The foregoing discussion showed the important role of the electronic subsystem in forming the structure of the alloy. The treatment was based, as far as possible, on simple approximations for the sake of clearness. In the real metallic alloys some other effects are present in the electronic structure which may cause the appearance of new minima in $V(\mathbf{k})$. The most essential of them are listed below.

First, the shape of the Fermi surface in the real alloys prove to be more complicated than in the cases considered above. In transition metal alloys, there may occur the Fermi surfaces resembling "monsters", crimped cylinders, needles, chains, "dog bones", pockets, etc. The general theory of singularities for such surfaces has not yet been devised but preliminary investigations show that we can make quite interesting observations here. For example, a straightforward complication of the Fermi surface, namely, the occurrence of two flat sections on it, gives rise to the singularities of the following kind:

$$\chi(q_1, q_Z) \sim -\pi \frac{4}{q_1^2} \left\{ \frac{1}{2}\left(\frac{q_Z}{2} \pm k_0\right)\sqrt{\left(\frac{q_Z}{2} \pm k_0\right)^2 - \frac{q_1^2}{4q^2}(4k_F^2 - q^2)} - \right.$$
$$- \frac{q_1^2}{8q^2}(4k_F^2 - q^2)\ln\left|\left(\frac{q_Z}{2} \pm k_0\right) + \sqrt{\left(\frac{q_Z}{2} \pm k_0\right)^2 - \frac{q_1^2}{4q^2}(4k_F^2 - q^2)}\right| -$$
$$\left. - \frac{q_1}{4q}\sqrt{4k_F^2 - q^2} - \frac{q_1^2}{8q^2}(4k_F^2 - q^2)\ln|4k_F^2 - q^2| \right\} +$$
$$- \frac{\pi}{q_1}\left\{ \frac{q_Z}{2q_1}\left[-\frac{q_Z}{2} \pm \frac{q_1}{2q}\sqrt{4k_F^2 - q^2} \right]^2 + \right.$$
$$\left. + \frac{q^2}{2q_1}\left(-\frac{q_Z}{2} \pm \frac{q_1}{2q}\sqrt{4k_F^2 - q^2} \right) \right\}. \tag{5.3.18}$$

The interaction of the Fermi surface with the Brillouin zone boundaries [389] modifies the structure of singularities in the susceptibility. The number of such examples can be significantly extended due to the variety of structures of the Fermi surface in alloys.

Secondly, the simplest approximations for the susceptibility of the electronic subsystem described in this section virtually corresponds to noninteracting electrons. Account of the electron interaction can be performed in two ways. The first consists of substituting $\chi_0(\mathrm{q}), \Lambda_0^{(3)}$, etc. for $\chi(\mathrm{q}), \Lambda^{(3)}$, etc., the quantities for a gas of interacting electrons. Another way is connected with the direct incorporation into the Hamiltonian of the alloy the terms corresponding to the many-fermion interaction. There exist serious grounds to assert (see, for example, [390]) that new singularities will emerge in this case in the Fourier transform of the ordering energy.

The third reason for the appearance of additional singularities in $V(\mathbf{K})$, is the departure from the random phase approximation and account of the nonlinear response of the electronic system to external perturbation.

Before closing this section, we want to make one essential remark. The appearance of the described singularities is caused by the sharp decrease in the energy distribution of electrons, i.e., the above formulas are valid only for an ideal alloy at T = 0 K. The Fermi level will be smeared out at non-zero temperatures due to the presence of short-range order and fluctuations in alloys, electron-phonon and electron-magnon interactions, etc. It can be shown [383] that, although the singularities are smeared out, the minima of the above-considered type in $V(\mathbf{K})$ will remain.

Thus, the appearance of the minima in the Fourier transform of the ordering energy and, respectively, of the particular structure of short-range and long-range order is determined, to a great extent, by the electronic structure of the alloy in the disordered state. The main factor in the formation of atomic order and its structure are instabilities in the electronic subsystem determined by the state and the properties of small groups of electrons which are located near the specific sections of the Fermi surface or have resonanse energies. In this connection, we can see the similarity between the driving forces in the order-disorder transformation and in other phase transformations in solids. The driving forces have a common origin, though the paths of transformations, i.e., their nature, are different. Their choice is dictated by the trend of the system to the minimum of the Gibbs free energy and by specific kinetic features of the transformation conditions.

5.4 The nature of the ordering phenomenon in the Cu-Au system

Various factors of the electron structure which can give rise to the formation of the ordered state have been considerd from the general standpoint in Sections 5.1 - 5.3. The present section is devoted to the calculations aimed at the check of the efficiency of the theoretical formalism developed in the preceeding sections and, simultaneously, at the quantitative analysis of the nature of forces responsible for atomic ordering. We chose as an example the Cu - Au system for which detailed calculations have been carried out.

The Fourier transform of the ordering energy and the contributions into it [391 - 395]. The appropriate formulas are given in Section 5.2. We employed in the calculations the form-factors of Moriarty for Cu and Au [345]. The calculations have been performed for the stoichiometric Cu_3Au, CuAu, and $CuAu_3$. The required experimental parameters were taken from the literature. The differences in effective valencies after renormalization are given in Table 26. The Fourier transform of the ordering energy was calculated for the cross section of 1/8 of the Brillouin zone assuming the spherical Fermi surface for the electronic susceptibility $\chi(q)$. The Fourier transform has a form of a cupola which is highly pronounced at l = 0, but for each alloy is smoothed out at greater l values; the height of the cupola rapidly decreases with l. The top of the cupola for $CuAu_3$ is more acute than that for Cu_3Au. As was established previously, the coordinates of the minimum of $V(\mathbf{K})$ determine the type of the superstructure under formation and the structure of short-range order only within a certain approximation. Nevertheless, they remain an interesting characteristic of the alloy.

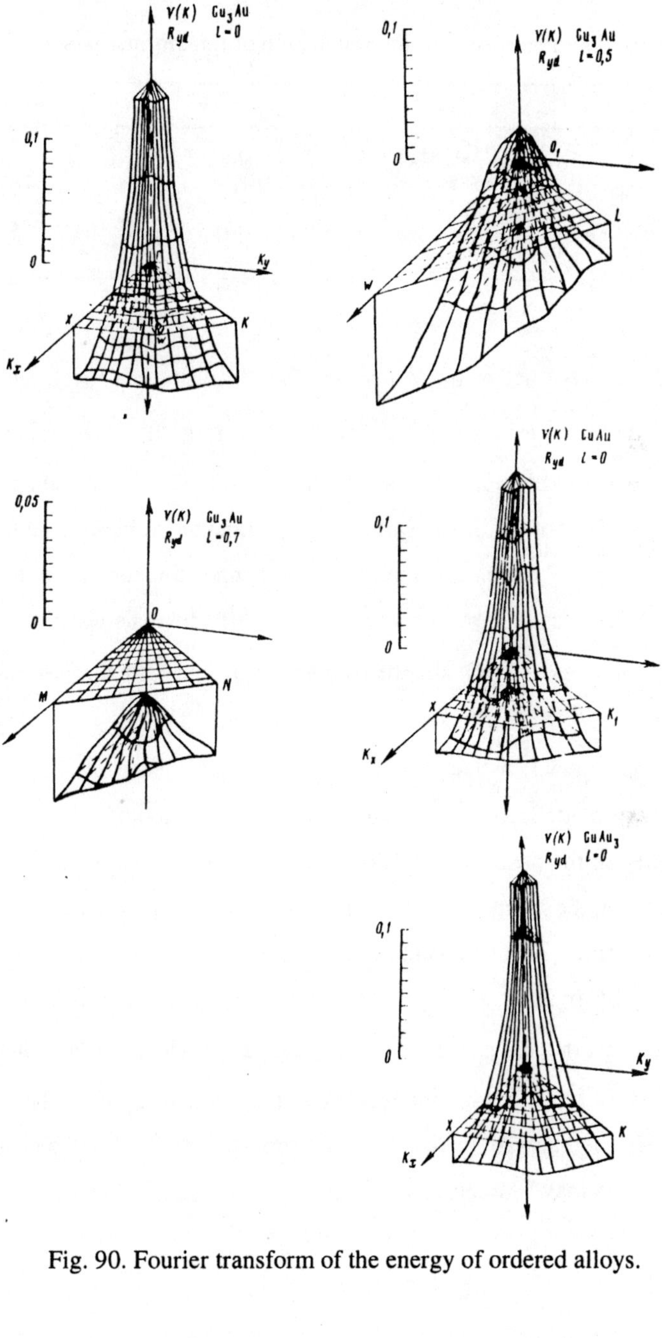

Fig. 90. Fourier transform of the energy of ordered alloys.

Table 26. Parameters used in the calculation of transformation temperature.

Alloy.	Cu_3Au	CuAu	Au_3Cu
$V(001)$	−0,0432	−0,0299	−0,0209
T_c, (theory), K;	1188	968	675
T_c (exp.), K.	663	683	477
Δz^*	0,2079	0,1897	0,1739

It should be noted that many relative minima form on the $V(\mathbf{K})$ surface, being only slightly different in the $V(\mathbf{K})$ values. Therefore, the determination of the coordinates of the absolute minimum is not a trivial task. The calculated coordinates of relative minima do not coincide with the experimental values of coordinates of superstructure vectors for ordinary and long-period superstructures near stoichiometric compositions. There exist three reasons for the difference between the calculated coordinates of minima of $V(\mathbf{K})$ and the coordinates of superstructure vectors estimated experimentally.

First, as was shown in Section 3.2, the exact coincidence of these coordinates should not be expected. Second, the approximate representation of the Fermi surface by a sphere adopted in the present calculation also does not promote such coincidence. In this approximation, the formation of long-range order is due mainly to the difference in the atomic properties of the components, whereas the finer features of the band structure of the disordered alloy are not taken into account. Third, in calculating $V(\mathbf{K})$, the fact that the alloys of the Cu-Au system are made up from noble metals has been accounted for only in an indirect way: the resonance terms have not been calculated separately but were included into the form-factors. We shall show below that the last two reasons may be important.

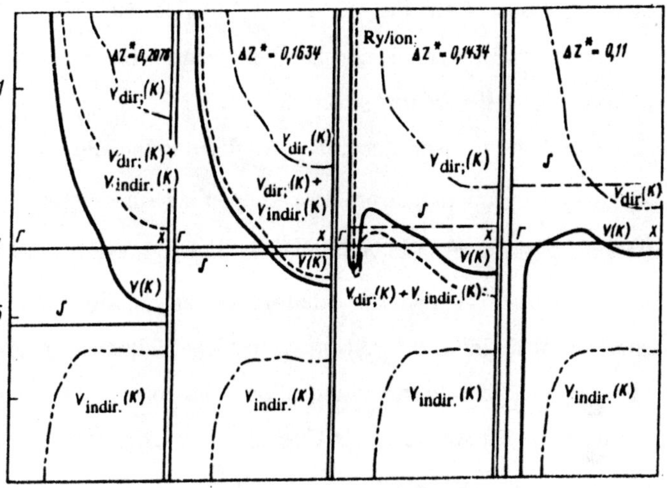

Fig. 91. $N(\kappa)$ along the ΓX direction and different contributions in it (direct interaction $V_{dir;j}(\kappa)$, indirect interaction $V_{indir.}(\kappa)$ and self-interaction $\int$) for the Cu_3Au alloy.

The $V(\mathbf{K})$ values in high-symmetry points of the Brillouin zone and the order-disorder transformation temperatures estimated with the use of these values in the simplest (GBW) approximation are given in Table 26. The order of T_c is predicted exactly but in order to reach a better agreement between the theory and experiment, the influence of the band structure factor upon $V(\mathbf{K})$ should be taken into account and the formulas for T_c should be based on more sophisticated statistical treatment. However, it is worth pointing out that the quantum-mechanical calculation performed in this section for the three compositions has predicted the possibility of forming in the Cu-Au system of both the high-symmetry and long-period superstructures with close energies.

It would be of interest to examine the role of the main contributions into the formation of $V(\mathbf{K})$. For this purpose, they are presented in Fig. 91 for the main directions in the Brillouin zone of the Cu_3Au alloy. According to (5.2.9) and (5.2.4), $V(\mathbf{K})$ includes three contributions: the direct interaction (the first term in (5.2.4)), the indirect interaction (the second term in (5.2.4)), and the self-interaction (the last term in

(5.2.9)). As may be seen from Fig. 91, the ordering is determined by the indirect interionic interaction via the conduction band electrons, wheras the direct interaction impedes the formation of the ordered structure.

The self-interaction (or, to be exact, its part related to the alloy) plays an important role. Although its contribution does not depend upon the wave vector K and therefore does not affect the coordinates of the minimum of $V(\mathbf{K})$, it exerts a decisive influence on the sign of V(0 0 1) thereby determining the possibility of forming the superstructure. To be definite, the self-interaction contribution is negative and, as a result, the long-range order forms in the alloy. The role of the self-interaction is important for the entire Cu-Au system. In $CuAu_3$, the ordering is also due to this effect [394]. Thus we have emphasized once more the important part of the structure-independent volume contributions in atomic ordering [396]. The self-interaction represents to some extent such a contribution.

Many ordering parameters including the coordinates and the depth of the minimum of $V(\mathbf{K})$ are influenced by the the difference in the effective valencies ΔZ^*. For the Cu_3Au alloy, the value of ΔZ^* varied in the range 0.2 to 0.08. As $V(\mathbf{K})$ decreases, the contribution of the direct interaction is lowered, but at the same time the indirect contribution also decreases. Starting from ΔZ^* 0.14, the self-interaction changes the sign and begins to impede the formation of order. Nevertheless, the long-range order persists all over the range of ΔZ^*, but the transformation temperature decreases with decreasing ΔZ^*(Fig. 92). Therefore, as ΔZ^*rises, the long-range order becomes more stable. The calculation showed that a similar behavior was observed also for the CuAu and $CuAu_3$ alloys. Note that despite the fact that the ordering is hindered by the contribution from the direct interaction and the latter increases with ΔZ^*, the stability of the ordered state also increases with ΔZ^*. This apparent contradiction will be discussed and resolved further.

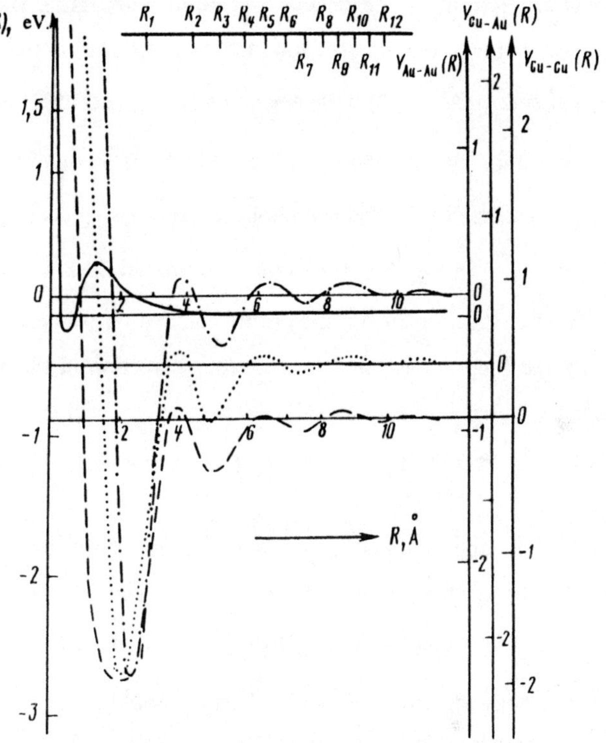

Fig. 92. Effect of the difference in effective valencies on the ordering temperature: Cu_3Au (1); Au_3Cu (2).

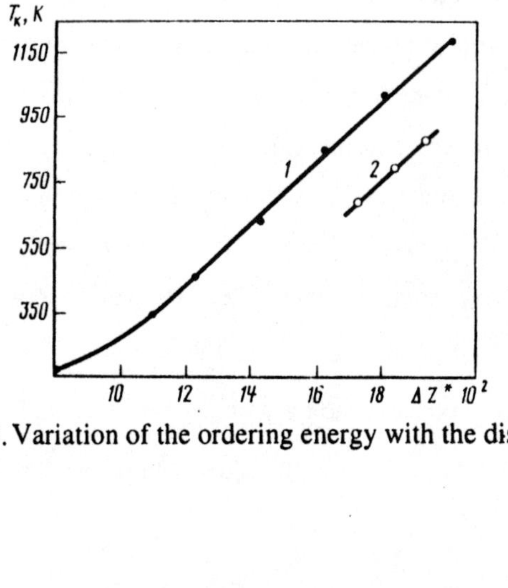

Fig. 93. Variation of the ordering energy with the distance.

The role of the long-range interaction in atomic ordering. Based on the same approximation as in the foregoing section, the calculations of the ordering energy $V(\mathbf{R})$ in the real space [397, 398] have been carried out with the use of the formulas derived in Section 5.2. The pair interaction potentials $V_{Cu-Cu}(R)$, $V_{Au-Au}(R)$ and $V_{Cu-Cu}(R)$ representing different contributions to the ordering energy and the function $V(\mathbf{R})$ itself are shown in Fig. 93. The radii of coordination shells are indicated. It is evident that the interaction in Cu-Au alloys (the calculation was carried out for T = 0 K) is essentially the long-range one. The indirect interaction in these alloys does not compensate for the direct interaction. The ordering energy oscillates in a complicated manner, the contributions from the distant shells, at least up to the fifth one, are significant.

In the three-shell approximation, the data concerning the ordering energy were placed on the stability diagram taken from Section 3.2. Only for the Cu_3Au alloy, the corresponding point falls into the region of stability of the $L1_2$ phase; its position, however, does not coincide with the place where it should be in accordance with the experimental data on the APB energy. For the remaining alloys, the point on the stability diagram does not lie in the range of stability of the $L1_2$ phase. We can see some reasons for that. First, the calculation of the interatomic interaction in the real space is connected with a mutual compensation of two contributions (associated with the direct and indirect interactions) whose values are two orders of magnitude greater than the required interaction. This calculation is quite a subtle issue. Second, the approximation of the Fermi surface by the sphere adopted in the present and preceding sections was shown to be insufficiently correct when treating some particular features of the interaction. For example, there is a discrepancy between the experimentally observed and calculated coordinates of the Cu–Au, minimum. The fact that the point characterizing the alloy falls beyond the stability range of the $L1_2$ superstructure is also connected with the spherical approximation. Although this approximation predicts correctly the ordering in the Cu-Au system and discloses the nature of the main contribution to the driving force of the phenomenon (and describes the structure of short-range order as well), it is inadequate for describing finer details of the ordering depending upon the behavior of $V(\mathbf{k})$ or $V(\mathbf{R})$ in the particular points of the reciprocal or real space. The interaction in this approach turns out to be spherically symmetric though the determination of the superstructure type requires a certain kind of anisotropy.

Account of topology of the real Fermi surface [399, 312, 221]. To do this account, one must, first, calculate the fine geometric features of the Fermi surface, at least near the sections where a small curvature is expected. Near the point K =(3/4, 3/4, 0) in the f.c.c. lattice (all the distances are given henceforth in the units 2π/a, while the energies and the values of the matrix elements -- in the units $(2\pi/a)^2$), there exists the degeneracy of the three states: $|\mathbf{k}\rangle$ =(3/4, 3/4, 0), $|\mathbf{k}+\mathbf{k}_1\rangle$=(--1/4, --1/4, 1) and $|\mathbf{k}+\mathbf{k}_2\rangle$=(--1/4, --1/4, 1), where $\mathbf{k}_1$ =(--1,--1,1) and $\mathbf{k}_2$ = (--1,--1,--1) are the vectors in the reciprocal lattice. By employing the perturbation technique for the degenerate states, which is connected with the restriction of the secular equation [57], one can show that the shape of the Fermi surface near the point K is determined in the three-wave approximation by solving the equation

$$\begin{vmatrix} \frac{3}{\sqrt{2}}x-\lambda & W_1^{(3)} & W_1^{(3)} \\ W_1^{(3)} & -\frac{x}{\sqrt{2}}+2Z-\lambda & W_2^{(3)} \\ W_1^{(3)} & W^{(3)} & -\frac{x}{\sqrt{2}}-2Z-\lambda \end{vmatrix} = 0, \tag{5.4.1}$$

where W_1 = W(111), W_2 = W(2 0 0), $\lambda=E_F - 9/8 - x^2 - y^2 - z^2$; the coordinates x, y, z denote the componens of the deviation from the point K. The restriction of the secular equation of the infinite order leads, as shown in [57], to the necessity of using in the calculations of the "renormalized" pseudopotential

$$W^{(n)}(\mathbf{k}_1) = \langle \mathbf{k}+\mathbf{k}_i \mid W \mid \mathbf{k}\rangle + \sum_{\mathbf{k}_n} \frac{\langle \mathbf{k}+\mathbf{k}_i \mid W \mid \mathbf{k}+\mathbf{k}_n\rangle\langle \mathbf{k}+\mathbf{k}_n \mid W \mid \mathbf{k}\rangle}{k^2 - |\mathbf{k}+\mathbf{k}_n|^2}, \quad (5.4.2)$$

where the summation extends over all vectors in the reciprocal lattice which are not taken into account in the basic approximation and n is the multiplicity of degeneracy. In our case the vectors $\mathbf{k}_1$ and $\mathbf{k}_2$ are excluded and we obtain

$$\langle \mathbf{k}+\mathbf{k}_n \mid W \mid \mathbf{k}\rangle = \frac{3}{4} W_{\omega}(\mathbf{k}_n) + \frac{1}{4} W_{\mathrm{Au}}(\mathbf{k}_n).$$

The calculations yield the values $W_1^{(3)} = 0.1374$ and $W_2^{(3)} = 0.0808$. The corrections $\delta E_F^{(111)}$ and $\delta E_F^{(200)}$ accounting for the interaction of the Fermi surface with the planes (111) and (200) of the first Brillouin zone are needed to be included in the calculation of the Fermi energy E_F. For the plane (111) we have $(k^2 - q_1^2/4) < -|W(111)|$, and near point L of the Brillouin zone the Fermi surface touches the plane (111) without crossing it. So, we obtain

$$\delta E_F^{(111)} = \frac{q_1}{4k_{F_0}}\left(\frac{q_1}{2} - k_{F_0}\right)^2 - \frac{(W_1^{(2)})^2}{2q_1 k_{F_0}}\left(\frac{1}{2} - \ln\left|\frac{W_1^{(2)}}{2q_1}\right|\right), \quad (5.4.3)$$

For the plane (200), according to [56],

$$\delta E_F^{(200)} = -\frac{W_2^{(2)}}{4q_3 k_{F_0}} \ln\left|\frac{q_2 + 2k_{F_0}}{q_2 - 2k_{F_0}}\right|. \quad (5.4.4)$$

Here q_1 and q_2 are the moduli of the vectors (111) and (200). For the twofold degenerate state near points L and X, the corrections are $\delta E_F^{(111)} = -0.0076$ and $\delta E_F^{(200)} = -0.0014$, and $E_F = 0.5456$. With the use of the calculated values $W_1^{(3)}$, $W_2^{(3)}$ and E_F, the shape of the Fermi surface was estimated from equation (5.4.1); its cross sections by the planes (000) and (110) are shown in Fig. 94. The so-called "truncation factor" t [52] defining the deviation of the Fermi surface from the sphere, can be deduced from the solution of the equation (5.4.1); in our case, we have $t = k_0/k_F = 0.9475$.

It is clear from Fig. 94 that the Fermi surface for the alloy Cu_3Au along the direction <110> has the plane section with k_0=0.7409. In this case, as was shown above, the Fourier transform of the ordering energy will contain the minima whose positions are determined by the condition $|\mathbf{k} + \mathbf{k}_n| = 2k_0$. The construction made by Kohn and Moss (Fig. 100) shows that the coordinates of the four minima of $V(\mathbf{k})$ near the site (110) will be determined by the value of $(10^{1/2}M)$ where M is the antiphase period and

$$\frac{1}{2M} = \sqrt{\left(\frac{12Z}{\pi}\right)^{1/2} t^2 - 1} - 1. \tag{5.4.5}$$

Putting the calculated t factor into this equation gives M = 5.405, which well agrees with the experimental value M = 5 – 6 [400, 401].

Thus, the account of the real form of the Fermi surface ensures much better agreement between the theory and experiment. The coordinates of the minimum of $V(\mathbf{k})$ alter and the use of the Kohn-Moss construction allows one to predict the antiphase period in accordance with the experimental data. At the same time, we find that the parameters of the forming long-period structure are stabilized by the indirect interionic interaction which will be screened weaker in a definite direction because of the singularity in the dielectric properties of the collectivized electrons due to the occurrence of the nearly plane section on the Fermi surface.

The nature of the ordering energy did not change but the cause for the formation of a particular superstructure changed after the real geometry of the Fermi surface has been accounted for. The results of the present sections allow one to arrive at another conclusion. As a matter of fact, the position of the Fermi surface, namely, its proximity

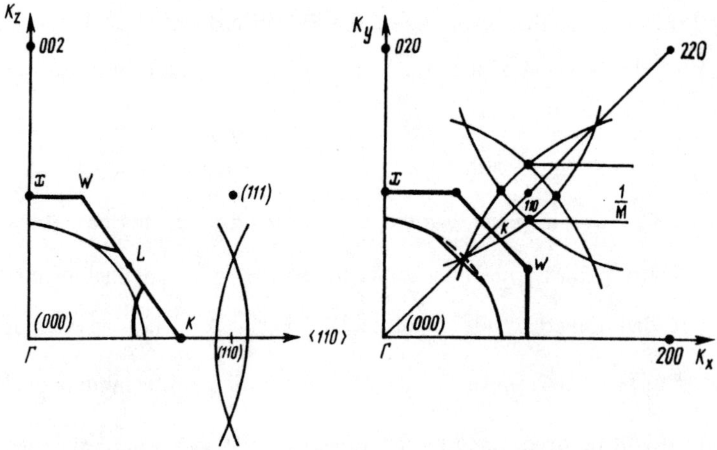

Fig. 94. Theoretical cross sections of the Fermi surface and Kohn--Moss diagram for the Cu_3Au alloy.

to tops and edges of the Brillouin zone determines the contribution of the many-particle interaction. The calculation made here shows that the three-particle interaction may be, at least, as important for ordering as the pair interaction on which the theory has been relied upon so far. We shall discuss this problem in more detail further. In this section we restricted ourselves to the treatment of the ordering problem in the approximation which did not take explicitly the contribution from the resonant d states into account. There has been a certain ground for this in the case of the Cu-Au system. Therefore, the conclusions formulated here may be referred not only to the alloys of this system, but also to alloys of simple and polyvalent metals.

The calculation of the short-range order structure in the Cu-Au system. Knowledge of the Fourier transform of the ordering energy $V(\mathbf{k})$ enables the structure of short-range order to be calculated in alloys of the Cu-Au system [392 - 395]. At the temperatures far above T_C, the Fourier transform of the short-range order parameter $\alpha(\mathbf{k})$ has the form

$$\alpha(\mathbf{k}) \sim \frac{C_A(1-C_A)}{1+C_A(1-C_A)\,V(\mathbf{k})/kT_K}. \tag{5.4.6}$$

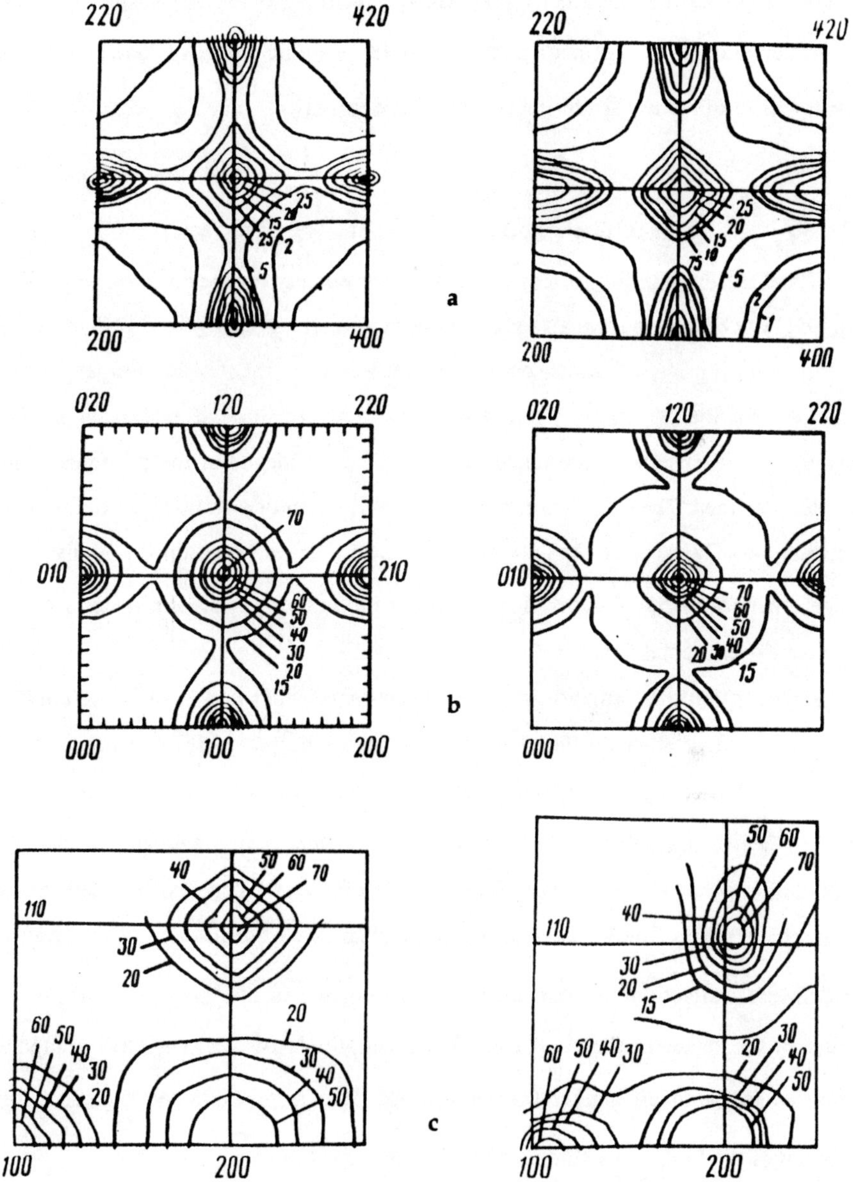

Fig. 95. Theoretical and experimental diffuse scattering patterns: Cu_3Au (a); CuAu (b); $CuAu_3$ (c). Left side -- theory, right side -- experiment.

It is most convenient to directly compare the intensity of diffuse scattering with the experimental data. The former can be calculated using the values of $\alpha(\mathbf{k})$ and $V(\mathbf{k})$. The intensity of diffuse scattering is proportional to $\alpha(\mathbf{k})$:

$$I(\mathbf{q}) = N\left|f_A - f_B\right|^2 \alpha(\mathbf{k}), \tag{5.4.7}$$

where $f_A - f_B$ is the difference between the atomic scattering factors.

The diffuse scattering pattern calculated theoretically for the (100) plane is given in Fig. 95. For comparison, we also present the experimental results from [258, 402 - 404]. With the exception of some particular features, the theoretical and experimental patterns coincide for all the three alloys. Not only qualitative agreement in diffuse scattering patterns should be noted but also the coincidence of the positions of many isointensity lines. The calculation predicts the formation of a short-range order near the compositions Cu_3Au and CuAu, while near the $CuAu_3$ composition not only a short-range order forms but a short-range decomposition takes place. The same is observed with the experimental data.

It means that the formation of atomic short-range order in the whole system Cu-Au is related, first of all, with the difference in the atomic properties of Cu and Au. The point is that in changing the concentration from Cu_3Au to $CuAu_3$, the band structure remains nearly unchanged, i.e., all variations are due to the difference in the form-factors of Cu and Au and to the change in the effective valencies. It is interesting to compare the alloys Cu_3Au and $CuAu_3$. The appearance of diffuse scattering around the fundamental reflections is connected with the change in the form of $\alpha(\mathbf{k})$ near the center of the band. In the Au_3Cu alloy, the amplitude of the long-wave components turns out to be much greater than in Cu_3Au. It can be explained by the different behavior of the long-wave components of the Fourier transform of $V(\mathbf{k})$.

The distinctions in the details of the theoretical and experimental scattering patterns are due, first, to the effects determined by the shape of the Fermi surface and the many-particle interaction which have not been taken into account in the present work and, second, to the fact that the experimental data have been obtained at temperatures not much higher than T_C , in which case relationship (5.4.6) resulted from the linearization is not quite correct.

It should be noted in conclusion that the nonparametrical calculation by the pseudopotential method has provided good agreement with the experimental diffuse scattering pattern. This pattern and, respectively, the structure of short-range order above T_C are determined by the behavior of $V(\mathbf{k})$ throughout the Brillouin zone. It is evident that these properties can be successfully described on the assumption that the Fermi surface is spherical. On the other hand, as was shown earlier, the calculation of the properties highly dependent on the behavior of $V(\mathbf{k})$ in the particular points of the Brillouin zone requires more accurate approaches.

The calculation of the total energy. Here we present the results of the calculations of the total energies and their structure-dependent constituents for different superstructures formed in this system [405]. We use the same approximation, namely, the spherical Fermi surface in the pseudopotential method. In the present section, the form-factors for pure Cu and Au calculated by Animalu [406] are employed, with the appropriate renormalization for the alloy. The formulas for the $L1_2$ superstructure were given in Section 5.1.

The electrostatic energy was calculated in the approximation of the point ions

$$E_{e.e} = -\frac{\alpha \bar{Z}^2}{R_0} - \beta\frac{(\Delta Z)^2}{R_0}\eta^2 + \frac{2\pi}{\Omega_0}(\Delta Z)^2 \sum_{\rho \neq 0} \alpha(\rho)n(\rho)\int j_0(q\rho)\frac{1}{q^2}\,dq, \tag{5.4.8}$$

where α is the Madelung constant, $\bar{Z} = C_A Z_A + C_B Z_B$ is the mean valency of the alloy, $R_0 = (3\Omega_0/4\pi)^{1/3}$; $\Delta Z = Z_A - Z_B$, $\alpha(\rho)$ is the short-range order parameter, and $j_0(k)$ is the Bessel function. For the $L1_2$ superstructure, $\alpha = 1.79172$, $\beta = 0.881160$ [356] and for the system Cu - Au, $\Delta Z = 0$.

The second order of the perturbation theory. The $L1_2$ superstructure. By using formulas (5.1.32), (5.1.33) and (5.1.34), the total energies can be represented by

$$E_{tot} = E_{e.e.} + E_{f.e.} + E^{(1)} + E_{ex,cor} + E_{b.s.}, \tag{5.4.9}$$

where

$$E_{f.e.} + E^{(1)} = \bar{Z}\left[\frac{3}{5}k_F^2 + \lim_{q\to 0}\left(\overline{\langle k|w_0|k+q\rangle} + \frac{8\pi Z}{\Omega_0 q^2}\right.\right. \tag{5.4.10}$$

is the energy of free electrons;

$$E_{ex,\,cor} = \bar{Z}\left(-0{,}115 + 0{,}031\ln r_s - \frac{0{,}916}{r_s}\right) \tag{5.4.11}$$

is the exchange-correlation energy of free electrons with the density r_S [407];

$$E_{b.s.} = E_1 + E_2 + E_3 + E_4 + E_5. \tag{5.4.12}$$

The successive terms in (5.4.12) are:

$$E_1 = \sum_{q = K_{OCH}} |W(q)|^2 \chi(q)\epsilon(q) \tag{5.4.13}$$

is the contribution of the band structure into the energy of mixing;

$$E_2 = \frac{1}{16}\eta^2 \sum_{q = K_{CC}} |\Delta w(q)|^2 \chi(q)\epsilon(q) \tag{5.4.14}$$

is the contribution into the ordering energy due to the appearance of the new Brillouin zones;

$$E_3 = C_A C_B \frac{1}{N} \sum_{q \neq 0} |\Delta w(q)|^2 \chi(q)\epsilon(q) \tag{5.4.15}$$

is the electron contribution into the energy of the concentration fluctuations;

$$E_4 = -\frac{3}{16}\eta^2\frac{1}{N}\sum_{\mathbf{q}\neq 0}|\Delta w(\mathbf{q})|^2\chi(\mathbf{q})\epsilon(\mathbf{q}) \tag{5.4.16}$$

is the band-structure contribution into the energy of fluctuations of the long-range order parameter; and

$$E_5 = \frac{1}{N}\sum_{\mathbf{q}\neq 0}\sum_{\rho\neq 0}\epsilon(\rho)\cos(\mathbf{q}\rho)|\Delta w(\mathbf{q})|^2\chi(\mathbf{q})\epsilon(\mathbf{q}) \tag{5.4.17}$$

is the electronic component of the correlation energy.

The calculations were based on different approximations for the exchange and correlation effects in the dielectric screening $\epsilon(\mathbf{q})$. We chose the following approximations [407, 408]:

(a) random phase approximation (RPA) with $G(\mathbf{q}) = 0$;

(b) the approximation of Hubbard-Sham with $G(\mathbf{q}) = \dfrac{q_2}{2(q^2 + k_{F_0}^2 + k_s^2)}$;

(c) the approximation of Singwi, with $G(\mathbf{q}) = A[1 - \exp(-Bq^2/K_{F_0})]$.

The results of the second-order perturbation theory calculations for the alloys Cu_3Au and $CuAu_3$ with the $L1_2$ superstructure are represented in the generalized form in Table 27. In this table,

$$E_C = E_2 + E_4 + E_5 \tag{5.4.18}$$

is the structure-dependent constituent of the alloy energy and

$$\Delta E_{tot} = E_{tot}(\eta) -- E_{tot}(\eta = 0) \tag{5.4.19}$$

is the contribution characterizing the total energy as a function of the long-range order parameter and correlation.

Consider first the results for the Cu_3Au alloy. The first conclusion regards a considerable influence of the exchange and correlation effects on the results of the calculation. For example, in the Singwi approximation a complete order proves to be energetically unfavorable in comparison to a perfect disorder. However, the RPA and the Hubbard-Sham treatments give the correct relationship between the energies in the states of perfect order and full disorder. The second, somewhat unexpected result, is that the energy of an alloy rises as the short-range order forms in the disordered state. We used in the calculation the short-range order parameters derived from experimental data.

It is worthwhile noting in this context that the calculation of the alloy with short-range order is somewhat inadequate since the static displacements have not

Table 27. Analysis of stability of the ordered state: parameters @[delta]E and @[delta]Q.

Alloy;	Model;	ΔE and ΔQ values, kal/gr			
		$\Delta E_m^{(2)}(\epsilon = 0)$	$\Delta E_n(\epsilon \neq 0)$	$\Delta E_c(\epsilon = 0)$	$\Delta E_c(\epsilon \neq 0)$
Cu_3Au	RPA	−4,5	−10,6	−1,0	−7,0
	Hubbard−Sham	−3,5	−9,28	−1,9	−7,7
	Singwi	–	−1,9	–	–
Au_3Cu	RPA	−4,39	–	−1,35	–
	Hubbard−Sham	−4,58	–	−1,73	–
	Singwi	–	–	−2,1	–

Alloy;	Model;	ΔE and ΔQ values, kal/gr.			
		ΔQ_n	ΔQ_c	ΔE экс
пер.	ΔQ exper.
Cu_3Au	RPA	6,1	3,5	2,6−6,5	1,1−2,3
	Hubbard−Sham	7,0	5,4	2,6−6,5	1,1−2,3
	Singwi	2,2	–	2,6−6,5	1,1−2,3
Au_3Cu	Au_3Cu	3,42	0,95	–	1,0−1,5
	Hubbard−Sham	3,98	1,35	–	1,0−1,5
	Singwi	1,5	1,7	–	1,0−1,5

been taken into account. The Animalu form-factors coupled with the spherical approximation for the Fermi surface seem to produce another structure of short-range order than the Moriarty form-factors. As compared with short-range order, the long-range order is predicted by all three approximations used in the calculation of the total energy. It gives evidence of the important part the volume effects play in the ordering energy. The fact is that in the Singwi approximation the structure contribution alone does not yield the ordering and if it had been the case, a long-range order would have been stabilized only by virtue of the volume effects. In general, we must underline the sensitivity of the total energy to the value of the lattice constant (to verify that we did not minimized in this calculation the total energy by varying the lattice parameter but determined the energy only for the values corresponding to the experimental lattice constant). In the RPA and Hubbard-Sham approximation, the ordering is provided by the structure-dependent contribution into the energy ($\sim \eta^2$).

As for the heat of transition and the difference in the total energies, the results of the calculation not involving any fitted parameters should be recognized as quite good. In our opinion, only the sign of the short-range order contribution slightly violates a general agreement between the theory and experiment. We can say therefore that the calculation based on the RPA or the Hubbard-Sham approach gives the best fit to the experimental differences in the total energies whereas the Singwi approximation proves to be most successful in predicting heats of transition. The calculation shows that the structure-dependent part gives the main contribution into the energy variation on ordering. However, the volume-dependent contribution is also significant. The latter promotes the atomic ordering in the Cu-Au system. In Table 27 we give the ultimate results of the corresponding calculation for the alloy $CuAu_3$. The conclusions concerning this alloy are virtually the same as for the Cu_3Au alloy.

The third order of the perturbation theory. The $L1_2$ superstructure. We have already emphasized the important role of the higher orders of perturbation theory and many-particle interionic interactions in alloys. Such calculations are quite laborious, however, and have been carried out in some particular cases [371, 374]. In this section we give the results of such calculation for the Cu_3Au alloy.

The electron energy in the third PP perturbation order may be written as

$$\langle E^{(3)}\rangle = \Big\{ \sum_{q_1 q_2 q_3 = K_{\text{осн}}} W(q_1)W(q_2)W(q_3) +$$
$$+ \frac{1}{16}\eta^2 \sum_{\substack{q_1 q_2 = k_{\text{осн}} \\ q_3 = k_{\text{осн}}}} \Delta w(q_1)\Delta w(q_2)[3w_B(q_3) + c_A \Delta w(q_3)] +$$
$$+ \frac{1}{64}\eta^3 \sum_{q_1, q_2, q_3 = k_{CC}} \Delta w(q_1)\Delta(q_2)\Delta w(q_3) +$$
$$+ \frac{1}{8} C_A \eta^2 \cdot \sum_{\substack{q_2, q_3 = k_{CC} \\ q_1 = k_{\text{осн}}}} \Delta w(q_1)\Delta w(q_2)\Delta w(q_3)\Big\} \Lambda^{(3)}(q_1, q_2, q_3)\Delta(q_1 + q_2 + q_3) +$$
$$+ \frac{3}{N} \sum_{q_1 \neq 0} \Big\{ \sum_{q_2 = k_{\text{осн}}} \Big[\Big(C_A C_B - \frac{3}{16}\eta^2\Big) + \sum_{\rho \neq 0} \epsilon(\rho) e^{-iq_1\rho}\Big] \times$$
$$\times \Delta W(q_1)\Delta W(q_2 - q_1)\Delta W(-q_2) +$$
$$+ \sum_{q_2 = k_{CC}} \Big[\frac{1}{16}(1 - 2C_A)\eta^2 - \frac{1}{32}\eta^2\Big] \Delta W(q_1)\Delta W(q_2 - q_1)\Delta W(-q_2)\Big\} \times$$
$$\times \Lambda^{(3)}(q_1, q_2 - q_1, -q_2) +$$
$$+ \frac{1}{N^2} \sum_{q_1, q_2 = 0} \Big\{\Big(C_A - 3C_A^2 + 2C_A^3 + \frac{9}{4} C_A \eta^2 - \frac{9}{16}\eta^2 + \frac{3}{16}\eta^3\Big) +$$
$$+ 3(1 - 2C_A) \sum_{\rho = 0} \epsilon(\rho) e^{-i(q_1 + q_2)\rho} +$$
$$+ \sum_{\rho_1, \rho_2 \neq 0} \epsilon(\rho_1 \rho_2) e^{-iq_1\rho_1} e^{-iq_2\rho_2} \Big\} \Delta W(q_1)\Delta W(q_2)\Delta W(q_1 - q_3) \times$$
$$\times \Lambda^{(3)}(q_1, q_2, -q_1 - q_2). \qquad (5.4.20)$$

The first sum corresponds to the contribution into the energy of mixing that appears if all three vectors q_1, q_2, q_3 are basic ($k_{\text{осн}}$). The second sum describes the contribution into the ordering energy due to the renormalization procedure: two vectors are basic and one is the superstructure vector. The third term has no analogs in the second order of the perturbation theory: it corresponds to the specific third-order contribution to the ordering energy. The same meaning has the fourth term. The remaining terms are due to the energy of fluctuations and the correlation energy. The fifth term gives the renormalized contribution from the energy of fluctuations and pair correlation which is analogous to the second-order contribution. The sixth term takes account of new types of fluctuations. And finally, the seventh term describes the

Table 28. Contributions into total energy in 3rd order in pseudopotential (Ry/ion).

model.	η	$E_n^{(2)}$	$E_n^{(3)}$	$E_n^{(2)}+E_n^{(3)}$	$E_c^{(2)}$	$E_c^{(3)}$	$E_c^{(2)}+E_c^{(3)}$
RPA	0,998	−0,8133	0,0919	−0,7214	0,0003	−0,0035	−0,0038
	0,000	−0,8119	0,0960	−0,7159	–	–	–
Hubbard–	0,998	−0,8213	0,1075	−0,7138	−0,0006	−0,0049	−0,0055
Sham	0,000	−0,8202	0,1122	−0,6080	–	–	–

energy of the three-particle correlations. Of the above third-order contributions, only the first sum gives the volume-dependent contribution, all the others are structure-dependent, i.e., they are related to the order parameter and correlations. A large number of contributions depending on the structure and having no analogs in the case of pure metals shows the important role of the higher perturbation order in calculating the energy of alloys. Formulas (5.4.20) were used to calculate the volume and the structure contributions to the energy of a solid solution. Final results of the calculations presented in Table 28 are obtained in the RPA and Hubbard-Sham approximations.

The third order contribution into the volume-dependent part of the total energy is small compared with that from the second order contribution. This gives evidence that the perturbation theory series for the volume-dependent part of the energy converges well. A different situation takes place for the contributions depending on the order parameters. Though quite small in absolute values, these contributions, not violating a common convergence of the pseudopotential series, point out simultaneously to the essential role of the higher orders of the perturbation theory in the problem of atomic ordering. The fact is that the energy difference in ordering is of the order of 10^{-3} Ry/ion.

The contribution of the third-order terms into the structure energy $E_c^{(3)}$ exceeds this value; therefore, in order to complete the quantum-mechanical calculation, the terms up to, at least, those of the sixth order should be included into consideration. The situation is completely analogous to that encountered in the calculation of the phonon spectra of polyvalent metals [371]. It should also be noted that in the third order of the perturbation theory, like in the second one, the ordering is provided by the contribution from the structure-dependent terms but the volume-dependent terms also give a significant contribution into the transformation energy.

Long-period superstructures [409]. In the Cu - Au system near the Cu_3Au and $CuAu_3$ stoichiometries, the long-period superstructures, namely, one-dimensional, $L1_2(M)$, or even two-dimensional, $L1_2(MM)$, are formed. Consider the superstructure of the $L1_2$ type with one-dimensional periodic antiphase boundaries (APB). For periodic APBs, the structure factor $S^{(\alpha)}$ can be written in the form

$$S^{(\alpha)}(\mathbf{q}) = \frac{1}{2\mu M} \sum_{j(\alpha)} e^{-i\mathbf{q}\mathbf{R}_j}. \tag{5.4.21}$$

Here μ is the number of atoms per unit cell of the alloy not containing APBs; the summation extends over all atoms of the α kind in the unit cell.

Now, for the planes of the new Brillouin zones specified by the vector $G(n_1, n_2, n_3)$ we obtain using the condition $S_0^{(\alpha)}(\mathbf{q}) = 0$:

$$W_G^B = \frac{1}{2\mu M} \sum_{k=0}^{2M_1-1} \psi(k) e^{-jn_1 k}, \tag{5.4.22}$$

where

$$\psi(k) = \left\{ W_G^B + W_G^A \left[e^{-\frac{1}{2}i(n_1+n_2)} + e^{-\frac{1}{2}i(n_2+n_3)} + e^{-\frac{1}{2}i(n_3+n_1)} \right] \right\} \times$$

$$\times\, \theta(M-k-\alpha) + \left\{ W_G^B + W_G^A \left[e^{-\frac{1}{2}i(n_1+n_2)} + e^{-\frac{1}{2}i(n_2+n_3)} + \right.\right.$$

$$\left.\left. + e^{-\frac{1}{2}i(n_3+n_1)} \right] \right\} \theta(k+\alpha-M),$$

$$\theta(\kappa) = \begin{cases} 1, & \kappa > 0 \\ 0, & \kappa < 0. \end{cases} \tag{5.4.23}$$

For the plane of the Brillouin zone that coincides with the former plane, we find

$$W_G^B = \frac{1}{2\mu M} \sum_{k=M}^{2M_1-1} \left(W_G^B - W_G^A\right)\left[e^{-\frac{1}{2}i(n_1+n_2)} + e^{-\frac{1}{2}i(n_2+n_3)} + \right.$$

$$\left. + e^{-\frac{1}{2}i(n_1+n_3)}\right] e^{-in_1 k}. \tag{5.4.24}$$

Obviously, with increasing M, the planes of the Brillouin zone and, therefore, the energy gaps will move toward the origin. It is seen from (5.4.22) and (5.4.24) that the gap becomes narrower as M increases. This fact explains the drop in intensity of satellites on the electron diffraction pattern with increasing M. Additional energy gaps, as may be seen from (5.4.23), appear on the planes whose indices n_1, n_2, n_3 have the same parity. If n_1, n_2 and n_3 have different parity, $\Delta E = 0$.

In a similar way one can obtain the expression for W_G^B in the case of two-dimensional periodic APBs with the period $2M_1$ in one direction and $2M_2$ -- in the other:

$$W_G^B = \frac{1}{2\mu(M_1+M_2)} \sum_{k=0}^{M_1-1} \sum_{l=0}^{M_2-1} e^{-i(kn_1+ln_2)} \left\{W_G^B + W_G^A\left[e^{-\frac{1}{2}i(n_1+n_2)} + \right.\right.$$

$$\left.\left. + e^{-\frac{1}{2}i(n_2+n_3)} + e^{-\frac{1}{2}i(n_3+n_1)}\right]\right\} + \sum_{k=M_1}^{2M_1-1} \sum_{l=M_2}^{2M_2-1} e^{-i(kn_1+ln_2)} \times$$

$$\times \left\{W_G^A + W_G^B\left[e^{-\frac{1}{2}i(n_1+n_2)} + e^{-\frac{1}{2}i(n_2+n_3)} + e^{-\frac{1}{2}i(n_1+n_3)}\right]\right\}. \tag{5.4.25}$$

For the planes that are coincident with the former planes, we have

$$W_G^B = \frac{1}{2\mu(M_1+M_2)} \sum_{k=M_1}^{2M_1-1} \sum_{l=M_2}^{2M_2-1} e^{-i(kn_1+ln_2)} \left(W_G^B - W_G^A\right) \times$$

$$\times \left[e^{-\frac{1}{2}i(n_1+n_2)} + e^{-\frac{1}{2}i(n_2+n_3)} + e^{-\frac{1}{2}i(n_3+n_1)}\right]. \tag{5.4.26}$$

Here, the additional energy gaps appear for the two directions. Now, we can write down the expression for the total energy of the alloy with the long-period superstructure. On changing the $L1_2$ superstructure for the $L1_2(M)$, the form of all contributions into E_n

given by formulas (5.4.8)–(5.4.17) remains unchanged except for that of the E_2 term. For some superstructure vectors, namely those for which n_1, n_2, n_3 have different parities and n_1+n_2 is odd, E_2 takes the form

$$E_2 = \frac{\eta^2}{16} \frac{1}{M^2 \sin^2 \frac{\pi n}{2M}} \sum_{q = k_{CC}} |\Delta W(q)|^2 \chi(q) \epsilon(q). \tag{5.4.27}$$

This expression describes the change of the band structure energy due to the occurrence of the satellite boundaries of the Brillouin zones.

Using formulas (5.4.8)–(5.4.17) and (5.4.27), the calculation has been carried out of the energies of the one-dimensional long-period superstructures $L1_2(M)$ in the alloys Cu_3Au and $CuAu_3$. The calculation showed that the energy of the $L1_2(M)$ superstructure in the corresponding alloys is greater than the energy of the $L1_2$ superstructure. This means that the long-period superstructure on the phase diagram would lie at higher temperature than $L1_2$. Such situation agrees with experiment. The heat of disordering for the long-period superstructures predicted by the theory proved to be 1 – 2 kal/atom which is also in accord with experimental data.

The role of the electrostatic interaction in atomic ordering. This issue is worth of a special discussion because a controversy may seem to exist in it. In the above analysis of the Fourier transform of the ordering energy, the conclusion was made that the electrostatic contribution impedes the formation of long-range order. The opposite follows from expression (5.4.8). Let us consider this question in more detail. Formula (5.4.8) comprises the full electrostatic contribution into the energy of a metallic crystal, while the Fourier transform $V(\mathbf{k})$ includes the indirect unscreened interaction.

It can be shown [56, 410] that the total electrostatic energy includes two main contributions: the unscreened Coulomb interaction and the interaction between ions and uniformly distributed conduction electrons (representing the so-called background energy). The first contribution is structure-dependent, the second depends only on the volume. The second term renormalizes the first one and may change its sign. As a consequence, analysis of the role of the electrostatic energy in ordering can be made only if all the contributions in this energy are taken into consideration. Given the renormalization of the structure-dependent terms by the volume-dependent ones is set up, the electrostatic energy will promote the ordering.

The main result of the present section is that the formation of the heat (and the energy) of the order-disorder transition occurs in a rather complicated way. The overall change of the energy is determined both by the structure-dependent and volume-dependent contributions; also of importance are electron and electrostatic energies. Many points in the Brillouin zone and the exchange-correlation effects play an essential role in forming the electron contribution to the energy. Also important are the higher orders of the perturbation theory. To date, the quantum theory of ordered solid solutions is capable of explaining the origin and the role of different contributions into the transformation energy and of calculating them without fitted parameters in fairly good agreement with experiment. That is what has been done in this section of the book.

Therefore, the type of short-range order and the crystal type of the superstructures formed in the considered alloys are determined by the shape of the Fermi surface. The existence of the plane sections in the direction <110> is connected with specific singularities in the indirect ion-ion interaction via the conduction electrons. It becomes quite clear from the above consideration that these results could not be obtained if the Fermi surface is approximated by a sphere.

Therefore, the material of the present chapter gives evidence of a complex nature of chosing the superstructure into which the alloy is ordered and especially of the stabilization of the ordered state [217, 403]. Atomic ordering connected with the transition to a particular superstructure can take place due to the difference in the atomic properties of the components, the occurrence of the plane sections of low curvature on the Fermi surface, the resonance character of the d states, and some peculiarities in many-particle interactions. The choice of a given superstructure is also

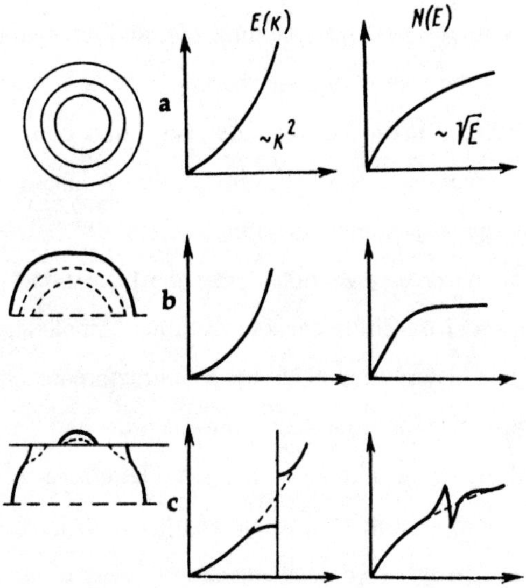

Fig. 96. Effect of the shape of the Fermi surface on the $E(\kappa)$ and N(E) (schematically). Fermi sphere (a), Fermi surface with plane section (b); appearance of new face of the Brillouin zone.

dictated by the specific features of the long-range interatomic interaction. It may be due to the collective Jan-Teller effect as well. In most cases, the choice of a particular superstructure is determined to the contribution of small groups of collectivized electrons.

The ordering is always related to the appearance of a new gap in the spectrum of electronic states. This gap, or the boundary of a new Brillouin zone, appears in contact with the Fermi surface, and their interaction yields substantial contribution to the superstructure formation energy [403] (Fig. 96).

Among other factors concurrently determining the possibility of establishing a long-range order in an alloy, very important is the spectrum of correlations and fluctuations.

The possibility of the formation of the phase with a long-range order, its stability and, therefore, T_c are determined by the total energy of the alloy in which, besides the configurational contributions, are also important the volume-dependent ones. While the former is connected with the necessary condition, i.e., with the formation of the long-range order, the account of the volume-dependent contributions is needed to formulate the sufficient condition.

Conclusion

Closing the survey of properties of ordered phases, we would like to point out once again their place in phase diagrams. A series of the five main classes of metallic compounds includes successively: primary solid solutions, ordered phases, electron compounds, size compounds, and compounds with normal valency. The significance of studying the properties of ordered phases is dictated by several reasons:

(1) The class of ordered phases is one of the most abundant in the line of compounds formed.

(2) A certain order in the atomic arrangement is typical of all metallic structures. Electron and size compounds are characterized by a long-range order, whereas primary solid solutions -- by a short-range order.

(3) A close packing typical of the ordered phases is also the main feature of other metallic compounds.

(4) A large number of metallic phases are characterized by the reduction of the sizes of the component atoms upon forming their compound. In other words, the phenomenon related to the structure contraction is typical of other metallic phases.

(5) It is customary for ordered phases to have wide ranges of solid solutions in which, as the composition deviates from stoichiometry, some specific features or defects characteristic of the nonstoichiometric compositions can be revealed. This is also a common property of metallic compounds.

(6) Ordered phases undergo martensitic transformations in the vicinity of which one can observe instability of the crystal lattice, short-range order in atomic displacements, or intermediate shear structures. The ordered phases can be transformed to an amorphous state. Several other metallic phases also display these properties.

(7) Stability of ordered phases, as well as of other metallic compounds, is due to the interaction between electron and size factors and to the state of atomic and topological order. The ordered phases possessing all the properties of metallic compounds, differ from electron and size compounds by simpler crystal lattices and relative simplicity of the phase transitions causing their formation. Hence, they represent the most convenient object for studying the properties of metallic compounds.

Close resemblance of the properties of metallic compounds can also be traced in some other aspects. It has been established that the decrease in symmetry of the superstructure reduces the concentration range in which the ordered phase exists. Because ordered phases are intermediary between solid solutions and electron compounds, a decrease in symmetry brings those phases closer to electron compounds. Common to ordered phases and other metallic compounds is the contact of the Fermi surface with the Brillouin zone boundaries. Also common is the fact that the regions of their existence are subject to certain limitations imposed on the electronic concentration and atomic radii.

It should be noted that both methodical approach to the study of ordered phases and the general formalism of the quantum theory of ordered solid solutions can be applied to the investigation of other metallic compounds. It should be born in mind, however, that in changing from primary solid solutions to Hume-Rothery and Lavez phases, the role of the configurational contributions declines, whereas the role of the nonconfigurational ones is enhanced and a pure metallic type of bonding will be complicated by the interatomic electron transfer and by the formation of directed orbitals.

Little attention in the monograph was paid to specific defects of the ordered phases. For example, the conception of the antiphase boundaries was merely introduced and the role of APBs was demonstrated in the mutual transformations of the crystal lattices of superstructures. Meanwhile, the antiphase boundaries of thermal origin are always present in the structure of ordered solid solutions. Near the APBs the degree of long-range order decreases, local atomic concentration is changed, appear specific directed and random displacements of atoms not inherent in the structure of a solid solution, and arise local atomic vibrations.

A considerable density of APBs greatly affects the electronic structure, physical and mechanical properties of ordered phases and characteristic features of their formation by means of the order-disorder phase transition. The structure of APBs becomes much more complicated in the nonstoichiometric phases and the role of APBs increases. The complication of the crystal structure of ordered solid solutions is also accompanied by the complication of the structure of individual APBs. Experimental evidence and methods described in this monograph form the basis for a deeper insight into the structure of individual APBs and the properties of their ensembles.

Crystal and electronic structure of ordered solid solutions, their composition and a set of defects present influence upon conductivity, thermal and halvanomagnetic properties, magnetism and magnetic structure, elastic behavior, phonon spectrum, etc. The formalism of quantum and statistical theory of ordering developed in this book and the experimental data on the order-disorder transformations form a reliable basis for constructing the theory of physical properties of ordered solid solutions.

And, finally, the problem of dislocations in ordered phases is still waiting for its solution in many regards. While a geometric theory of superdislocations in various superstructures can be considered as completed, the structure of dislocation core and, particularly, their mobility in connection with more complex crystallography of superstructures is virtually not investigated. In a rigorous form it can be developed on the grounds of conceptions put forward in this monograph concerning fine structure, the state of order, binding forces and stability of crystal lattices of ordered phases. It is expected that such work will be made in future.

It should be noted, in conclusion, that whilst the main trends and statistical physics of the order-disorder transformation are considered here in much detail, we presented only one of the possible versions of the quantum theory of solid solutions.

REFERENCES

1. Belov, N.V., Structura Ionnykh Kristallov i Metallicheskikh Faz (Structures of Ionic Crystals and Metallic Phases), Moscow, USSR Academy of Sciences, 1943, 236 pp.
2. Hume-Rothery, B. and Reynor, G.V., Structures of Metals and Alloys [Russian translation], Moscow, Metallurgizdat, 1959, 391 pp.
3. Schubert, K., Crystal Structures of Two-Component Phases [Russian translation], Moscow, Metallurgia, 1971, 531 pp.
4. Smirnova, N.L., Kristallografiya, v. 1, p. 165, 1956.
5. Smirnova, N.L., ibid., p. 502.
6. Smirnova, N.L., Kristallografiya, v. 3, p. 232, 1958.
7. Smirnova, N.L., ibid., p. 362.
8. Smirnova, N.L., Kristallografiya, v .4, p. 20, 1959.
9. Gladyshevskii, E.I., Zalutskii, I.I. and Tyvanchuk, A.T., Vestnik L'vovskogo Universiteta, Khimiya, no. 14, p. 9, 1973.
10. Zalutskii, I.I. and Tyvanchuk, A.T., Vestnik L'vovskogo Universiteta, Khimiya, no. 15, p. 3, 1974.
11. Gladyshevskii, E.I., Kristallochimiya Silitsidov i Germanidov (Crystallochemistry of Silicides and Germanides), Moscow, Metallurgiya, 1971, 296 pp.
12. Iziumov, Yu.A., Naish, V.E. and Ozerov, R.P., Neitronografiya Magnetikov (Neutronography of Magnetics), Moscow, Atomizdat, 1981, v. 2, 311 pp.
13. Syromyatnikov, V.N., Magnitnoe uporyadochenie v kubicheskikh kristallakh. Magnitnye struktury granatov, in: Simmetriinyi Analiz v Neitronografii Magnitnykh Struktur (Symmetry Analysis in Magnetic Structure Neutronography), Sverdlovsk, Ural Sci. Center, USSR Acad. Sci., 1980, p. 18.

14. Gufan, Yu.M., Strukturnye Fazovye Perekhody (Structural Phase Transitions), Moscow, Nauka, 1982, 304 pp.

15. Gufan, Yu.M., Dmitriev, V.P., Popov, V.I. and Chechin, G.M., Fiz. Met. Metalloved., v. 46, no. 6, p. 1134, 1978.

16. Gufan, Yu.M., Dmitriev, V.P., Popov, V.I., and Chechin, G.M., Fiz. Tverd. Tela (Leningrad), v. 21, no. 2, p. 554, 1979.

17. Lifshits, E.M., Zh. Eksp. Teor. Fiz. , v. 11, p. 255, 1941.

18. Villan, J., J. Phys. Chem. Sol., v. 11, p. 303, 1959.

19. Khachaturyan, A.G., Teoriya Fazovykh Prevrashchenii i Struktura Tverdykh Rastvorov (Theory of Phase Transformations and Structure of Solid Solutions), Moscow, Nauka, 1974, 384 pp.

20. Khachaturyan, A.G., Fiz. Tverd. Tela (Leningrad), v. 5, p. 26, 1963.

21. Khachaturyan, A.G., Zh. Eksp. Teor. Fiz. , v. 63, p. 1421, 1972.

22. Kozlov, E.V. and Shtern, D.M., Izv. Vyssh. Uchebn. Zaved., Fyzika, v. 27, no. 6, p. 26, 1985.

23. Bir, G.L. and Pikus, G.E., Simmetriya i Deformatsionnye Effekty v Poluprovodnikakh (Symmetry and Deformation Effects in Semiconductors), Moscow, Nauka, 1972, 584 pp.

24. Landau, L.D. and Lifshits, E.M., Statisticheskaya Fizika (Statistical Physics), Moscow, Nauka, 1976, Part 1, 583 pp.

25. Friedel, J., Transition metals. Electron structure of d band. Its role in crystalline and magnetic structures. In: Physics of Metals. v. 1. Electrons, ed. J.M. Ziman, London, Cambridge Univ., 1969, pp. 373 - 461.

26. Jones, H., Theory of Brillouin Zones and Electronic States in Crystals, Amsterdam, North-Holland, 1962, 264 pp.

27. Luehrmann, A.W., Adv. Phys., v. 17, no. 65, p. 1, 1968.

28. Golubenko, T.Ya. and Kozlov, E.V., Simmetriinye izmeneniya pri obrazovanii dal'nego poryadka v splave CuAu, in: Uporyadochenie Atomov i Svoistva Splavov

(Atomic Ordering and Properties of Alloys), Kiev, Naukova Dumka, 1979, pp. 224 - 227.

29. Bersuker, I.B., Elektronnoe Stroenie i Svoistva Koordinatsionnykh Soedinenii (Electronic Structure and Properties of Coordination Compounds), Leningrad, Khimiya, 1976, 349 pp.

30. Sviridov, D.T., Sviridova, R.K. and Smirnov, Yu.F., Opticheskie Spektry Ionov Perekhodnykh Metallov v Kristallakh (Optical Spectra of Transition Metal Ions in Crystals, Moscow, Nauka, 1976, 267 pp.

31. Vonsovskii, S.V., Grum-Grzhimailo, S.V., Cherepanov, V.I., et al., Teoriya Kristallicheskogo Polya i Opticheskie Spektry Primesnykh Ionov s Nezapolnennoi d-Obolochkoi (Crystal Field Theory and Optical Spectra of Impurity Ions with Unfilled d Shell), Moscow, Nauka, 1969, 179 pp.

32. Ziman, J.M., Principles of the Theory of Solids, Cambridge, Cambridge Univ., 1964.

33. Harrison, W.A., Solid State Theory, New York, McGraw-Hill. , 1970, 616 pp.

34. Heine, V., Electronic structure of metals, in: Physics of Metals, v. 1. Electrons, ed. J.M. Ziman, London, Cambridge Univ., 1968, pp. 11- 74.

35. Kozlov, E.V., Dement'ev, V.M., Emel'yanov, V.N., et al., Symmetry, pair interatomic interactions and electronic states in ordering alloys, in: Order-Disorder Transformations in Alloys, ed. H. Warlimont, Heidelberg, Springer, 1974, pp. 58 - 79.

36. Knox, R.S. and Gold, A., Symmetry in the Solid State, New York, Benjamin, 1964, 424 pp.

37. Kozlov, E.V., Kormin, N.M., and Matveeva, N.M., Analiz uslovii obrazovaniya plotno-upakovannykh sverkhstruktur sostava AB , in: Proc. IV All-Union Conf. on Atomic Ordering and Its Influence on the Properties of Alloys, Tomsk, Tomsk Univ., 1974, Part.1, 146 - 152.

38. Kozlov, E.V., Kormin, N.M. and Matveeva, Izv. Akad. Nauk SSSR, Metally, no. 5, p. 192, 1977.

39. Elliott, R.P., Constitution of Binary Alloys, First Suppl., v. 1, New York, McGraw-Hill. , 1965, 455 pp.

40. Elliott, R.P., Constitution of Binary Alloys, First Suppl., v.2, New York, McGraw-Hill. , 1965, 472 pp.

41. Shunk, F.A., Constitution of Binary Alloys, Second Suppl., New York, McGraw-Hill, 1965, 760 pp.

42. Hansen, M.and Anderko, K., Constitution of Binary Alloys, v. 1, New York, McGraw-Hill. , 1958, pp. 1 - 607.

43. Hansen, M. and Anderko, K., Constitution of Binary Alloys, v.2, New York, McGraw-Hill. , 1958, pp. 613 - 1488.

44. Vol, A.E., Stroenie i Svoistva Dvoinykh Metallicheskikh Sistem (Constitution and Properties of Binary Metallic Systems), Moscow, Fizmatgiz, 1959, v. 1, 755 pp.

45. Vol, A.E., Stroenie i Svoistva Dvoinykh Metallicheskikh Sistem (Constitution and Properties of Binary Metallic Systems), Moscow, Fizmatgiz, 1962, v. 2, 982 pp.

46. Vol, A.E. and Kagan, I.K., Stroenie i Svoistva Dvoinykh Metallicheskikh Sistem (Constitution and Properties of Binary Metallic Systems), Moscow, Nauka, 1976, v. 3, 814 pp.

47. Vol, A.E and Kagan, I.K., Stroenie i Svoistva Dvoinykh Metallicheskikh Sistem (Constitution and Properties of Binary Metallic Systems), Moscow, Nauka, 1979, v. 4, 576 pp.

48. Taylor, K.N.R. and Darby, M.I., Physics of Rare Earth Solids, London, Chapman and Hall, 1972, p. 374.

49. Pearson, W.B., Handbook of Lattice Spacings and Structures of Metals and Alloys, London, Pergamon, 1967, v. 1, 939 pp.

50. Pearson, W.B., Handbook of Lattice Spacings and Structures of Metals and Alloys, London, Pergamon, 1967, v. 2, 1446 pp.

51. Umanskii, Ya.S., Finkel'shtein, B.N., Blanter, M.E., et al., Fizicheskoe Metallovedenie (Physical Metallurgy), Moscow, Metallurgizdat, 1955, 724 pp.

52. Sato, H. and Toth, R.S., Long-period superlattices in alloys, in: Alloying Behavior and Effects in Concentrated Solid Solutions, New York, Academic, 1963, no. 9, pp. 295 - 419.

53. Jones, H., J. Phys. F: Metal Phys., v. 2, p. 733, 1969.

54. Inglesfield, J.E., J. Phys. F: Metal. Phys., v. 1, p. 149, 1968.

55. Sato, H. and Toth, R.S., Phys. Rev., v. 124, no. 6, p. 1833, 1961.

56. Harrison, W.A., Pseudopotentials in the Theory of Metals, New York, Benjamin, 1966, 366 pp.

57. Heine, V., Cohen, M. and Weaire, D., in: Solid State Physics, v. 24, eds. H. Ehrenreich, F. Seitz, and D. Turnbull, New York, Academic, 1970.

58. Williams, A.R. and Weaire, D., J. Phys. C: Condens. Matter, v. 3, p. 387, 1970.

59. Kozlov, E.V., Dement'ev, V.M. and Kormin, N.M., Predstavlenie ob effektivnom parnom mezhionnom vzaimodeistvii v binarnykh metallicheskikh fazakh, in: Stroenie, Svoistva i Primenenie Metallidov (Constitution, Properties and Application of Metallides), Moscow, Nauka, 1974, pp. 31 - 35.

60. Kormin, N.M., Kristallokhimicheskii analiz stabil'nosti sverkhstruktur, in: Uporyadochenie Atomov i Ego Vliyanie na Svoistva Splavov (Atomic Ordering and Its Influence on the Properties of Alloys), Tomsk, Tomsk Univ., 1978, 26 - 29.

61. Pearson, W.B., Crystal Chemistry and Physics of Metals and Alloys, Part 1, New York, Wiley Interscience, 1972, 419 pp.

62. Pearson, W.B., Crystal Chemistry and Physics of Metals and Alloys, Part 2, New York, Wiley Interscience, 1972, 471 pp.

63. Massalski, T.B., Structure of solid solutions, in: Physical Metallurgy, ed. R.W.Cahn, Chs. II-V, XXI, Amsterdam, North-Holland, 1965, pp. 144 - 219.

64. Darken, L.S. and Gurry, R.W., Physical Chemistry of Metals, New York, McGraw-Hill, 1953, 583 pp.

65. Kormin, N.M., Issledovanie Roli Razlichnykh Kristallofizicheskikh Faktorov v Stabil'nosti Sverkhstruktur (The Study of the Role of Various Crystallophysical

Factors in Superstructure Stability), Cand. Sci. (Phys.-Mat.) Dissertation, Tomsk, 1980, 18 pp.

66. Brown, P. and Taylor, U., Experimental study of metal and alloy structure, in: Physics of Metals, v. 1. Electrons, ed. J.M. Ziman, Cambridge, Cambridge Univ., 1969, pp. 51 - 372.

67. Cracknell, A.P. and Wong, K.C., The Fermi Surface. Its Concept, Determination, and Use in the Physics of Metals, Oxford, Clarendon, 1973, 348 pp.

68. Batsanov, S.S. and Zvyagina, R.A., Integraly Perekryvaniya i Problema Effektivnykh Zaryadov (Overlap Integrals and the Problem of Effective Charges, v. 1, Novosibirsk, Nauka, 1966, 386 pp.

69. Ziman, J.M., Electrons and Phonons. The Theory of Transport Phenomena in Solids, Oxford, Clarendon, 1960, 488 pp.

70. Gray, D. and Brown, E., Phys. Rev., v. 160, no. 3, p. 567, 1967.

71. Airoldi, G., Asdente, M., and Drosi, M., Phil. Mag., v. 14, no. 129, p. 455, 1966.

72. Kozlov, E.V., Dement'ev, V.M., Emel'yanov, V.N., et al., Simmetriya, parnoe mezhatomnoe vzaimodeistvie i priroda sil, otvetstvennykh za atomnyi dal'nii poryadok (Symmetry, pair interatomic interaction and the nature of forces responsible for atomic long-range order), in: Proc. IV All-Union Conf. on Atomic Ordering and Its Influence on Properties of Alloys, Tomsk, Tomsk Univ., 1974, Part I, pp. 68 - 82.

73. Weiss, R.J., X-Ray Determination of Electron Distributions, Amsterdam, North-Holland; New York, Wiley, 1966, 196 pp.

74. Weiss, R.J., Solid State Physics [Russian translation], Moscow, Atomizdat, 1968.

75. Emel'yanov, V.N. and Kozlov, E.V., Elektronnye konfiguratsii i atomnyi faktor rasseyaniya rentgenovskikh luchei v metallicheskikh fazakh perekhodnykh metallov, in: Elektronnoe Stroenie i Fiziko-Khimicheskie Svoistva Splavov i Soedinenii na Osnove Perekhodnykh Metallov (Electronic Structure and Physicochemical Properties of the Alloys and Compounds on the Base of Transitional Metals), Kiev, Naukova Dumka, 1976, pp. 30 - 34.

76. Krivoglaz, M.A., Primenenie rasseyaniya rentgenovskikh luchei i teplovykh neitronov dlya issledovaniya nesovershenstv v kristallakh, in: Elektrony, Metally, Reshetka (Electrons, Metals, Crystal Lattice), Kiev, Naukova Dumka, 1975, pp. 355 - 388.

77. Van Tendenloo, G. and Amelinckx, S., Acta Crystallogr., v. 30, no. 2, 431, 1974.

78. Marcincovski, M.J., Electron Microscopy and Strength of Crystals, New York, Wiley Interscience, 1963, pp. 215 - 320.

79. Beeler, Joe, R. (Jr.), Antiphase boundaries and domains, in: Intermetallic Compounds, ed. J.H. Westbrook, New York, Wiley, 1967, pp. 233 - 268.

80. Amelinckx, S., Surface Sci., v. 31, p. 296, 1972.

81. Amelinckx, S., J. Less-Common Metals, vol. 28, 325, 1972.

82. Cowley, J.M., Short-range ordering in crystals, in: Proc. 30th Ann. EMSA Meet., Baton Rouge, 1975, pp. 35 - 85.

83. Kornilov, I.I., Matveeva, N.M., Pryakhina, L.I., and Polyakova, R.S., Metallokhimicheskie Svoistva Elementov Periodicheskoi Sistemy (Metallochemical Properties of Elements of Periodic System), Moscow, Nauka, 1966, 351 pp.

84. Kornilov, I.I., Titan. Istochniki, Sostavy, Svoistva, Metallokhimiya i Primenenie (Titanium: Sources, Compositions, Metallochemistry and Application), Moscow, Nauka, 1975, 308 pp.

85. Dvoinye i Mnogokomponentnye Sistemy na Osnove Medi (Binary and Multicomponent Copper-Base Systems), Handbook, Moscow, Metallurgiya, 1986, 439 pp.

86. Diagrammy Sostoyaniya Dvoinykh i Mnogokomponentnykh Sistem na Osnove Zheleza (Phase Diagrams of Binary and Multicomponent Systems), Handbook, Moscow, Metallurgiya, 1986, 439 pp.

87. Diagrammy Sostoyaniya Metallicheskikh Sistem (Phase Diagrams of Metallic Systems), Moscow, VINITI, 1955 - 1986, nos. 1 - 31.

88. Kornilov, I.I. and Boriskina, N.G., Dokl. Akad. Nauk SSSR, v. 108, no. 6, p. 1083, 1956.

89. Chernov, D.B., Paskal, Yu.I., Gyunter, V.E., and Monasevich, L.A., Izv. Vyssh. Uchebn. Zaved, Fizika, v. 24, n. 3, p. 93, 1981.

90. Kornilov, I.I., Belousov, O.K., and Kachur, E.V., Nikelid Titana i Drugie Splavy s Effektom "Pamyati" (Titanium Nickelide and Other Shape Memory Alloys, Moscow, Nauka, 1977, 180 pp.

91. Kovneristii, Yu.K., Belousov, O.K., Matveeva, N.M., et al., Termodinamicheskie i strukturnye aspekty issledovaniya splavov na osnove nikelida titana s effektom pamyaty formy, in: Splavy Titana s osobymi svoistvami (titanium alloys with special properties), Moscow, Nauka, 1982, pp. 4 - 10.

92. Matveeva, N.M. and Shcherbakova, I.E., Diagrammy sostoyaniya i splavy sistem zheleza i titana, proyavlyayushchie effect pamyati formy, in: Diagrammy Sostoyaniya v Materialovedenii (Phase Diagrams in Materials Science), 1984, Kiev, Naukova Dumka, pp. 61 - 72.

93. Eremenko, V.N. and Shtepa, T.D., Stroenie diagramm sostoyaniya dvoinykh sistem - metall VIII gruppy periodicheskoi sistemy elementov, in: Obshchie Zakonomernosti v Stroenii Diagramm Sostoyaniya Metallicheskikh Sistem (General Trends in the Constitution of the Phase Diagrams of Metallic Systems), Moscow, Nauka, 1973, 41 - 45.

94. Donkersloot, H. and Van Vucht, J., J. Less-Common Metals, v. 20, no. 2, p. 83, 1970.

95. Sivokha, V.P., Khachin, V.N., and Matveeva, N.M., Sverkhuprugost', effekt pamyaty formy i ikh primenenie v novoi tekhnike, in: All-Union Sci. Conf., Abstracts, Tomsk, 1985, pp.111- 112.

96. Savvinov, A.S., Khachin, V.N., and Matveeva, N.M., Izv. Akad. Nauk SSSR, Metally, no. 5, p. 118, 1985.

97. Sivokha, V.P., Martensitnye Prevrashcheniya i Effekt Pamyati Formy v TiNi(Pd), TiNi(Pt), TiNi(Au), Cand. Sci. (Phys.-Mat.) Dissertstion, Tomsk, 1986, 17 pp.

98. Matveeva, N.M., Khachin, V.N., and Sivokha, V.P., Diagrammy martensitnykh prevrashchenii v sisteme TiNi-TiPd, in: Stabil'nye i Metastabil'nye Fazovye Ravnovesiya v Metallicheskikh Sistemakh (Stable and Metastable Phase Equilibria in Metallic Systems), Moscow, Nauka, 1985, pp. 25 - 29.

99. Matveeva, N.M., Kovneristyi, Yu.K., Savvinov, A.S., Sivokha, V.P., and Khachin, V.N., J. de Phys. C4, vol. 43, no. 12, p. 249, 1982.

100. Matveeva, N.M., Khachin, V.N., and Sivokha, V.P., Vliyanie platiny na martensitnoe prevrashchenie i effekt pamyati formy nikelida titana, in: Splavy Redkikh Metallov s Osobymi Svoistvami. Redkozemel'nye i Blagorodnye Metally (Alloys of Non-Common Metals with Special Properties. Rare Earth and Noble Metals), Moscow, Nauka, 1983, pp. 163 - 165.

101. Beck, P.A., Darvy, J.B., and Arora, O.P., Trans. AIME, v. 206, p. 148, 1956.

102. Philip, T.W. and Beck, P.A., Trans. AIME, v. 209, p. 1269, 1957.

103. Vitting, L.M., Zh. Neorg. Khim., v. 2, p. 845, 1957.

104. Koster, W. and Godecke, T., Z. Metallk., v. 71, no. 12, p. 765, 1980.

105. Warlimont, H.A., Z. Metallk., v. 60, p. 195, 1969.

106. Schuzman, E.H., Arch. Eisenh@uttenw., v. 311, p. 325, 1980.

107. Hiromichi, S. and Kensuke, O., J. Jap. Inst. Metals, vol. 43, no. 7, p. 667, 1979.

108. Beandry, B. and Deane, A., Trans. AIME, v. 218, no. 5, p. 854, 1960.

109. Svechnikov, V.N., Shurin, A.K., Dmitrieva, G.P., and Alfintseva, R.A., Diagrammy Sostoyaniya Metallicheskikh Sistem (Phase Diagrams of Metallic Systems), Moscow, Nauka, 1968, p. 153-156.

110. Friedrich, K., Jonsson, S., Aldinger, F. et al., Z. Metallk., v. 71, n. 9, 584 - 587, 1980.

111. Budurov, S., Wassilev, G., and Nentschew, N., Z. Metallk., v. 65, no. 11, p. 681, 1974.

112. Taylor, A. and Doyle, N., J. Appl. Crystallogr., v. 5, no. 3, p. 210, 1972.

113. Litvinov, V.S. and Arkhangelskaya, A.A., Martensitnye prevrashcheniya v sisteme Ni-Al (Martensitic transformations in the Ni-Al System), in: Proc. Int. Conf. "Martensitic Transformations", Kiev, 1977, p. 75.
114. Sigli, C. and Sanchez, J.M., Acta Met., v. 33, no. 6, p. 1097, 1985.
115. Feschotte, P. and Eggimann, P., J. Less-Common Metals, v. 63, no. 1, p. 15, 1979.
116. Best, K. and Godecke, T., Z. Metallk., v. 60, no. 8, p. 659, 1969.
117. Markiv, V.Ya., Gavrilenko, I.S., Pet'kov, V.V., and Belyavina, N.N., in: Metallofizika (Metal Physics) , Republ. Interdiscipl. Collation, Kiev, no. 73, p. 39, 1978.
118. Stocks, G. and Winter, H., Proc. NATO Adv. Study Inst., Ghent, July 12-23, 1982; New York, 1984, pp. 464 - 579.
119. Kaufmann, L. and Tanner, L., CALPHAD, v. 8, no. 2, p. 121, 1984.
120. Warlimont, H. and Diley, L., Martensitic Transformations in Copper- , Silver- and Gold-Based Alloys [Russian translation], Moscow, Nauka, 1980, 205 pp.
121. Ljungstromer, l., Stoperitidende, v. 40, no. 1, p. 5, 1974.
122. Kaufman, L. and Nesor, H., CALPHAD, v. 2, no. 4, p. 325, 1978.
123. Naumkin, O.P., Terekhova, V.T. and Savitskii, E.M., Izv. Akad. Nauk SSSR, Metally, no. 4, p. 176, 1965.
124. Koster, W. and Wachtel, E., Z. Metallk., v. 51, no. 5, p. 271, 1960.
125. Tokayama T., Wey Mylong Yong, and Nishizawa, T., Trans. Jap. Inst. Metals, v. 22, no. 5, p. 315, 1981.
126. Kaufman, L. and Nesor, H., Met. Trans. A, v. 6, no. 11, p. 2115, 1975.
127. Obrowski, W., Metall., v. 17, no. 2, p. 108, 1963.
128. Obrowski, W., Naturwissenschaft., v. 47, p. 14, 1960.
129. Eremenko, V.N., Buyanov, Yu.I., and Prima, S.B., Poroshkovaya Metallurgiya, no. 6, p. 77, 1966.
130. Karlson, H., J. Inst. Metals, no. 79, p. 391, 1951.
131. Matveeva, N.M. and Kostyanaya, O.V., Temperature dependence of the structure of rapidly cooled alloys of the TiCu-TiNi system, in: Fyzikokhimiya Amorfnykh

Metallicheskikh Sistem (Physics and Chemistry of Amorphous Metallic Systems, Moscow, Nauka, 1987, pp. 117 - 122.

132. Kaufman, L., CALPHAD, v. 2, n. 2, p. 117, 1978.
133. Eremenko, V.N., Buyanov, Yu.I., and Panchenko, N.M., Poroshkovaya Metallurgiya, no. 7, p. 55, 1969.
134. Gros, J., Ansara, L., Allibert, M., et al., Mem. et etud. sci. rev. met., v. 83, no. 9, p. 448, 1986.
135. Grossley, F., Trans. Met. Soc. AIME, v. 236, no. 8, p. 1174, 1966.
136. Potzschke, M. and Schubert, K., Z. Metallk., v. 53, no. 7, p. 474, 1962.
137. Hashimoto, S., Morita, H., Fujinaga, Y., et al., KEK Progr.Rept., no. 3, p. 168, 1989.
138. Gruzin, P.L., Rodionov, Yu.A., Bannatov, K.T., et. al, Magnitnye Svoistva Kristallicheskikh i Amorfnykh Materialov (Magnetic Properties of Crystalline and Amorphous Materials), Irkutsk, 1983, pp. 137 - 141.
139. Tsiuplakis, K. and Kneller, E., Z. Metallk., v. 60, no. 5, p. 433, 1969.
140. Budurov, S., Wassilew, G., and Nentschew, N., Z. Metallk., v. 65, no. 11, p. 681, 1974.
141. Pianelli, A. and Faivre, R., Compt. Rend. Acad. Sci., v. 245, no. 18, p. 1537, 1957.
142. Fontaine, D., Kikuchi, R., et al., US Dept. Commerce, NBS Spec. Publ., no. 496/2, p. 999, 1978.
143. Raub, E. and Roschel, E., Z. Metallk., v. 57, no. 7, p. 546, 1966.
144. Eremenko, V.N. and Shtepa, T.D., Izv. Akad. Nauk SSSR, Metally, no. 6, p. 198, 1970.
145. Heine, W. and Zwicker, U., Z. Metallk., v. 53, no. 6, p. 380, 1962.
146. Anderko, K., Z. Metallk., v. 49, no. 4, p. 165, 1958.
147. Van Deen, J.K. and van der Wonde, F., Acta Met., v. 29, no. 7, p. 1255, 1981.
148. Gruzin, P.A., Rodionov, Yu.L., and Pryakhin, V.A., Dokl. Akad. Nauk SSSR, v. 251, no. 6, p. 1384, 1980.

149. Dunne, D. and Wayman, C., Met. Trans., v. 4, p. 137, 1973.

150. Kachi, S., Bando, V., and Higuchi, S., Jap. J. Appl. Phys., v. 1, no. 6, p. 307, 1962.

151. Haush, J. and Warlimont, H., Phys. Lett. A, v. 36, p. 415, 1971.

152. Owen, E. and Liu, V., J. Iron and Steel Inst., v. 163, no. 2, p. 132, 1949.

153. Marchand, A. and Paulere, J., Compt. Rend., v. 257, p. 2987, 1963.

154. Matveeva, N.M. and Kolyada, A.A., Issledovanie metodom messbauerovskoi spektroskopii struktury pri martensitnykh prevrashcheniyakh v splavakh zheleza s 25 i 30% nikelya, in: Kristallicheskaya Struktura i Svoistva Metallicheskikh Splavov (Crystal Structure and Properties of Metallic Alloys), Moscow, Nauka, 1978, pp. 86 - 93.

155. Nakagawa, Y., Tanji, Y., Morita, H., et al., J. Magn. and Magn. Mater., v. 10, no. 2/3, p. 145, 1979.

156. Umemoto, M. and Wayman, C., Met. Trans. A, v. 9, p. 891, 1978.

157. Kajiwara, S. and Owen, W.S., Met. Trans., v. 5, p. 2047, 1974.

158. Tadaki, T. and Simizu, K., Scripta Met., v. 9, p. 771, 1975.

159. Focs, M. and Frants, M., Proc. Int. Conf. Martensitic Transformations ICOMAT, Oxford (USA), 1979, p. 485.

160. Tailashev, A.S., Matveeva, N.M., and Ganzina, L.L., Splavy Redkilh i Tugoplavkikh Metallov s Osobymi Fizicheskimi Svoistvami (Rare Earth and Refractory Metal Alloys with Special Physical Properties), Moscow, Nauka, 1979, pp. 304 - 307.

161. Focs, M., Frants, C., and Gantois, M., Shape Memory Effects in Alloys, New York, Plenum, 1975, pp. 342 - 358.

162. Matveeva, n.M., Demensky, G.K., Petrov, L.A., et al., J. phys. C4, v. 43, no. 12, p. 377, 1982.

163. Koster, W. and Godecke, T., Z. metallk., v. 68, no. 10, p. 582 - 589, 661 - 666, 1977.

164. Predel, B. and Frebel, M., Z. Metallk., v. 63, no. 7, p. 393, 1972.

165. Kaufman, L. and Bernstein, H., Computer Calculation of Phase Diagrams, Man Lahs, Cambridge, Massachusets, Academic, 1970.

166. Kaufman, L., CALPHAD, v. 2, no. 2, p. 117, 1978.

167. Sanchez, J., Barefoot, J., Jarrett, R., and Tien, J., Acta Met., v. 32, no. 9, p. 1519, 1984.

168. Pearson, W. and Rimek, D., Canad J. Phys., v. 35, no. 10, p. 1228, 1957.

169. Best, K. and Godecke, T., Z. metallk., v. 60, no. 8, p. 659, 1969.

170. Ellner, M., Heinrich, S., and Schubert, K., J. Less-Common Metals, v. 66, no. 2, p. 163, 1979.

171. Lesailis, S., Hamar-Thibauet, S., Z. Metallk., v. 75, no. 10, p. 764, 1984.

172. Gasselton, R. and Hume-Rothery, W., J. Less-Common Metals, v. 7, no. 3, p. 212, 1964.

173. Kang, S., Song, Y., Kayssar, W., et al., Z. metallk., v. 75, n. 1, p. 86, 1984.

174. Gabriel, A., Lukas, H., Allibert, C., et al., Z. Metallk., v. 76, no. 9, p. 589, 1985.

175. Stolz, E. and Schubert, K., Z. metallk., v. 53, no. 7, p. 433, 1962.

176. Koster, W. and Nordskog, H., Z. Metallk., v. 51, no. 9, p. 501, 1960.

177. Plukiger, R., Susz, Ch., Heiniger, F., et al., J. Less-Common Metals, v. 40, no. 1, p. 103, 1975.

178. Koster, W. and Hummel, R., Z. metallk., v. 55, no. 4, p. 175, 1964.

179. Massalski, T. B. and Okamoto, H., Bull. Alloy Phase Diagr., v. 6, no. 5, p. 434, 1985.

180. Khachin, V.N., Izv. Vyssh. Uchebn. Zaved., Fizika, no. 5, p. 88, 1985.

181. Savvinov, A.S., Khachin, V.N., and Matveeva, N.M., Dokl. Akad. Nauk SSSR, v. 270, no. 3, p. 609, 1983.

182. Khachin, V.N., Martensitnaya Neuprugost' B2 Soedinenii Titana (Martensitic Inelasticity of B2 Titanium Compounds), Dr. Sci. (Physics) Dissertation, Tomsk, 1987.

183. Matveeva, N.M., Kovneristyi, Yu.K., Savvinov, A.S., et al., Sverkhuprugost', effekt pamyati formy i ikh primenenie v novoi tekhnike (Superelasticity, Shape Memory Effect and Their Application in Modern Technique), in: Proc. All-Union Conf., Abstracts of papers, 1982, pp. 66 - 67, 85.

184. Savvinov, A.S., Khachin, V.N., and Sivokha, V.P., Izv. Vyssh. Uchebn. Zaved., Fizika, v. 7, p. 34, 1983.

185. Khachin, V.N., Voronin, V.P., and Matveeva, N.M., Some trends in development of martensitic B2 - B19 transformation in TiFe-TiPd, in: Proc. 4th Conf. on Aging of Metallic Alloys, Sverdlovsk, 1984, pp. 133 - 134.

186. Kester, V., Short-range ordering and clustering in solid solutions, in: Tonkaya Struktura i Svoistva Tverdykh Rastvorov (Fine Structure and Properties of Solid Solutions), Moscow, Metallurgiya, 1968, pp. 196 - 220.

187. Iveronova, V.I. and Katsnel'son, A.A., Blizhnii Poryadok v Tverdykh Rastvorakh (Short-Range Order in Solid Solutions), Moscow, Nauka, 1977.

188. Smirnov, A.A., Molekulyarno-Kineticheskaya Teoriya Metallov (Molecular-Kinetic Theory of Metals), Moscow, Nauka, 1966.

189. Krivoglaz, M.A. and Smirnov, A.A., Teoriya Uporyadochivayushchikhsya Splavov (Theory of Ordering Alloys), Moscow, Fizmatgiz, 1958.

190. Clapp, P.C., Phys. Rev., v. 164, no. 3, p. 1018, 1967.

191. Stratanovich, R.L., Teoriya Informatsii (Information Theory), Moscow, Sov. Radio, 1975.

192. Niks, F.Ch. and Shockly, V., Usp. Fyz. Nauk, v. 20, no. 3, p. 344; no. 4, p. 536, 1938.

193. Muto, T. and Takagi, Yu., Theory of Order-Disorder Transitions in Alloys, Solid State Phys., v. 1, Eds. F. Seitz and H. Turnbull, pp. 193 - 282, 1954.

194. Guttman, L., Solid State Phys., v. 3, Eds. F. Seitz and H. Turnbull, pp. 145 - 223, 1956.

195. Warren, B.E. and Averbach, B.L., Diffuse scattering of X-rays, in: Modern Research Techniques in Physical Metallurgy, ASM, Cleveland, Ohio, 1952, pp. 109 - 149.

196. Cowley, J.M., Phys. Rev., v. 120, n. 5, p.1648, 1960.

197. Cramer, H., Mathematical Methods of Statistics, Univ. of Stokholm, 1946, 648 pp.

198. Janoshi, L., Teoriya i Praktika Obrabotki Resultatov Izmerenii (Theory and Practice of Measurement Results Processing), Moscow, Mir, 1968.

199. Malakhov, A.I., Kumulantnyi Analiz (Cumulant Analysis), Moscow, Sov. Radio, 1978.

200. Lukach, E., Kharacteristicheskie Funktsii (Characteristic Functions), Moscow, Nauka, 1979.

201. Kitaigorodskii, A.I., Teoriya Strukturnogo Analiza (Theory of Structural Analysis), Moscow, USSR Acad. Sci., 1957.

202. Mott, N.F. and Massey, H.S.F., The Theory of Atomic Collisions, Oxford, Clarendon, 1965.

203. James, R.W., The Optical Principles of Diffraction of X-Rays, Ed. by Sir Lawrence Bragg, London, 1950.

204. Krivoglaz, M.A., Teoriya Rasseyaniya Rentgenovskikh Luchei i Teplovykh Neitronov Real'nymi Kristallami (The Theory of X-Ray and Thermal Neutron Scattering by Real Crystals), Moscow, Nauka, 1967.

205. Iveronova, V.I. and Revkevich, G.P., Teoriya Rasseyaniya Rentgenovskikh Luchei (The Theory of X-Ray Scattering), Moscow, Moscow State Univ., 1978.

206. Clapp, P.C., J. Phys. Chem Solids, v. 30, p. 2589, 1969.

207. Clapp, P.C., Phys. Rev. B, v. 4, no. 2, p. 255, 1971.

208. Schwartz, L.H. and Cohen, J.B., J. Appl. Phys., v. 36, p. 598, 1965.

209. Bardhan, P. and Cohen, J.B., Acta Crystallogr. A, v. 32, p. 597, 1976.

210. Gehlen, P.C. and Cohen, J.B., Phys. Rev., v. 139, no. 1, p. 844, 1965.

211. Iveronova, V.I. and Katsnel'son, A.A., Izv. Vyssh. Uchebn. Zaved., Fizika, no. 3, p. 40, 1976.

212. Popov, L.E., Koneva, N.A., and Tereshko, I.V., Deformatsionnoe Uprochnenie Uporyadochennykh Splavov (Deformation Hardening of Ordered Alloys), Moscow, Metallurgiya, 1979.

213. Golosov, N.S. and Dudka, B.V., phys. status solidi B, v. 66, no. 1, p. 439, 1974.

214. Golosov, N.S. and Dudka, B.V., phys. status solidi B, v. 59, no. 1, p. 361, 1973.

215. Kozlov, E.V. and Shtern, D.M., Izv. Vyssh. Uchebn. Zaved., Fizika, no. 6, p. 43, 1977.

216. Fisher, I.Z., Statisticheskaya Teoriya Zhidkostei (Statistical Theory of Liquids), Moscow, Fizmatgiz, 1961, 280 pp.

217. Kozlov, E.V., Dement'ev, V.M., and Shtern, D.M., Quantum-statistical theories of atomic ordering, in: Struktura i Plasticheskoe Povedenie Splavov (Structure and Plasticity of Alloys), Tomsk, Tomsk Univ., 1983.

218. Kozlov, E.V., Shtern, D.M., and Kormin, N.M., Izv. Vyssh. Uchebn. Zaved., Fizika, no. 9, p. 22, 1980.

219. Richards, M.J. and Cahn, J.W., Acta Met., v. 19, p. 1263, 1971.

220. Allen, S.M. and Cahn, J.W., Acta Met., v. 20, p. 423, 1972.

221. Kozlov, E.V., Izv. Vyssh. Uchebn. Zaved., Fizika, no. 8, p. 82, 1976.

222. De Fontaine, Acta Met., v. 23, p. 553, 1975.

223. Kozlov, E.V., Shtern, D.M., Kormin, N.M., and Gudin, O.A., Zh. Strukt. Khim., v. 18, no. 1, p. 80, 1977.

224. Bokii, G.G., Vvedenie v Kristallokhimiyu (Introduction to Crystallochemistry), Moscow, Moscow State Univ., 1954.

225. Kozlov, E.V. and Shtern, D.M., Izv. Vyssh. Uchebn. Zaved., Fizika, no. 3, p. 63, 1976.

226. Ogawa, S., On the long-period ordered alloys, in: Order-Disorder Transformations in Alloys, Ed. H. Warlimont, Berlin, Springer, 1974, pp. 240 - 264.

227. Kozlov, E.V., Semukhin, B.S., Rubinovich, L.M. and Shtern, D.M., Fiz. Met. Metalloved., v. 56, p. 547, 1983.

228. Kozlov, E.V. and Starenchenko, S.V., Phase transitions in binary alloys with long-period structures, in: Stabil'nye i Mezhstabil'nye fazy v Materialakh (Stable and Interstable Phases in Materials), Kiev, Naukova Dumka, 1987, pp. 143 - 153.

229. Kozlov, E.V., Klopotov, A.A., Legotina, N.D. and Tailashev, A.S., Atomic long-range order and stability of crystal lattice with respect to fcc - bcc transformation in the CuPd alloy, in: Martensitnye Prevrashcheniya (Martensitic Transformations), Kiev, Naukova Dumka, 1977, pp. 51- 53.

230. Klopotov, A.A., Tailashev, A.S., and Kozlov, E.V., Kinetics of the B1- A1phase transition in a Cu - 40 at.% Pd alloy, in: Uporyadochenie Atomov i Svoistva Splavov (Atomic Ordering and Properties of Alloys), Kiev, Naukova Dumka, 1979, pp. 115 - 118.

231. Kozlov, E.V., Tailashev, A.S., Popov, L.E., et al., Atomic displacements in the ordered alloys NiAl and CuAu, in: Proc. 4th All-Union Conf. on Atomic Ordering and Its Influence on the Properties of Alloys, Tomsk, Tomsk Univ., 1974, pp. 286 - 292.

232. Klopotov, A.A., Tailashev, A.S., and Kozlov, E.V., Izv. Vyssh. Uchebn. Zaved., Fizika, v. 31, no. 6, p. 67, 1988.

233. Kozlov, E.V., Meisner, L.L., Klopotov, A.A., and Tailashev, A.S., Izv. Vyssh. Uchebn. Zaved., Fizika, v. 28, no. 5, p. 118, 1985.

234. Kozlov, E.V., Meisner, L.L., and Tailashev, A.S., Fiz. Met. Metalloved., v. 58, no. 3, p. 498, 1984.

235. Hill, F.L., Statistical Mechanics. Principles and Selected Applications, New York, McGraw-Hill, 1956.

236. Rolov, B.N., Razmytye Fazovye Perekhody (Smeared Phase Transitions), Riga, Zimatne, 1972.

237. Parsonidge, N. and Stavely, L., Disorder in Crystals [Russian translation], Moscow, Mir, Part I, 1982.

238. Semenchenko, V.K., Izbrannye Glavy Teoreticheskoi Fiziki (Selected Chapters in Theoretical Physics), Moscow, Prosveshchenie, 1966.

239. Cowley, J.M., J.Appl. Phys., v. 21, no. 1, p. 24, 1950.

240. Warren, B.E., Trans. AIME, v. 233, p. 1802, 1965.

241. Keating, D.T. and Warren, B.E., J. Appl. Phys., v. 22, no. 3, p. 286, 1951.

242. Goman'kov, V.I., Puzei, I.M., and Rukosuev, M.N., Ordering in NiFe, in: Uporyadochenie Atomov i Ego Vliyanie na Svoistva Splavov (Atomic Ordering and Its Influence on Properties of Alloys), Kiev, Naukova Dumka, 1968, pp. 105 - 108.

243. Goman'kov, G.I., Puzei, I.M., and Loshmanov, A.A., Kristallografiya, v. 10, no. 3, p. 416, 1964.

244. Goman'kov, G.I., Nogin, N.I., and Kozis, E.B., Fiz. Met. Metalloved., v. 55, p. 125, 1983.

245. Marcinkowski, M.J. and Brown, N., J. Appl. Phys., v. 32, no. 3, p. 375, 1961.

246. Beers, J.C. and Guttman, L., Phys. Rev. B, v. 9, no. 9, p. 3941, 1974.

247. Vintaikin, E.Z., Litvin, D.F., and Udovenko, V.A., Fiz. Met. Metalloved., v. 33, no. 1, p. 77, 1972.

248. Berg, H. and Cohen, J.B., Met. Trans., v. 3, no. 7, p. 1797, 1972.

249. Kozlov, E.V., Tailashev, A.S., Shtern, D.M., and Klopotov, A.A., Uzv. Vyssh. Uchebn. Zaved., Fizika, no. 5, p. 32, 1977.

250. Kozlov, E.V., Tailashev, A.S., Sazanov, Yu.A., et al., Study of the ordered state of the CuPt alloy with periodic antiphase boundaries, in: Strukturnyi Mekhanizm Fazovykh Prevrashchenii Metallov i Splavov (Structural Mechanism of Phase Transitions in Metals and Alloys), Moscow, Nauka, 1976, pp. 146 - 149.

251. Kozlov, E.V. and Starenchenko, S.V., Izv. Vyssh. Uchebn. Zaved., Fizika, v. 23, no. 3, p. 70, 1980.

252. Tailashev, A.S., Starenchenko, S.V., Kushnarenko, V.M., et al., Order-disorder phase transition in binary alloys based on an fcc lattice, in: Uporyadochenie

Atomov i Ego Vliyanie na Svoistva Splavov (Atomic Ordering and Its Influence on Properties of Alloys), Tomsk, Tomsk Univ., 1988, pp. 22 - 25.

253. Kozlov, E.V. and Tailashev, A.S., Fiz. Met. Metalloved., v. 43, p. 610, 1977.

254. Ganzina, L.L., Tailashev, A.S., and Kozlov, E.V., Structural changes in the order-disorder phase transition in the FePt alloy, in: Uporyadochenie Atomov i Svoistva Splavov (Atomic Ordering and Properties of Alloys), Kiev, Naukova Dumka, 1979, pp. 158 - 160.

255. Feder, R., Mooney, M., and Nowick, A.S., Acta Met., v. 6, no. 4, p. 266, 1958.

256. Lyashenko, B.G., Litvin, D.F., Puzei, I.M., and Abov, Yu.G., Kristallografiya, v. 2, p. 64, 1957.

257. Fournet, J., Ordering and solid solutions, in: Phase Stability in Metals and Alloys, Eds. P.S. Kudman, J. Stringer, and R.I. Jaffee, McGraw-Hill. , New York, 1967, pp. 270 - 291.

258. Roberts, B.W., Acta Met., v.2, no. 7, p. 597, 1954.

259. Iwasaki, H. and Ogawa, S., J. Phys. Soc. Japan, v. 22, no. 1, p. 158, 1967.

260. Scott, R.E., J. Appl. Phys., v. 31, no. 12, p. 2112, 1960.

261. Iakkel, H.L., J. Appl. Phys., v. 33, no. 8, p. 2439, 1962.

262. Starenchenko, S.V. and Kozlov, E.V., Order-disorder phase transformation in alloys with periodic antiphase domain structure, in: Uporyadochenie Atomov i Ego Vliyanie na Svoistva Splavov (Atomic Ordering and Its Influence on Properties of Alloys), Kiev, Naukova Dumka, 1979, pp. 105 - 107.

263. Kozlov, E.V., Emel'yanov, V.N., Tailashev, A.S., and Shtern, D.M., Atomic interaction, correlation and atomic long-range order in binary metallic alloys, in: Fizika Tverdogo Tela i Metallofizika (Solid State Physics and Metal Physics), Alma-Ata, Nauka, 1979, pp. 105 - 107.

264. Sato, H. and Toth, R.S., Phys. Rev., v. 127, no. 2, p. 469, 1962.

265. Gangulee, A. and Moss, S.C., J. Appl. Crystallogr., v. 1, p. 61, 1968.

266. Popov, S.N., Sazonov, Yu.A., Shtern, D.M., and Kozlov, E.V., Order-disorder phase transition in the AgMg alloy, in: Uporyadochenie Atomov i Ego Vliyanie na

Svoistva Splavov (Atomic Ordering and Its Influence on Properties of Alloys), Kiev, Naukova Dumka, 1979, pp. 323 - 326.

267. Kozlov, E.V. and Starenchenko, S.V., Fiz. Met. Metalloved., v. 48, p. 1220, 1979.

268. Podkovka, V.P., Kormin, N.M., Shtern, D.M., and Kozlov, E.V., Study of atomic ordering in DO superstructure alloys, in: Uporyadochenie Atomov i Ego Vliyanie na Svoistva Splavov (Atomic Ordering and Its Influence on Properties of Alloys), Kiev, Naukova Dumka, 1979, pp. 112 - 115.

269. Ermolaev, L.A., Starenchenko, S.V., and Kozlov, E.V., Physical properties and order-disorder transition in Au-base alloys, in: Uporyadochenie Atomov i Ego Vliyanie na Svoistva Splavov (Atomic Ordering and Its Influence on Properties of Alloys), Kiev, Naukova Dumka, 1979, pp. 276 - 280.

270. Kozlov, E.V., Kushnarenko, V.M., and Pushkareva, G.V., Study of ordering kinetics and properties of the NiMo alloy, in: Proc. 4th All-Union Conf. on Atomic Ordering and Its Influence on Properties of Alloys, Tomsk, Tomsk Univ., 1974, Part I, pp. 232 - 237.

271. Kozlov, E.V. and Kushnarenko, V.M., Fiz. Met. Metalloved., v. 46, no. 2, p. 320, 1978.

272. Kozlov, E.V., Kushnarenko, V.M., and Shtern, D.M., Izv.Vyssh. Uchebn. Zaved., Fizika, no. 4, p. 92, 1978.

273. Chipman, D. and Warren, B.E., J. Appl. Phys., v. 21, no. 7, p. 696, 1950.

274. Muldawer, L., J. Appl. Phys., v. 22, no. 5, p. 663, 1951.

275. Goman'kov, V.I., Litvin, D.F., Loshmanov, A.A., et al., Ukr. Fiz. Zh., v. 8, p. 268, 1963.

276. Vintaikin, E.Z. and Urushadze, G.G., Fiz. Met. Metalloved.,v. 27, no. 4, p. 895, 1969.

277. Vintaikin, E.Z. and Loshmanov, A.A., Fiz. Met. Metalloved., v. 24, no. 4, p. 754, 1967.

278. Girifalko, L.A., Statistical Physics of Materials, New York, Wiley Interscience, 1973, 382 pp.

279. Kubo, R., Statistical Mechanics, Amsterdam, North-Holland, 1965.

280. Huang Kerson, Statistical Mechanics, New York, Wiley, 1963, 520 pp.

281. Liboff, R., Introduction to the Theory of Kinetic Equations, New York, Wiley, 1969, 371 pp.

282. Isihara, A., Statistical Physics, New York, Academic, 1971, 471 pp.

283. Mayer, J. E. and Goeppert Mayer, M., Statistical Mechanics, New York, 1946, 477 pp.

284. Bragg, W.L. and Williams, E.T., Proc. Roy. Soc. London, A, v. 145, p. 699, 1934.

285. Fowler, R.H. and Guggenheim, E.A., Statistical Thermodynamics, Cambridge, Cambridge Univ., 1939.

286. Yang, C.N., J. Chem. Phys., v. 13, p. 66, 1945.

287. Yang, C.N. and Li, Y.Y., Chin. J. Phys., v. 7, p. 59, 1974.

288. Kikuchi, R., J. Chem. Phys., v. 60, p. 1071, 1974.

289. Cowley, T.M., Phys. Rev., v. 77, p. 669, 1950.

290. Christy, D.O. and Hall, G.L., Phys. Rev., v. 132, p. 1958, 1963.

291. Kirkwood, T.G., J. Chem. Phys., v. 6, p. 70, 1938.

292. Gurov, K.P., Metallofizika, no. 20, p. 68, 1968.

293. Zernike, F., Physica, v. 1, p. 565, 1940.

294. Rushbrooke, J., Equilibrium theory of liquid state, in: Physics of Simple Liquids, Eds. H.N.J. Temperley, J.S. Rowlinson, and G.S. Rushbrooke, Amsterdam, North-Holland, 1968, pp. 30 - 62.

295. Bethe, H.A., Proc. Roy. Soc. London A, v. 150, p. 552, 1935.

296. Peierls, R., Proc. Roy Soc. London A, v. 154, p. 207, 1936.

297. Clapp, P.C. and Moss, S.C., Phys. Rev., v. 142, p. 418, 1966; v. 171, p. 754, 1968.

298. Tahir-Khelli, R.Q., Phys. Rev., v. 188, p. 1142, 1969.

299. Golosov, N.S., Izv.Vyssh. Uchebn. Zaved., Fizika, no. 8, p. 64, 1976.

300. Gahn, U., phys. status solidi A, v. 29, p. 529, 1975.

301. Gahn, U., phys. status solidi A, v. 40, p. 463, 1977.

302. Domb, C., Adv. Phys., v. 9, p. 148, 1960.

303. Green, H.S. and Hurst, C.A., Order-Disorder Phenomena, London, Interchange, 1964.
304. Pines, B.Ya., Outlines of Metal Physics, Khar'kov, Khar'kov Univ., 1961.
305. Born, M. and Huang, K., Dynamical Theory of Crystal Lattices, Oxford, Clarendon, 1954.
306. Aptekar', I.L. and Finkel'shtein, B.N., Zh. Eksp. Teor. Fiz., v. 21, p. 900, 1951.
307. Geichenko, V.V., Kanyuka, A.K., and Smirnov, A.A., Ukr. Fiz. Zh., v. 24, p. 721, 1979.
308. Clapp, P.C., Phys. Rev., v. 164, p. 1018, 1967.
309. Zubarev, D.N., Non-Equilibrium Statistical Thermodynamics, Moscow, Nauka, 1971.
310. Kozlov, E.V. and Gudin, O.A., Fiz. Met. Metalloved., v. 37, p. 973, 1974.
311. Kozlov, E.V. and Rubinovich, L.M., Izv. Vyssh. Uchebn. Zaved., Fizika, no. 10, p. 30, 1982.
312. Kozlov, E.V., Gudin, O.A., Dement'ev, V.M., et al., Crystal structure of ordered alloys and order-disorder phase transitions, in: Uporyadochenie Atomov i Ego Vliyanie na Svoistva Splavov (Atomic Ordering and Its Influence on Properties of Alloys), Tomsk, Tomsk Univ., 1978, pp. 11 - 17.
313. Kozlov, E.V. and Shtern, D.M., Fiz. Met. Metalloved., v. 47, no. 3, p. 464, 1979.
314. Kozlov, E.V., Shtern D.M., Kashirin, V.P., and Gudin, O.A., Application of updating methods of statistical physics to the theory of order-disorder phase transitions, in: Termodinamicheskie Svoistva Tverdykh Metallicheskikh Splavov (Thermodynamic Properties of Solid Metal Alloys), Minsk, Belarus' Univ., 1976, pp. 6 - 8.
315. Kozlov, E.V. and Rubinovich, L.M., Izv. Vyssh. Uchebn. Zaved., Fizika, no. 2, p. 104, 1982.
316. Kashirin, V.B. and Kozlov, E.V., Theory of atomic ordering based on entropy expansion into a series in correlation potentials, in: Uporyadochenie Atomov i

Svoistva Splavov, Kiev, Naukova Dumka (Atomic Ordering and Alloy Properties), Kiev, Naukova Dumka, 1979, pp. 95 - 97.

317. Kozlov, E.V. and Kashirin, V.B., Fiz. Met. Metalloved., v. 51, no. 4, p. 702, 1981.

318. Shtern, D.M. and Kozlov, E.V., Izv. Vyssh. Uchebn. Zaved., Fizika, no. 10, p. 40, 1982.

319. Solov'eva, M.I., Shtern, D.M., Rubinovich, A.M., and Kozlov, E.V., Izv. Vyssh. Uchebn. Zaved., Fizika, no. 12, p. 25, 1986.

320. Kozlov, E.V. and Shtern, D.M., Statisticheskaya Teoriya Fazovykh Perekhodov, Tomsk, Tomsk Univ., 1989.

321. Shtremel', M.A., Characteristic features of the AB superstructure (of the Pt Mo type) in the face-centered cubic lattice, Fiz. Met. Metalloved., v. 20, no. 6, p. 829, 1965.

322. Golosov, N.S., Pudan, L.Ya., Golosova, G.S., et al., The dependence of atomic ordering in fcc solid solutions on the ratio of the interaction energies in the first and second coordination spheres, Fiz. Tverd. Tela (Leningrad), v. 14, no. 5, p. 1494, 1972.

323. Golosov, N.S., Popov, L.E., and Pudan, L.Ya., J. Phys. Chem. Solids, v. 34, no. 7, p. 1149, 1976.

324. Rudman, P.S. and Overbach, B.L., Acta Met., v. 5, no. 2, p. 65, 1957.

325. Greenholz, M., Kidron, Q., and Shimony, J. Appl. Crystallogr., v. 7, p. 83, 1974.

326. Lesnik, A.G., Modeli Mezhatomnogo Vzaimodeistviya v Statisticheskoi Teorii Splavov (Interatomic Interaction Models in Statistical Theory of Alloys), Moscow, Fizmatgiz, 1962.

327. Rudman, P.S., Acta Met., v. 13, no. 4, p. 387, 1965.

328. Leibfried, G., Gittertheorie der Mechanischen und Thermischen Eigenschaften der Kristalle, Band VII, Teil 2, Berlin, Springer, 1955.

329. Kozlov, E.V., Popov, L.E., and Starostenkov, M.D., Izv. Vyssh. Uchebn. Zaved., Fizika, no. 3, p. 107, 1972.

330. Moss, S.C., Phys. Rev., v. 171, p. 764, 1968.

331. Matysina, Z.A., Molekulyarno-Kineticheskaya Teoriya Uporyadochivayushchikhsya Tverdykh Rastvorov (Molecular-Kinetic Theory of Ordering Solid Solutions), Dnepropetrovsk, Dnepropetrovsk Univ., 1978, 119 pp.

332. Kozlov, E.V. and Ginzburg, A.E., Izv. Vyssh. Uchebn. Zaved., Fizika, no. 9, p. 7, 1971.

333. Kozlov, E.V. and Ginzburg, A.E., Izv. Vyssh. Uchebn. Zaved., Fizika, no. 3, p. 7, 1972.

334. Kozlov, E.V., Ginzburg, A.E., and Popov, L.E., Izv. Vyssh. Uchebn. Zaved., Fizika, no. 10, p. 79, 1972.

335. Warren, P.E., Averbach, B.L., and Roberts, B.W., J. Appl. Phys., v. 22, no. 12, p. 1493, 1951.

336. Chipman, D.R., J. Appl. Phys., v. 27, no. 7, p. 739, 1956.

337. Schwartz, L.H. and Cohen, J. Appl. Phys., v. 36, no. 2, p. 598, 1965.

338. Sykes, C. and Jones, F.W., Proc. Roy. Soc. London A, vol. 157, pp. 213 - 233, 1936.

339. Ziman, J.M., The Calculation of Bloch Functions, in: Solid State Physics, v. 26, Eds. H. Ehrenreich, F. Seitz, and D. Turnbull, New York, Academic, 1971.

340. Callaway, J., Energy Band Theory, New York, Academic, 1964.

341. Kozlov, E.V., Starostenkov, M.D., and Popov, L.E., Primenenie potentsialov parnogo vzaimodeistviya v teorii atomnogo dal'nego poryadka, in: Stroenie, Svoistva i Primenenie Metallidov (Constitution, Properties, and Application of Metallides), Nauka, Moscow, 1974, pp. 35 - 39.

342. Smirnov, V.A., Podkorytov, S.I., Kiselev, Yu.V., and Kozlov, E.V., Fiz. Met. Metalloved., vol. 47, p. 261, 1979.

343. Smirnov, V.A., Podkorytov, S.I., Kiselev, Yu.V., and Kozlov, E.V., Izv. Vyssh. Uchebn. Zaved., Fizika, no. 8, p. 39, 1979.

344. Harrison, W.A., Phys. Rev., v. 181, p. 1036, 1969.

345. Moriarty J.A., Phys. Rev. B, v. 1, no. 4, p. 1363, 1970.

346. Moriarty J.A., Phys. Rev. B, v. 5, no. 6, p. 2066, 1972.

347. Moriarty J.A., Phys. Rev. B, v. 6, no. 4, p. 1239, 1972.

348. Fuks, D.L. Zhorovkov, M.F., and Panin, V.E., phys. status solidi B, v. 70, no. 1, p. 793, 1975.

349. Fuks, D.L., Zhorovkov, M.F., and Panin, V.E., Fiz. Met. Metalloved., no. 7, p. 80, 1975.

350. Fuks, D.L., Panin, V.E., and Zhorovkov, M.F., Fiz. Met. Metalloved., v. 39, no. 2, p. 884, 1975.

351. Zhorovkov, M.F., Fuks, D.L., and Panin, V.E., phys. status solidi B, v. 68, no. 2, p. 767, 1975.

352. Hayes, T., Brooks, H., and Bienenstock, A., Phys. Rev., v. 175, no. 3, p. 699, 1968.

353. Inglesfield, J.E., J. Phys. C: Condens. Matter, v. 2, p. 1285, 1969.

354. Inglesfield, J.E., Acta Met., v. 17, p. 1395, 1969.

355. Kogachi, M. and Matsuo, J. Phys. Chem. Solids, vol. 32, p. 2393, 1971.

356. Kogachi, M., J. Phys. Chem. Solids, vol. 34, p. 67, 1973.

357. Kogachi, M., J. Phys. Chem. Solids, vol. 35, p. 109, 1974.

358. Matysina, Z.A., Fiz. Met. Metalloved., v. 32, p. 699, 1971.

359. Krasko, G.L. and Makhnovetskii, A.B., Fiz. Tverd. Tela (Leningrad), v. 15, p. 3114, 1973.

360. Krasko, G.L. and Makhnovetskii, A.B., phys. status solidi B, v. 65, p. 869, 1974; v. 66, p. 349, 1974.

361. Taut, M. and Paasch, G., phys. status solidi B, v. 51, p. 295, 1972.

362. Dement'ev, V.M. and Kozlov, E.V., The total electron energy of binary phases of transition metals, in: Elektronnoe Stroenie i Fiziko-Khimicheskie Svoistva Splavov i Soedinenii na Osnove Perekhodnykh Metallov (Electronic Structure and Physical and Chemical Properties of Alloys and Compounds Based on Transition Metals), Kiev, Naukova Dumka, 1976, pp. 34 - 39.

363. Kozlov, E.V. and Dement'ev, V.M., Izv. Vyssh. Uchebn. Zaved., Fizika, v. 24, no. 12, p. 63, 1982.

364. Kozlov, E.V., Dement'ev, V.M., Emel'yanov, et al., The Origin of bonds in binary metallic phases, in: Metallofizika (Metal Physics), Kiev, Naukova Dumka, 1974, no. 52, pp. 35 - 40.

365. Hirschfelder, J.O., Curtiss, Ch.F., and Bird, R.B., Molecular Theory of Gases and Liquids, Wiley, New York; Chapman and Hall, London, 1954.

366. Morse, P.M. and Feshbach, H., Methods of Theoretical Physics, Part 2, Chs. 9 - 13, New York, McGraw-Hill. , 1953.

367. Krasko, G.L., Pis'ma Zh. Eksp. Teor. Fiz., v. 13, p. 218, 1971.

368. Moriarty, J.A., J. Phys. F: Metal Physics, v. 5, p. 873, 1975.

369. Dement'ev, V.M. and Kozlov, E.V., Izv. Vyssh. Uchebn. Zaved., Fizika, no. 8, p. 21, 1973.

370. Dement'ev, V.M. and Kozlov, E.V., Applications of pseudopotential method, in: Proc. 4th All-Union Conf. on Atomic Ordering and Its Influence on the Properties of Alloys, Tomsk, Tomsk Univ., 1974, part I, pp. 142 - 145.

371. Brovman, E.G. and Kagan, Yu.M., Usp. Fiz. Nauk, v. 112, no. 3, p. 369, 1974.

372. Shtern, D.M., Dement'ev, V.M., and Kozlov, E.V., Theory of order-disorder phase transitions involving many-body contributions, in: Uporyadochenie Atomov i Svoistva Splavov (Atomic Ordering and Properties of Alloys), Kiev, Naukova Dumka, 1979, p. 68 - 70.

373. Dement'ev, V.M., Shtern, D.M., and Kozlov, E.V., Effect of many-body interactions on stability of ordered phases, in: Uporyadochenie Atomov i Svoistva Splavov (Atomic Ordering and Properties of Alloys), Kiev, Naukova Dumka, 1979, pp. 70 -73.

374. Katsnel'son, A.A. and Yastrebov, L.I., Psevdopotential'naya Teoriya Kristallicheskikh Struktur (Pseudopotential Theory of Crystal Structures), Moscow, Moscow Univ., 1981.

375. Yastrebov, L.I. and Katsnel'son, A.A., Osnovy Odnoelektronnoi Teorii Tverdogo Tela (Principles of One-Electron Theory of Solids), Moscow, Nauka, 1981.

376. Dement'ev, V.M. and Kozlov, E.V., Izv. Vyssh. Uchebn. Zaved., Fizika, no. 6, p. 30, 1974.

377. Teramoto, K. and Tachiki, M., Phys. Rev., v. 166, p. 656, 1968.

378. Krivoglaz, M.A. and Tyu Khao, On the theory of X-ray diffuse scattering by alloys, in: Metallofizika (Metal Physics), Kiev, Naukova Dumka, 1973, no. 24, p. 84 - 96.

379. Krivoglaz, M.A., Application of X-ray and neutron scattering to the investigation of imperfections in crystals, in: Metally, Elektrony, Reshetka (Metals, Electrons, Crystal Lattice), Kiev, Naukova Dumka, 1975, p. 355.

380. Moss, S.C., Phys. Rev. Lett., v. 22, no. 21, p. 1108, 1969.

381. Moss, S.C. and Walker, R.H., J. Appl. Crystallogr., v. 8, p. 96, 1974.

382. Khanannov, Sh.Kh., Fiz. Met. Metalloved., v. 43, no. 2, p. 431, 1976.

383. Krivoglaz, M.A., Smearing out of Pecularities in Frequency Dependences of Elementary Excitations in Crystals, in: Voprocy Fiziki Metallov i Metallovedeniya (Problems of Metal Physics and Physical Metallurgy), 1962, no. 15, p. 100 - 116.

384. March, N.H., Effective ion-ion interaction in liquid metals, in: Physics of Simple Liquids, Eds. N.V. Temperley, J.S. Rowlinson, and G.S. Rushbrooke, Amsterdam, North-Holland, 1968, p. 241- 286.

385. Afanas'ev, A.M. and Kagan, Yu.M., Zh. Eksp. Teor. Fiz., v. 43, no. 4(10), p. 1456, 1962.

386. Roth Laura, M., Zeiger, H.J., and Kaplan, T.A., Phys. Rev., v. 149, no. 2, p. 519, 1966.

387. Lloyd, P. and Sholl, C.A., J. Phys. C: Condens. Matter, v. 2, p. 1620, 1968.

388. Brovman, E.G. and Kagan, Yu.M., Zh. Eksp. Teor. Fiz., v. 63, no. 5(11), p. 1937, 1972.

389. Williams, A.R. and Weaire, D., J. Phys. C: Condens. Matter., v. 3, no. 2, p. 387, 1970.

390. Veger, M. and Goldberg, I., Some lattice and electron properties of phases with @[beta]-W structure, in: Solid State Physics, New York, v. 28, pp. 2 - 172, 1973.

391. Kolodeznaya, S.F., Dement'ev, V.M., and Kozlov, E.V., Izv. Vyssh. Uchebn. Zaved., Fiz., v. 23, no. 10, p. 84, 1980.

392. Kolodeznaya, S.F., Dement'ev, V.M., and Kozlov, E.V., The Fourier-transform of ordering energy and properties of the CuAu Alloy, in: Uporyadochenie i Svoistva Splavov (Ordering and Properties of Alloys), Kiev, Naukova Dumka, 1979, pp. 201 - 204.

393. Kolodeznaya, S.F. and Kozlov, E.V., Fiz. Met. Metalloved., v. 53, no. 1, p. 191, 1982.

394. Kolodeznaya, S.F. and Kozlov, E.V., Izv. Vyssh. Uchebn. Zaved., Fizika, v. 25, no. 3, p. 118, 1982.

395. Kolodeznaya, S.F. and Kozlov, E.V., phys. status solidi B, v. 119, p. 131, 1983.

396. Kozlov, E.V. and Kolodeznaya, S.F., Izv. Vyssh. Uchebn. Zaved., Fizika, no. 7, p. 103, 1984.

397. Kozlov, E.V., Dement'ev, V.M., and Kolodeznaya, S.F., Mater. Res. Soc., v. 39, p. 65, 1985.

398. Kozlov, E.V. and Kolodeznaya, S.F., Calculation and use of pair interatomic long-range potentials in CuAu alloy, in: Voprosy Atomnoy Nauki i Tekhniki. Fizika Radiatsionnykh Povrezhdenii i Radiatsionnoe Materialovedenie (Problems of Atomic Science and Technology. Physics of Radiation Damage, Radiation Materials Science), 1983, no. 5, pp. 86 - 91.

399. Dement'ev, V.M. and Kozlov, E.V., Symmetry of Fermi surface and structure of the ordered state of metallic alloys, in: Metody Rascheta Energeticheskoi Struktury i Fizicheskikh Svoistv Kristallov (Methods of Calculations of Energy Structure and Physical Properties of Crystals), Kiev, Naukova Dumka, 1977, pp. 38 - 44.

400. von Raether, H., Z. Angew. Phys., v. 4, no. 1, p. 53, 1952.

401. Singlair, R., Schneider, K., and Thomas, G., Acta Metall., v. 23, no. 7, p. 873, 1975.

402. Cowley, J.M., J. Appl. Phys., v. 21, no. 1, p. 24, 1950.

403. Kozlov, E.V., Theoretical and experimental problems of order-disorder phase transitions, in: Termodinamicheskie Svoistva Intermetallicheskikh Faz (Thermodynamic Properties of Intermetallic Phases), Kiev, Inst. Materials Science Problems, Ukrainian Academy of Sciences, 1982, pp. 99 - 111.

404. Batterman, B.W., J. Appl. Phys., v. 28, no. 5, p. 556, 1954.

405. Kozlov, E.V., Emel'yanov, V.N., and Antonov, V.A., phys. status solidi B, v. 81, no. 2, p. 719, 1977.

406. Animalu, A.O.E., Phys. Rev. B, v. 8, no. 8, p. 3542, 1973.

407. Pines, D. and Nozi@eres, Ph., The Theory of Quantum Liquids, v. 1, New York, Benjamin, 1966.

408. Dolgov, O.V. and Maksimov, E.G., Usp. Fiz. Nauk, v. 135, no. 3, p. 441, 1981.

409. Kozlov, E.V. and Dement'ev, V.M., Fiz. Met. Metalloved., v. 35, no. 3, p. 500, 1973.

410. Smirnov, V.A., Kiselev, Yu.V., and Kozlov, E.V., Izv. Vyssh. Uchebn. Zaved., Fizika, vol. 25, p. 28, 1982.

SUBJECT INDEX